AF710071

ANIMAL COGNITION AND SEQUENTIAL BEHAVIOR

Behavioral, Biological, and Computational Perspectives

ANIMAL COGNITION AND SEQUENTIAL BEHAVIOR

Behavioral, Biological, and Computational Perspectives

edited by

Stephen B. Fountain, Michael D. Bunsey, Joseph H. Danks, and Michael K. McBeath

Kent State University

KLUWER ACADEMIC PUBLISHERS
Boston / Dordrecht / London

Distributors for North, Central and South America:
Kluwer Academic Publishers
101 Philip Drive
Assinippi Park
Norwell, Massachusetts 02061 USA
Telephone (781) 871-6600
Fax (781) 681-9045
E-Mail <kluwer@wkap.com>

Distributors for all other countries:
Kluwer Academic Publishers Group
Distribution Centre
Post Office Box 322
3300 AH Dordrecht, THE NETHERLANDS
Telephone 31 78 6392 392
Fax 31 78 6546 474
E-Mail <services@wkap.nl>

Electronic Services <http://www.wkap.nl>

Library of Congress Cataloging-in-Publication Data

A C.I.P. Catalogue record for this book is available
from the Library of Congress.

Copyright © 2002 by Kluwer Academic Publishers

All rights reserved. No part of this publication may be reproduced, stored in a retrieval system or transmitted in any form or by any means, mechanical, photo-copying, recording, or otherwise, without the prior written permission of the publisher, Kluwer Academic Publishers, 101 Philip Drive, Assinippi Park, Norwell, Massachusetts 02061

Printed on acid-free paper. Printed in the United States of America

The Publisher offers discounts on this book for course use and bulk purchases. For further information, send email to <michael.williams@wkap.com>.

For Maria, Stephanie, and Natalie

CONTENTS

CONTRIBUTORS

Michael Bunsey (Deceased)
Department of Psychology
Kent State University
Kent, Ohio 44242

E. John Capaldi
Department of Psychological Sciences
Purdue University
West Lafayette, Indiana 47907

Stephen B. Fountain, Douglas G. Wallace, & James D. Rowan
Department of Psychology
Kent State University
Kent, Ohio 44242

Stewart H. Hulse
Department of Psychology
Johns Hopkins University
Baltimore, Maryland 21218

Raymond P. Kesner
Department of Psychology
University of Utah
Salt Lake City, UT 84112

Brendan McGonigle & Margaret Chalmers
Laboratory for Cognitive Neuroscience and Intelligent Systems
The University of Edinburgh
Edinburgh, Scotland
EH8 9YL

Warren H. Meck
Department of Psychological and Brain Sciences
Duke University
Durham, NC 27708

William A. Roberts
Department of Psychology
University of Western Ontario
London, Ontario, Canada
N6A 5C2

Herbert L. Roitblat
Department of Psychology
University of Hawaii
Honolulu, Hawaii 96822

Karyl B. Swartz & Sharon A. Himmanen
Department of Psychology
Lehman College of the City University of New York
Bronx, New York 10468

H. S. Terrace
Department of Psychology
Columbia University
New York, New York 10027

Anthony A. Wright
Department of Neurobiology and Anatomy
University of Texas Medical School at Houston
Houston, Texas 77227

Participants of the Eighth Kent Psychological Forum. From left to right: Warren H. Meck, J. Gregor Fetterman, Karyl B. Swartz, Stewart H. Hulse, Herbert L. Roitblat, Stephen B. Fountain, Michael Bunsey, Brendan McGonigle (kneeling), Raymond P. Kesner, William A. Roberts, E. J. Capaldi, Sarah T. Boysen, H. S. Terrace, Anthony A. Wright, F. Robert Treichler.

FOREWORD

The Inn at Honey Run is not easy to find; it is barely visible up the hill and through the woods from the narrow country road that provides access to it. The Inn is situated in the hilly Amish country of central Ohio an uncomfortable distance from the nearest major airport. Yet, as it turns out, it seems to have been the perfect setting—relaxed, scenic, secluded, even contemplative—for the Eighth Kent Psychological Forum. In April, 1996, some of the leading researchers in the field of animal learning, memory, and cognition gathered at the Inn to discuss recent advances and new perspectives on some of the central problems in animal cognition, namely, how sequential and temporal behavior are organized. Each of the major participants delivered an hour-long formal presentation at the conference. Equally important, though, were the informal and often spirited discussions that took place between sessions, during meals, and late into the evening. This edited volume was stimulated by the Forum, and many of the chapters echo research and ideas that were discussed in that setting.

Recent years have witnessed the successful development of model systems for studying simple behavior such as Pavlovian conditioning from a mechanistic perspective. The most profitable approaches have been interdisciplinary and have included developing psychological and neurobiological model systems using lower animals and using sophisticated connectionist approaches for computer simulation of simple behavioral systems. The result of this effort has been tremendous advances in our understanding of a number of fundamental psychobiological systems relevant to normal behavior and health. Complex cognitive behaviors, however, simply because of their greater complexity, are just beginning to yield to similar analytical approaches. Nevertheless, developments in this area are eagerly awaited because of their potential for broadening our understanding of human and animal cognitive abilities and for their implications for understanding and treating cognitive dysfunction. The Eighth Kent Psychological Forum focused on bringing together psychologists studying sequential and temporal organization in animal behavior who, therefore, have a common interest in understanding function and dysfunction in the realm of complex cognitive behavior.

This volume is divided into an introductory chapter and three broad sections that together, we hope, will provide an overview of current directions in the field. In Part 1, Stew Hulse sets the stage for the remainder of the volume by providing both a historical context and some prognostications for research in the field. The chapter draws heavily on his presentation at the Forum. His comments on the historical development of the field and his sentiments and presentiments regarding the current status and future direction of the field stimulated lively discussion at the Forum. That chapter provides a reference point for thinking about developments in the field that are represented in the remainder of the volume. Part 2 focuses on current approaches to the study of sequential learning and memory. Part 3 considers counting and timing behaviors. Part 4 focuses on cognitive structure, then we close, in Part 5, with some comments on research and ideas presented at the Forum and in this volume.

In addition, it should be mentioned that three issues were at the center of many lively discussions at the Forum, both during sessions and in informal conversations, though some are not explicitly addressed in chapters. First, one recurring question, almost a leitmotif of the conference, was how best to characterize complex animal behavior of the sort discussed at the Forum. Can sequential behavior, timing, and counting be understood as complex emanations of simple processes? Do these putatively complex behaviors demand explanation in terms of complex cognitive processes? In the first presentation of the Forum, Stew Hulse admonished us all to be wary of anthropomorphism as explanation. Hulse thus adds his voice to others recently warning of the dangers of yielding too easily to the siren calls of "animal minds" or to the flawed logic of the "argument from design." On the positive side, one can hope that the chapters of this volume will provide blueprints for successfully approaching complex behavior with empirical and theoretical self-restraint.

A second topic that was treated both in Stew Hulse's chapter as well as in discussion throughout the Forum was the issue of the relative value of basic learning research versus neurobiology research. It is clear which way the wind blows—funding for behavioral neuroscience is up, whereas funding for pure behavioral work is scarce. Although all agreed that behavioral neuroscience approaches provide added power and critical converging evidence, it was also agreed that purely behavioral approaches are also necessary. Without pure behavioral work we cannot know enough about behavior to support behavioral neuroscience. Those in the behavioral neurosciences need a behavioral literature out there at which to direct their neuroscientific tools. It is too much to expect them to derive understanding of the mechanisms of behavior *and* to probe the neurobiological substrates—if anything, the Forum and this volume prove that difficult behavioral questions remain unanswered despite the undivided efforts of purely behavioral researchers.

Finally, some participants at the Forum, Sally Boysen in particular, emphasized the need for vigilance regarding the activities of animal rights groups. It was agreed that animal researchers need to take a more proactive approach to issues of animal experimentation. Currently, it is the animal rights groups that are going into schools and promoting their cause, whereas animal researchers are generally reacting, speaking up only in response to attacks on a case-by-case basis. It was agreed that we, as animal researchers, need to make our case more aggressively. For example, we need to go to schools and discuss the importance of animal research so that children are exposed to these ideas from the earliest age possible. As Sally Boysen noted, the other controversies ultimately pale in comparison if animal rights groups block our very ability to conduct research.

ACKNOWLEDGMENTS

The Eighth Kent Psychological Forum, like each annual Forum before and since, was sponsored by the Kent State University Applied Psychology Center (APC). The APC, under the direction of Stevan E. Hobfoll, has been remarkably successful in fulfilling its mission to facilitate applied psychological research in all its varieties at Kent. As the organizer of the Eighth Forum, I am especially grateful and think it to the APC's credit that the deeper meanings and broader implications of "fundamental" research are properly appreciated for their applied significance. As the Forum organizer and principal editor of this volume, I am especially indebted to the Forum participants for their remarkably enthusiastic support of the Forum and preparation of this volume. Special thanks indeed go to Sadhana Moneypenny of the APC. Her organizational prowess and unrelenting good humor in the face of crisis overcame most, if not all, of the organizer's limitations in this regard. The Forum was a memorable success in large measure because of her efforts. I also thank my co-editors, Mike Bunsey, Joe Danks, now Dean of Arts and Sciences at Kent, and Mike McBeath, now at Arizona State University, for their contributions, both organizational and editorial. Special thanks in this regard go to Mike Bunsey, whose unexpected passing too early in his career deprived all who knew him of a truly gifted colleague and a kind and gentle friend. Thanks also to Sally Boysen (Ohio State Univerisity), Greg Fetterman (Indian University-Purdue University, Indianapolis), and Bob Treichler (Kent State University), who delivered stimulating presentations of their work at the Forum but did not contribute to this volume. I also wish to thank a group of Kent graduate students who read and provided feedback to the editors on most of the chapters: Steven Harrod, Brian Hock, Kevin Kaut, Cantey Land, Nannette Stempowski, and Douglas Wallace. Finally, on a personal note, I thank Maria S. Zaragoza—former Forum organizer, colleague, department chair, and, most importantly, spouse—for her advice and unconditional support.

Stephen B. Fountain
Kent, Ohio, August 2001

Part 1: Historical Perspectives

1

PERSPECTIVES ON COMPARATIVE COGNITION
Past, Present, And Future

Stewart H. Hulse
Johns Hopkins University

In this chapter, first, I would like to note that 1996 marks the 20th anniversary of an important event in the history of comparative cognition, an event that helps set our activities this weekend into historical context. After a few words about that, I will then turn to a problem with which I have had a certain measure of familiarity over the years, namely, the problem of serial order in behavior. I intend to examine that issue with an historical approach. History captures much that remains important in our field, not only regarding the problem of serial order, but also regarding the larger framework of comparative cognition as a whole. Finally, I would like to make a few observations about where I think the field of comparative cognition may be headed. We have made great progress in the past few decades, but I fear that serious pitfalls are either with us or lurk immediately ahead. Because history has a way of repeating itself, as the historian has said, we must heed some looming danger signals, the harbingers of which are already among us. Still, we have made progress, and there is much to look forward to.

THE DALHOUSIE CONFERENCE

Let me turn to the historical event to which I refer. The event was, to my knowledge, the first conference dealing by name with cognitive processes in nonhuman animals. Werner Honig, Harry Fowler, and I were discussing the current state of affairs in animal learning at a meeting of the Psychonomic Society in Boston in the Fall of 1975. We decided to organize a conference to be held at Dalhousie University in the spring of 1976—20 years less 2 months earlier than the Applied Psychology Forum at Kent State. The group assembled in Halifax, Nova Scotia in June and spent four intensive days exploring the idea that studies of animal learning and behavior might profit from a cognitive perspective. A book similar to this one resulted from that conference (Hulse, Fowler, and Honig, 1978).

It is interesting to note the parallel between the topics that were considered timely then, and those that comprise the chapters of this book. The topic of memory was certainly well represented, with Werner Honig and David Olton holding forth on that problem. To my knowledge, incidentally, that was the first public discussion of data obtained from the radial arm maze and the remarkable capacity of rats to negotiate the series of choices in that structure. Russell Church discussed the internal clock and some of the first data obtained on the timing behavior of rats. David Olton discussed spatial memory, of course, and Emil Menzel discussed cognitive mapping in chimpanzees. I spoke about serial learning and some of the initial data we had been getting from rats on this issue. David Premack spoke about abstraction in animals, with special emphasis on the role of language and language training on concept formation. Al Riley (with Herb Roitblat as co-author) presented a paper on selective attention in pigeons. The parallel is intriguing between the topics that were important then and the topics that appear in this book. Clearly certain issues have remained significant.

It is also interesting to note the topics discussed then that are not represented in this volume. There were several papers on Pavlovian conditioning, most of which were only tangentially cognitive in nature. Bob Rescorla, Harry Fowler, and Nick Mackintosh described current issues in that field, and Allan Wagner presented some of his initial modeling on the priming of short term memory, a topic that later became incorporated into his SOP model of classical conditioning. Bob Bolles discussed defensive behavior, and Eliot Hearst described data on autoshaping and sign tracking. But one had a sense that Pavlovian conditioning, although of importance for understanding associative structure, was not to be a significant part of studies in comparative cognition as that field was to develop. Clearly, something new was afoot, an enterprise that has flourished in the 20 years since 1976.

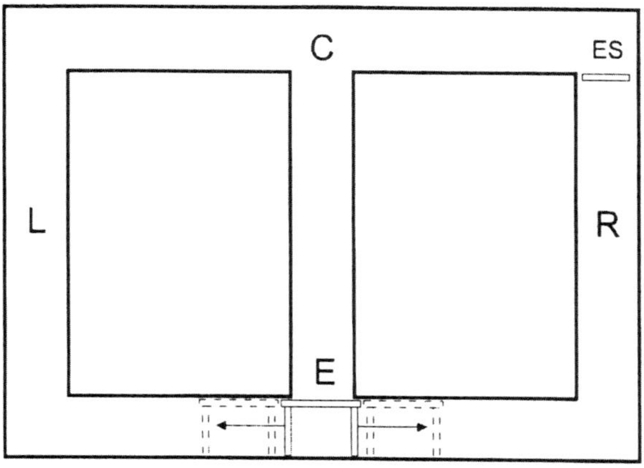

Figure 1. Hunter's temporal maze. After the experimenter inserts the rat into the maze at point E, the rat makes turns into the right (R) or left (L) alleys at choice point C. The moveable barrier at the bottom of the figure forced the rat up the center alley after selecting the R or L side alleys. The end stop (ES) was inserted into the right or left alley after the first or second choice on a single- or double-alternation trial to force the rat to the opposite alley as the animal looped through the maze.

THE PROBLEM OF SERIAL ORDER IN BEHAVIOR

I have always found it instructive to trace the history behind facts and theories that are important today. Historical perspective often stimulates future progress. There is one problem that encompasses many fundamental questions in psychology, questions ranging from memory to the perception of time and space. The problem interests many of those contributing to this symposium either directly or indirectly. I refer, as you might have guessed, to the problem of serial order in behavior.

That problem is fundamental. Without solving it, people—and nonhuman animals—could not learn to solve spatial and temporal mazes, could not learn to utter acoustic signals, including language, and could not learn to sing or to make instrumental music. Let me trace some important historical landmarks surrounding the problem of serial order in behavior.

Walter S. Hunter and the Temporal Maze

Certainly one of the earliest developments in the study of serial order in behavior was Walter Hunter's experiments with rats in the *temporal maze* (Hunter, 1920), shown in Figure 1. As you will recall, the maze is a series of four alleys arranged as a rectangle with a central alley running up the middle. The rat leaves the entrance point (E) and must decide to go right or left at the choice point labeled "C" in the figure. In a *single alternation* problem, the rat must go up the central alley and to the right at point C, then loop back to the start point. Here, the experimenter moves a block to force the rat to go up the center alley again. When it reaches the choice point the second time, the rat must go to the left and loop through that side of the maze.

In a *double alternation* problem, the rat must go right at the choice point, looping twice around the right side of the maze. Then the rat must turn left at the choice point, looping twice around that side of the maze. The end stop (ES) is placed in the maze so the rat cannot retrace through the right side of the maze after completing its journey there. Successful solution for both single and double alternation tasks requires that the rat remember two things: (1) the spatial location of the choice point, and (2) the proper temporal sequence of right and left turns.

Hunter's rats, a few of them at least, were able to solve a single alternation problem. However, no rat was ever able to solve the more complex spatial and temporal contingencies of a double alternation problem. Perhaps, in retrospect, that was because Hunter did not fully appreciate the implications of Thorndike's Law of Effect. Although Hunter sometimes used punishment for wrong responses in maze experiments, he generally did not reinforce his rats with food in the temporal maze until the entire daily session of 10 trials had been completed! This was perhaps a reward for cooperative social interaction between man and rat, but probably not useful for much else.

But why was the temporal maze important in Hunter's era? At the time, there was substantial controversy as to how animals like rats could find their way through mazes. What sensory systems did they require to navigate a complicated series of correct paths and blind alleys? One of the primary theoretical ideas of the time was that proprioceptive feedback from muscular movements provided a distinct train of stimulation as the rat ran its way through the maze. This conjecture was based on the fact that rats could do pretty well on mazes if they were blinded, if their vibrissae were cut off, and if they were otherwise deprived of their primary senses. Hunter's temporal maze addressed the issue of whether the rats could solve the puzzle of getting through the temporal maze without the cue provided by the unique kinesthetic "feel" of muscular movements associated with successive turns and choice points. As we have seen, Hunter's rats weren't very good at such tasks. He concluded, then, that kinesthesis by itself could not play a major role in serial problems like the temporal maze. If proprioception played a role at all, it did so only in conjunction with stimulation from other senses like vision and olfaction.

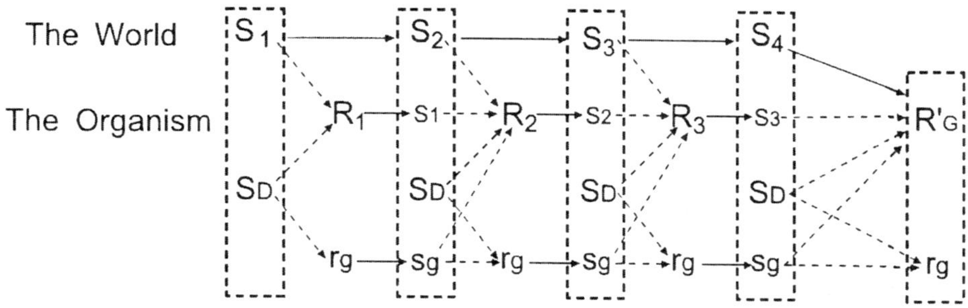

Figure 2. Hull's (1931) model for serial stimulus-response sequences in a maze learning task. This was the first appearance of the r_g–s_g mechanism. Note that four stimuli became associated with a response: the external stimulus, S, the proprioceptive stimulus from prior responses, S, the drive stimulus, S_D, and the fractional goal stimulus, s_g.

Hull, Spence, and the r_g–s_g Mechanism: Goal Directed Activity

The next link in our chain of events was an article published by Clark Hull in 1931 in the *Psychological Review* (Hull, 1931). Its title was "Goal Attraction and Directing Ideas Conceived as Habit Phenomena." Hull set out to develop a system that expressed "ideas" in purely physical terms, that is, in terms of stimulus, response, and drive stimulus. The complete model is shown in the next figure.

As you can see in Figure 2, The World was reduced to a series of stimuli, S_1 through S_4. The Organism was reduced to a series of responses, R_1 through R_3, the goal response R'G, and the drive stimulus, S_D. The drive stimulus was especially important because it provided a mediating stimulus that persisted throughout the response chain and assured the development of the r_g–s_g mechanism. The latter, if you will recall, was a mechanism by which rats "thought ahead" about the goal and represented, in fact, Hull's conception of an "idea." This model, or minor variations upon it, still stimulates research to this day. That is, theoretical systems of serial learning which depend upon stimulus-response chains are related to the principles espoused by Hull—at least insofar as the feedback from one response provides a component of the stimulus dynamics controlling the next response in the chain.

Skinner's Operant Response Chain

sS^D : *visual lever* . R : *lifting* →

sS^D : *tactual lever* . R : *pressing* →

sS^D : *tray* S^D : *sound of magazine* . R : *approach to tray* →

S : *food* . R *seizing*

Figure 3. A diagram of Skinner's (1938) response chain. Note the importance of feedback from a prior response and a discriminative stimulus for the next response in the chain.

Skinner

Hull's system was paralleled by Skinner's idea of the response chain. The next figure shows the representation of response chaining that Skinner presented in *The Behavior of Organisms* in 1938 (Skinner, 1938). As Figure 3 shows, Skinner's chain is based upon discriminative stimuli (S^Ds) followed by responses that are also associated with feedback from prior stimulation, such as the tactile sensation of the lever or from the rat's lifting its paws. The feedback stimuli provide part of the stimulus component of succeeding responses in the chain. Skinner, unlike Hull, did not postulate a drive stimulus as a critical component of the chain. But clearly, the two systems are highly related. Both claim that serial order in behavior is described in terms of an alternation between the stimulus components at one point in the chain and the response components that follow, the response components leading in turn to a new set of stimulus components, and so on. Both systems are clearly expressed in peripheralistic terms and framed to meet the criteria of the behaviorism that was so much part of the theoretical scene at the time.

Lashley

The next milestone in our story, indeed a watershed, was the appearance of Lashley's seminal paper concerning the problem of serial order in behavior (Lashley, 1951). The paper appeared in a collection of articles from the Hixon Conference at Princeton University and started a trend of theoretical thinking that remains important to this day. Essentially,

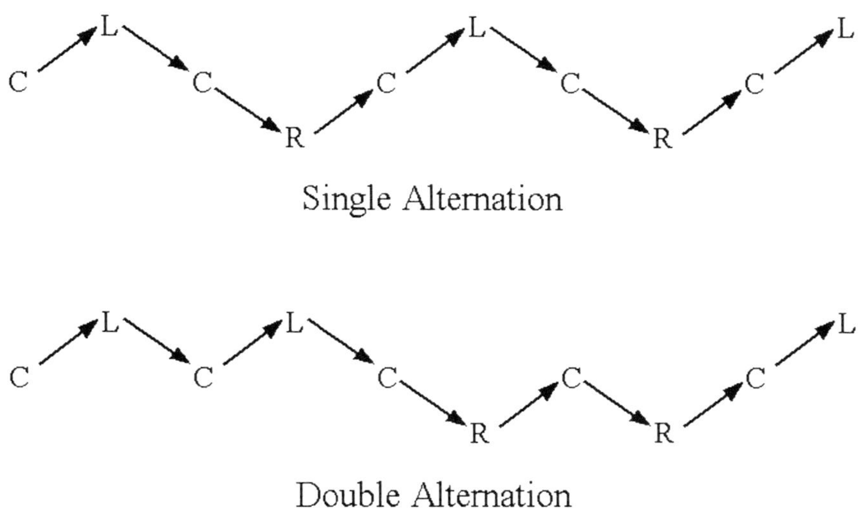

Figure 4. A diagram demonstrating the branching problem for a single alternation and a double alternation sequence in Hunter's temporal maze. Right (R) or left (L) turns must be made to the identical stimuli at the choice point, C, as the rat returns to the choice point following successive loops through the maze.

Lashley's article was a concise description of the shortcomings of associative chains in accounting for the serial order that so often appears in behavior. One important problem, he noted, was the problem of *branching* in an associative chain. Branching occurs when the same stimulus calls for different responses at different points in the chain. For example, in Hunter's temporal maze, the rat must learn to turn right on one occasion to the stimuli comprising the choice point, but to turn left to exactly the same stimuli at the same choice point on the next trial. The scheme appears in the next figure.

As you can see in Figure 4, the rat's appropriate choice at the decision point is determined strictly by the sequential location of the choice in the appropriate sequence of right and left turns. Of course, neither Hunter nor Lashley used the term "branching," because the computer metaphor was not part of the lexicon at that time. However, the idea was clearly stated in Lashley's article. For instance, one of his objections to associative chaining was based upon an analysis of speech production. He gives an example based on the use of the word "right" in a sentence. He says, and I quote (and paraphrase slightly), "The word right, for example, is noun, adjective, adverb and verb and has four spellings and at least ten meanings. In such a sentence, 'The mill right on my right thinks that some conventional rite ...symbolizes the right of every man to write as he pleases,' (...the proper...) arrangement is obviously not due to any direct association of the word *right* itself to other words but to meanings which are determined by some broader relations." And then later he says, "From such considerations, it is certain that any grammatical form which ascribes (...the arrangement...) to associative linkage of the words of the sentence overlooks the essential structure of speech.... The order is imposed by some other agent." He then goes on to point out that such considerations are true not only of language but also of other skilled movements. The gaits of a horse—trotting, pacing, and singlefooting—he says, involve essentially the same pattern of muscular contraction in the horse's legs, but the gait is imposed by some mechanism in addition to direct reciprocal innervation among the low-level sensory-motor nuclei controlling the leg.

The same is true for pianists regarding the order in which fingers fall on the keys. The order of finger placement does *not* come from the association of successive movements of the individual digits. Learning to play a musical composition, which is an extraordinarily complicated stimulus structure, involves breaking the composition into clumps of movements at various levels of hierarchical organization and learning a central representation of how finger movements are to be produced in time and space within each level in the hierarchy. The nervous system simply cannot act fast enough for response feedback to play any significant role in controlling this sequence. The control must come from some higher-level afferent source. The main point is that peripheral movement, and the learning of an associative chain, is merely the surface manifestation of a vastly more complex, centrally mediated, hierarchical organization. This is a familiar idea, but it bears repeating.

It is interesting to note that Lashley discusses virtually every topic that might be included in a symposium or a book on comparative cognition. These include not only the problem of serial order, but also timing, spatial organization, and the many other ways we register and recall past experience. Of course, the structure and function of the central nervous system overarches all these issues.

The Advent of the Computer Metaphor

The early theoretical history of the problem of serial order ends, perhaps, with a development that is only indirectly related to the problem of serial order itself. I refer to the advent of the modern digital computer as a metaphor for describing how the central nervous system works to produce behavior. This idea was ushered into psychology by, among others, Miller, Galanter and Pribram in their book, *Plans and the Structure of Behavior*

(1960). The computer metaphor stimulated by that and other writing of the time provided a lexicon that is now part of everyday discourse. We retrieve memories after we store them, we process information, and so on. The computer program as a metaphor for the unfolding of serial events has certainly played a profound role in the way we conceptualize the serial structure of behavior.

Modern Theoretical History

As an example of the computer metaphor at work, let us turn to some more recent history, coming primarily from human experimental psychology, that set the stage for further developments in understanding the problem of serial order. The first development of significance stems from the work of Simon and Kotovsky (1963),on the one hand, and Frank Restle's work on the same issues (e.g., Restle, 1970). Rule structure and computer models designed to simulate peoples' behavior when they were dealing with serial order were the new ideas that these people introduced. The proposition that people learned hierarchically-organized rules instead of associative chains in encoding serial response patterns was a fundamental premise of this newer approach. The idea is well expressed by performance on the *Thurstone letter completion* task that Simon described both in 1963 and in a more recent lecture delivered to the American Psychological Association in the late 1980's. For example, when asked to complete the letter string A-B-C-B-C-D-C-D-E-__ by filling in the blank most people would do so with a D (presumably). One is led to induce the rule structure of the sequence of letters and to supply the missing letter based on an induction and then an extrapolation of the rule system involved in the sequence. Simon pointed out that there were several principals involved in such induction. First of all, people tended to look for *periodicities* within the surface structure of the pattern. Then they search within those periodicities, or *chunks*, to find rules to describe how each element in the sequence was related to the next. Rules such as Next, Reverse, Complement, and so on, could be brought into play to produce serial surface structure of extraordinary complexity.

A significant advantage of this system was that a complex surface structure was not learned by memorizing an equally complicated associative chain, fastening the sequence of elements link by link. Rather the surface structure was learned by encoding the underlying rules, and then regenerating the pattern from the memorized rules when called upon to do so. For one thing, this solved the branching problem that we saw to be a potential weakness in the work of Hunter, Hull, Spence, and Skinner. According to the new rule-encoding approach, what an organism chose to do in response to a stimulus depended on where that stimulus appeared in a *hierarchical* context where the significance of the stimulus changed across different levels of the hierarchy.

It is important to recognize that Simon, Restle, and the others were working to describe a *psychological* process with their computer programs. They were not merely writing computer algorithms to generate formally-constructed serial strings. For example, from a formal standpoint, E-D-C-D-C-B-C-B-A is a perfectly permissible continuation of the A-B-C-B-C-D-C-D-E string we saw earlier. It is a human *predilection* to extrapolate the string based on the string's induced rule structure, but formally, a Reverse rule at the place where humans would normally extrapolate would be equally permissible.

Whether or not the principles of rule learning could be applied to the behavior of nonhuman animals was first proposed and tested by me and my students in the late 1970's and early 1980's. We provided evidence that rats behaved as if they could indeed encode rule systems. The rats found serial sequences that were formally complex to be more difficult to learn than serial systems that were formally simple (Hulse & Dorsky, 1977). For example, it was easier for a rat to anticipate (as measured by slow running) a 0 quantity of food in the a series of runs to different food quantities if the food quantities were ordered 14-7-3-1-0 than if the quantities were ordered 14-1-3-7-0.

Rats extrapolated patterns based on induced rule structure (Fountain & Hulse, 1981), learning to anticipate a 0 food quantity more accurately (even though they had never before experienced it) after training with a 14-7-3-1 sequence that after training with a 14-1-3-7 sequence of food quantities.

Learning stimulus sequences of food quantities was facilitated by temporal markers (Fountain, Henne, & Hulse, 1984). That is, in a repeating sequence of 14-7-3-1-0 food pellets, the 0 food pellet was anticipated with greater facility if the repetitions were separated (chunked) by a long inter-sequence interval.

Rats found spatial markers helpful in the formation of rule structures (Hulse & O'Leary, 1982). If the elements of a 14-6-1-0 sequence of different quantities of food were associated with unique places in space in a radial arm maze, a 0 quantity of food was easier to anticipate than if the elements of the sequence were associated with varying places in space.

Stephen Fountain has expanded this initial work in many new directions. For example, he discovered that brain stimulation turned out to be a very powerful tool for facilitating the learning of complex sequences of positions in many experiments modeled directly after the initial work of Restle on the learning of serial sequences by humans (e.g., Fountain & Rowan, 1995). The parallels he has shown between human and rat performance on such serial learning tasks are remarkable.

All this earlier work did not, however, seek to identify the means by which rats in fact, encoded such structures neurologically. Their behavior was well described and predicted by a rule-based system, but a more fundamental explanation, at a neurobiological level for example, remains to be specified. Given the significance of serial learning in the organization of behavior, a search for the biological substrate of serial organization should prove profitable indeed. Fountain and his colleagues have made some initial steps in that direction (e.g., Fountain & Rowan, 1992).

Of course, throughout all these later decades, the idea of the associative chain has continued to find favor. There is a continuing tension between those who believe rats and other nonhuman and nonverbal organisms can process serially organized events on the basis of some centrally-encoded scheme, and those who believe that more peripheralistically-founded theoretical models can still serve the problem well. This tension is expressed in some of the articles in this book. I hope that this brief historical reminiscence will help motivate resolutions to that tension both now and in the future.

FUTURE ISSUES IN COMPARATIVE COGNITION

Let me turn, finally, to what I perceive to be important problems that we who study comparative cognition will have to address in the future. We are already grappling with some of them. I want to devote my attention to just three issues that I perceive to be most troublesome. (Sarah Boysen raised a fourth issue at the conference, namely that of the perils represented by the animal rights movement, and that problem certainly remains with us as the animal rights activists become ever more clever at attacking animal research.)

Put the Comparison into Comparative Psychology

First of all, we need to continue to make comparative cognition truly comparative. Quite often, and I have been guilty of this myself, we do experiments that parallel those that are done with people in a test for the capacity of a nonhuman animal to match that of a human. Rarely, however, do we do this with strategies that encourage true comparisons. "Can this species do it, too?" often provides the primary motivating force behind the research. Instead, it seems to me, we need to make *direct* comparisons of two or more meaningfully- chosen species in the same way, adapting apparatus and procedures as necessary to facilitate experimentation. In so doing, of course, we must keep an evolutionary perspective in mind. Birds have different evolutionary histories than mammals, for example, and we must decide whether our choice of species is appropriate for the comparison at hand. Why is it inherently interesting, for example, to compare the memory of pigeons or starlings with the memory of nonhuman primates or human beings? No doubt, there are many appropriate comparative problems to be researched with those species. But as we design our research, does our choice of species get as much attention as our choice of problems? It ought to.

I am certainly not the first to make this point, which is really, of course, merely an expression of the distinction between the general-process and the ethologically-oriented approach to the study of behavior. As an ethologist, Sara Shettleworth has used evolution as a *tool* to study the role of memory in the behavior of food storing birds (e.g., Shettleworth, 1993). She has, for example, compared the food-storing behavior of closely related species of corvid jays. By thus holding genetic makeup mostly constant, she has then compared the behavior and, especially, the neurology of the species to determine those factors that are truly critical for the long-term memory involved in storing and later retrieving food. Alan Kamil, David Sherry, and John Krebs, among others, have adopted a similar approach. It is powerful.

The same thing may be said when it comes to comparing the behavior of verbal with closely-related nonverbal species (nonverbal, at least, in the human sense), like chimpanzees. There are many investigators who come to the study of such behavior from a comparative, perhaps ethological perspective, and we should all pay attention to what they have to say.

There is still another advantage of the comparative approach I espouse. It might draw the attention of our counterparts in human experimental psychology. Right now, as I have lamented elsewhere recently, I fear our colleagues in laboratories of human psychology tend to ignore the contributions we comparative psychologists can make to cognitive psychology more generally conceived. I well remember, for example, a symposium that Tony Wright organized on the topic of memory for the annual meeting of the Psychonomic Society some ten years ago. He had carefully gathered participants who worked with both human and nonhuman subjects. It was truly fascinating to watch the behavior of those attending the symposium—not the behavior of the speakers, but the behavior of the audience. The listeners working with animals stayed throughout and heard all the speakers. However, their counterparts working with humans came, and left, returned, and left again, listening only to those speakers discussing human memory. There is a lesson here that needs our attention. From the larger perspective of trying to understand memory as a general process, this episode was—and remains—lamentable. Comparative studies have a special advantage. They allow us not only to characterize any given species, but also to characterize any species uniquely. I suspect, for example, that the years of research on language in chimpanzees have taught us more about human language than about chimpanzee "language." We have learned more about the properties that make human language uniquely human from studies of what chimpanzees *cannot* say as from demonstrations of whatever capacity for human language chimpanzees may possess.

The same may be true for the study of behavior writ large. Comparative psychology affords *contrasts* that identify the capacities as well as the limitations of any species under scrutiny. Just as we psychologists who study animals have learned much from students of human behavior, students of human behavior have much to learn from us. They ought to pay more attention than they do, and we ought to encourage them in any way we can.

The Neurobiology of Cognition

I turn now to the second problem that is with us and promises to become more vexing in the future. In this decade of the brain, it goes without saying that the biology, physiology, and anatomy of the central nervous system are very important in furthering our understanding of behavior. Many of you are already embarked upon studies exploring the neurobiology of the cognitive behavior that your animals exhibit. That is a wise adaptive strategy if our field is to survive. For now, at least, the study of behavior for its own sake—cognitive or otherwise—will not attract much of the attention of our colleagues in behavioral neuroscience or in the neurosciences more generally. What's more, behavioral studies alone won't attract much attention from funding agencies, as some of you can surely attest. We must all face the practical exigencies of modern life in the research laboratory, a life that has been dependent primarily on government financial support.

However, that support is disappearing. It would be a tragic loss if psychological science—and the government agencies that support it—were to abandon the study of behavior per se. If that happens, we run the eventual risk of garnering masses of information about the brain while neglecting what the brain does.

The problem is not only insidious, it spreads subtly as graduate and postdoctoral programs turn more and more to neuroscience and away from the behavior that neuroscience is attempting to explain. The problem is worse in neuroscience programs themselves, but it also exists in psychology departments. In both cases, but especially the former, students' knowledge of the behavioral facts and principles underlying learning and memory, for example, are often woefully lacking. Studies of memory, for example, may involve elegant experiments providing detailed descriptions of neurobiological function at the level of the synapse, cell receptors—even gene expression—but often only a single, simple, behavioral assay of the psychological process of memory itself. How much of what we learn about the neurobiology and the pharmacology of learning and memory is to be based solely on poorly founded studies of passive avoidance learning? Or of classical conditioning of the rabbit's nictitating membrane? It is as if the massive amounts of knowledge about the measurement and control of behavior that we have acquired in the last 30 or 40 years has been neglected by the newer generation of neuroscientists and biopsychologists.

What are we to do in the face of this problem? I urge that instead of many complicated neurobiological manipulations measured by just one behavioral assay, we strive to use one simple neurobiological manipulation measured by many behavioral assays. An altered brain creates an entirely new animal for comparative study, and who knows how that alteration may manifest itself in behavior? We won't find out unless we look, and look from many angles. Psychology, including comparative cognition, has a unique perspective that we must keep alive. After all, we provide the questions that the neurosciences must answer.

I most emphatically do *not* mean to say that the neurobiology of comparative cognition, or of any psychological process, ought to be neglected in favor of a return to the study of behavior alone. That would be folly. Such regression would ignore the vast quantity of important information that the neurosciences have brought and will bring to the study of behavior. And a return to the past would blunt and misdirect the stimulating new ideas that behavioral studies such as those discussed at the Forum can bring to the neurobiology of behavior.

What I *am* saying is that there ought to be a true balance between the study of behavior and the biology that underlies it. Forty years of effort on the behavioral side should not be ignored. I speak to a drift away from an appropriate balance, and I argue merely for a return to equilibrium. Fortunately, as Ray Kesner and others have pointed out, leading neuroscientists and animal psychologists have recognized the problem, and the future promises circumstances where behavioral neuroscientists and comparative psychologists will work hand in hand.

The Dangerous "New Look" of Comparative Cognition

Finally, I have a serious concern about a drift into anthropomorphism in our field that is reminiscent of the late 19th century. Blumberg and Wasserman (1995) have summarized the issue very well in their recent article in the *American Psychologist*. They stress that there are certain comparative psychologists among us who are adopting a new mentalism based on an *argument from design*.

An argument from design, they remind us, consists essentially of three steps. First, we develop an awe of the complexity and functionality of nature. We cannot, second, understand how complexity and functionality could have arisen without an external Designer, and therefore, third, we turn to familiarity and analogical reasoning to prove the existence of the Designer. One can substitute for the term "Designer" the ideas of Mind and Thought.

Blumberg and Wasserman go on to point out how this way of thinking has crept into the way certain of us have written about cognitive behavior. The following book titles are cases in point. They are my selections, not Blumberg and Wasserman's.

How monkeys see the world: Inside the mind of another species (Cheney & Seyfarth, 1990)
The question of animal awareness: Evolutionary continuity of mental experience (Griffin, 1976)
Cognitive ethology: The minds of other animals (Ristau, 1991)

To my mind the attribution of the term "mind" to nonhuman animals in the titles of these books is both gratuitous and frightening. One can only hope that the authors mean to refer to a "mock" anthropomorphism, to borrow Blumberg and Wasserman's term, where one presumes that animals have mental capacities like humans as a heuristic in order to get ideas for research. I fear, however, that they refer to an "explicit" anthropomorphism where familiarity and analogical reasoning are embraced without reason or reservation.

I began with an appeal to history as a useful source of ideas for the future stimulation of our science. In the case of this emerging enterprise we might call comparative *pseudo*-cognition, I see history turning upon itself to the detriment of our scientific venture. My concern is that we will repeat the mistakes inherent in the anthropomorphism of the late 19th century which led us into a fruitless, anecdotal analysis of supposed animal "minds." And, if we do that, where will our science be then? And, who among us, a hundred years later, will be wise and brave enough to be our C. Lloyd Morgan and curb the illegitimate claims that were so easily made about the nature and power of animal "minds?" Furthermore, I suspect it will take heavy artillery, not a simple canon, to resolve this growing tension in our science. The development bears very close watching, otherwise our hard-won progress in comparative cognition will be swept away in another frothy wave of mentalism. We must prevent history from repeating itself.

Given this caution, I believe the future of comparative cognition can be luminous. In recent years, we have combined the techniques and principles of experimental psychology with the tenets of evolution and behavioral ecology. This melding joins the best of the European ethological tradition with the best of the American behaviorism. If we can avoid the pitfalls of pseudocongnitivism, and I believe we can, we have many exciting theories and discoveries to look forward to.

ACKNOWLEDGMENTS

Preparation of this article was supported by National Science Foundation Research Grant IBN 9317868. I thank Timothy Gentner, Scott MacDougall-Shackleton, and Amy Wisniewski for their thoughtful comments about earlier drafts of the work. Correspondence concerning the article should be addressed to Stewart H. Hulse, Department of Psychology, Johns Hopkins University, Baltimore, MD 21218. Electronic mail may be sent via Internet to hulse@jhu.edu.

REFERENCES

Blumberg M. S., & Wasserman, E. A. (1995). Animal mind and the argument from design. *American Psychologist*, **50**, 133-144.

Cheney, D. L., & Seyfarth, R. M. (1990). *How monkeys see the world: Inside the mind of another species*. Chicago: University of Chicago Press.

Griffin, D. R. (1976). *The question of animal awareness: Evolutionary continuity of mental experience*. New York: Rockefeller University Press.

Hulse, S. H., & Dorsky, N. P. (1977). Structural complexity as a determinant of serial pattern learning. *Learning and Motivation*, **8**, 488-506.

Hulse, S. H., Fowler, H., & Honig, W. K. (1978). *Cognitive processes in animal behavior*. Hillsdale, NJ: Erlbaum.

Hull, C. L. (1931). Goal attraction and directing ideas conceived as habit phenomena. *Psychological Review*, **38**, 4887-506.

Hunter, W. S. (1920). The temporal maze and kinaesthetic sensory processes in the white rat. *Psychobiology*, **2**, 1-17.

Fountain, S. B., & Hulse, S. H. (1981). Extrapolation of serial stimulus patterns by rats. *Animal Learning & Behavior*, **9**, 381-384.

Hulse, S. H., & O'Leary, D. K. (1982). Serial pattern learning: Teaching an alphabet to rats. *Journal of Experimental Psychology: Animal Behavior Processes*, **8**, 260-273.

Fountain, S. B., Henne, D., & Hulse, S. H. (1984). Phrasing cues and hierarchical organization in serial pattern learning by rats. *Journal of Experimental Psychology: Animal Behavior Processes*, **10**, 30-45.

Fountain, S. B., & Rowan, J. D. (1992). MK-801 impairs serial learning ut not retention for serial patterns in rats. *Society for Neuroscience Abstracts*, **18**, 1060.

Fountain, S. B., & Rowan, J. D. (1995). Sensitivity to violations of "run" and "trill" structures in rat serial-pattern learning. *Journal of Experimental Psychology: Animal Behavior Processes*, **21**, 78-81.

Lashley, K. S. (1951) The problem of serial order in behavior. In L. A. Jeffress (Ed.), *Cerebral mechanisms in behavior* (pp. 112-146). New York: Wiley.

Miller, G. A., Galanter, E., & Pribram, K. (1960). *Plans and the structure of behavior*. New York: Holt, Rinehart, & Winston.

Restle, F. (1970). Theory of serial pattern learning: Structural trees. *Psychological Review*, **77**, 481-495.

Ristau, C. A., (Ed.). (1991). *Cognitive ethology: The minds of other animals*. Hillsdale, NJ: Erlbaum.

Shettleworth, S. J. (1993). Varieties of learning and memory in animals. *Journal of Experimental Psychology: Animal Behavior Processes*, **19**, 5-14.

Simon, H. A., & Kotovsky, K. (1963). Human acquisition of concepts for sequential patterns. *Psychological Review*, **70**, 534-546.

Skinner, B. F. (1938). *The behavior of organisms*. New York: Appleton-Century-Crofts.

Part 2: Sequential Learning and Memory

2

THE COMPARATIVE PSYCHOLOGY OF CHUNKING

H. S. Terrace
Columbia University

The ability to learn arbitrary sequences is crucial for intelligent action, both verbal and non-verbal. For more than a century, psychologists have investigated the organization of such sequences in experiments on the memorization of nonsense syllables (Ebbinghaus, 1964) and the mastery of various types of mazes (Small, 1900). The results of both types of experiment gave rise the classic theory that serially organized behavior can be represented as a linear sequence of associations.

Ebbinghaus explained list learning by reference to associations between successive items and between a particular item and its list position. Hull offered a similar explanation of maze learning by rats (Hull, 1952). Thus, associative principles that were used to explain how a human adult memorizes a list of arbitrary items were used to explain how an experimentally naive rat learns a sequence of arbitrary responses, and vice versa (Osgood, 1953; Underwood, 1957).

After dominating psychological thinking for more than half a century, the validity of association theories of serially organized behavior was questioned on a variety of theoretical and empirical grounds (Lashley, 1951; Chomsky, 1957). In his classic analysis of serially organized behavior, Lashley rejected linear models because they could not explain knowledge of relationships between non-adjacent items (for example, between words before and after an embedded clause) and because inter-response times between successive responses are often shorter than the time that would be needed for feedback from one response to trigger the next (for example, playing a sequence of notes on a musical instrument). These and related arguments have been described in detail by others and will not be elaborated in this chapter (e.g., Anderson & Bower, 1974; Gardner, 1985). Instead our focus will be the concept of chunking (Miller, 1956), one of the most influential but, as we shall see, one of the most poorly understood concepts of modern cognitive psychology.

The significance of chunking derives from its ability to overcome objections to linear models of serially organized behavior association by augmenting linear structures with hierarchical structures. Although the concept of chunking was proposed to define the capacity of short-term memory, it has also been used to characterize such diverse phenomena as long-term memory, visual perception, and motor plans. The main purpose of this chapter is twofold: to examine some problems that arise when the concept of chunking is used uncritically and to distinguish between two basically different types of chunks: *input* and *output* chunks.

George Miller introduced the concept of chunking in his classic paper, "On the magical number 7 ± 2" (1956). Miller argued that a chunk was the basic unit for measuring the capacity of immediate memory [in current terminology, *short-term* or *working* memory; see Baddeley (1992) for a discussion of the taxonomy of different memory systems]. The idea was that subjects could retain a large number of discrete items of information if they were encoded as chunks before they were transferred to long-term memory. For example, the 12 digits 1-4-9-2-1-7-7-6-1-8-1-2 could be encoded as 3 historical dates. In contrast to the enormous capacity of *long-term memory* (LTM), Miller estimated the capacity of STM to be 7 ± 2 chunks and argued that the amount of information that is retained in STM is independent of the amount of information contained by each chunk.

The facilitory effect of chunking on human memory has been confirmed by a broad variety of experiments. Familiar examples include the enhancement of recall on lists on which subjects can assign items to verbally defined categories (Bousfield & Bousfield, 1966; Bower, 1972a) and on lists composed of temporally defined clusters of items (Bower & Winzenz, 1969). On a conceptual level, chunking is regarded as a basic cognitive process despite differences between theorists as to the functional and anatomical boundaries of STM and LTM (Atkinson & Shiffrin, 1971; Baddeley, 1981; Craik & Tulving, 1975; Mishkin & Petri, 1984; Squire, 1986; Weiskrantz, 1970), the types of evidence that have been used to distinguish STM and LTM (Bower, 1972b; Estes, 1972; Johnson, 1972; Murdock, 1993) and the actual capacity of STM (Broadbent, 1975; Crowder, 1976; Mandler & Dean, 1969; Wickelgren, 1964, 1967).

Chunking has also been used to explain the performance of animals on a variety of sequential tasks, e.g., detecting the temporal pattern of the amount of reinforcement that can be earned from trial to trial (Capaldi et al., 1986), learning a rule for responding to a row of response levers in a particular sequence (Fountain & Annau, 1984), learning to execute a simultaneous chain on the basis of qualitative similarities between list items (Terrace, 1987) and organizing response sequences in a radial maze on the basis of the different types of food reward used to bait each arm (Dallal & Meck, 1990). Dallal and Meck's study is of particular interest because it demonstrates chunking in the serially organized behavior of an animal that is free to respond in any order it chooses [cf. free recall studies performed on human subjects (Bousfield, 1953; Bousfield et al., 1964)].

Curiously absent from the literature on chunking, both human and animal, is an operational definition. As Miller noted perceptively at the end of his article: "...we are not very definite about what constitutes a chunk of information" (Miller, 1956). Forty years and dozens of published articles later, psychologists appear to be just as indefinite (Cowan, 2001). The few definitions that have been proposed are ad hoc and apply only to items that can be encoded verbally. Consider an instructive example (from a recent commentary on Miller's paper) in which a chunk is defined as "...a pronounceable label that may be cycled within short-term memory" (Shiffrin & Nosofsky, 1994). The capacity of STM is then defined as "... the number of labels pronounceable in 2 s".

Without an operational definition of chunking, it is difficult to decide whether the serial tasks on which animals have been trained are homologous to those used in experiments on human chunking. For example, the literature on human chunking is based on the performance of verbal subjects on tasks that require the retention of sequential information in STM. The literature on animal chunking is based entirely on the performance of non-verbal subjects who were trained on tasks that require the retention of sequential information in LTM. This problem is compounded by differences in the pre-experimental histories of human and animal subjects. Human subjects are typically college students who have had extensive experience at learning lists. Animal subjects typically have no prior training on serial tasks.

Verbal subjects have obvious advantages over non-verbal subjects. Verbal subjects can readily grasp spoken and written instructions and they can describe the experience of learning a list. Less obvious are certain disadvantages of using human subjects to investigate basic processes of memory. Foremost, is the implication that chunking presupposes linguistic knowledge and that subjects learn verbal associations while mastering a list, whether between adjacent and non-adjacent items or between an item and its ordinal position (e.g., "first", "fourth", "middle", "second-before-middle", etc). Reliance on verbal subjects also obscures the influence of previously acquired expertise at chunking. That expertise is the product of years of experience with spoken and written language and the memorization of innumerable rote lists (e.g., the alphabet, sequences of numbers, the planets, etc.).

The experiments I will describe in this chapter are concerned with long-term memory of sequences in animals and humans. In some instances, the experimenter provides a basis for organizing subsets of the required sequence during training. In others, the subjects organized the sequences they were asked to learn into smaller temporally defined subsets, without any requirement to do so. The spontaneous temporal organization of sequences will be used as a criterion for defining *output* chunks, as contrasted with *input* chunks, the type of chunk originally proposed by Miller (1956).

SIMULTANEOUS CHAINING PARADIGM

In most of the experiments I will review, sequences were trained as *simultaneous chains*, sequences that differ in many respects from those used in previous studies of serial learning in animals. The simultaneous chaining paradigm was first used in an experiment whose purpose was to question claims about the grammatical competence of apes, in particular, the claim that sequences trained by rote were sentences (Straub et al., 1979). To simulate the conditions used in ape language experiments, Straub et al. trained pigeons to respond to randomly configured arrays of four colors in a particular sequence: *red→green→yellow→blue*. All of the colors were presented simultaneously, in a different configuration on each trial. Random variation of the position of the colors from trial to trial insured that subjects could not learn the sequence as a specific motor program (i.e., as a successive chain). Correct responses to each item allowed a trial to continue. A correct response to the last item of the sequence produced food reward. An error terminated the trial (e.g., forward errors, such as B, A→B→D, C, A→D and backward errors, such as, A→B→A, A→B→C→A, etc.). Repetitive responses to the same item had no consequences (e.g., A→A→A→B→B→C→C→ C→C→D).

If training began with trials on which all of the items appeared simultaneously, subjects might stop responding because of the low probability of guessing the correct sequence by chance. On a simultaneous chain, subjects have to determine the identity of each item of a simultaneous chain by trial and error. The probability of guessing A on the first trial of a 4-item list is 1/4. Because repetitive responses to the same items are not considered errors, the probability of guessing B following a correct response to A is 1/3. Given the generous assumption that subjects are able to recall any of the items to which they've responded previously, the probability of guessing the entire sequence is p = 1/4 x 1/3 x 1/2 x 1 = .04. To avoid the risk of extinction at the start of training, each list was introduced by the *successive phase method.* A new item was added to the end of a partial list each time the subject satisfied an accuracy criterion. On a 4-item list, the successive phases of training were A, A→B, A→B→C, and A→B→C→D.

All of the subjects of the Straub, et al. experiment learned the 4-item list of colors on which they were trained. That result suggested a simpler interpretation of experiments purporting to demonstrate the grammatical ability of apes. Why interpret the sequences of plastic chips or lexigrams that apes were trained to produce as anything more complicated than rotely learned sequences whose function was to obtain some specific reward [e.g., *Mary→give→Sarah→apple* (Premack, 1976) or *Please→machine→give→ apple* (Rumbaugh, 1977)? As Terrace (1979) has noted, there is no evidence the first 3 symbols of these sequences had any meaning for the apes in question.

The sequences on which pigeons (and apes) were trained are also of interest because they cannot be explained as *successive chains,* the kind of sequence that has been used in previous experiments on serial learning (Terrace, 1984). As illustrated in the following comparison, successive and simultaneous chains are fundamentally different:

Successive Chain: S_A: R_A----> $S_B R_B$----> S_C: R_C----> S_D: R_D----> S^R
Simultaneous Chain: $S_A S_B S_C S_D$: R_A----> R_B----> R_C----> R_D---->S^R

Unlike the successive chaining paradigm, a simultaneous chaining paradigm presents all list items throughout each trial (e.g., the numbers on the face of a telephone). In a successive chain, the subject encounters each cue individually (e.g., the choice points in a maze). A second difference is the variation of the physical configuration of list items from trial to trial. This prevents subjects from using a particular physical sequence of responses to produce the required list (for example, when making a telephone call with a sequence of rotely learned movements on a number pad). To execute a simultaneous chain correctly, the subject has to respond to each item in a particular order, regardless of its spatial position.

A third distinguishing feature of the simultaneous chaining paradigm is the absence of differential feedback during the execution of a correct sequence. Following a correct response to $item_n$, no information is provided as to the identity of $item_{n+1}$. Consider, for example, the consequences of responding to item B on the 4-item list, A→B→C→D. After responding to A, subjects are given no information that the next response should be directed to C (as opposed to A or D).

Yet another important difference between successive and simultaneous chains is the order in which individual responses are trained. On simultaneous chains, the *first* response is trained first. On successive chains, the *last* response must be trained first. The backward training of a successive chain follows from theoretical analyses of the backward effect of reinforcement for each response of a successive chain (Hull, 1932) and the *Law of Chaining* (Skinner, 1938). The idea was that response sequences can be broken down into a series of linked responses: "The response of one reflex may constitute or produce the eliciting or discriminative stimulus of another" (Skinner, 1938, p. 32). Skinner illustrated the *Law of Chaining* with the method he used to train a rat to press a bar to obtain food. First the rat is trained to approach the food tray. That behavior consists of the 2-response sequence: S^D (tray): R_1 (approach) $\rightarrow$ S^R (food): R_2 (approach seize food). Subsequently, the rat is shaped to perform a longer sequence: S^D (visual lever): R_1 (approach (lifting body) $\rightarrow$ S^D (tactual lever): R_2 (pressing) $\rightarrow$ S^D (sound of magazine: R_3 (approach tray) $\rightarrow$ S^R (food): R_4 (seize food).

Simultaneous chains cannot be trained backwards because subjects have a strong bias for making forward errors. Consider, for example, the backward phases of training that would be needed to train a 4-item sequence: D, C→D, B→C→D, A→B→C→D. During the initial phase of training, each response to D is followed by reinforcement. When C & D are presented simultaneously, the subject persists in responding to D (a forward error). Since each response to D ends the trial without reinforcement, responding to D is eventually extinguished.

Numerous attempts to overcome the bias for responding to D have been unsuccessful. Training the sequence in a forward manner (A, A→B, A→B→C, A→B→C→D) avoids this problem. After learning to respond to A, the subject is trained to respond to displays of A and B in the sequence A→B. Since repetitive responses were not treated as errors, the only consequence of additional responses to A is to prolong the trial. Eventually, the subject responds to B and the trial ends with reinforcement. The same process is repeated when C & D are introduced.

LIST LEARNING BY PIGEONS

Straub & Terrace (1981) extended Straub et al.'s findings by evaluating subjects' knowledge of associations between *non-adjacent* items. First, Straub & Terrace (1981) showed that the number of sessions needed to satisfy the accuracy criterion increased progressively each time the list was

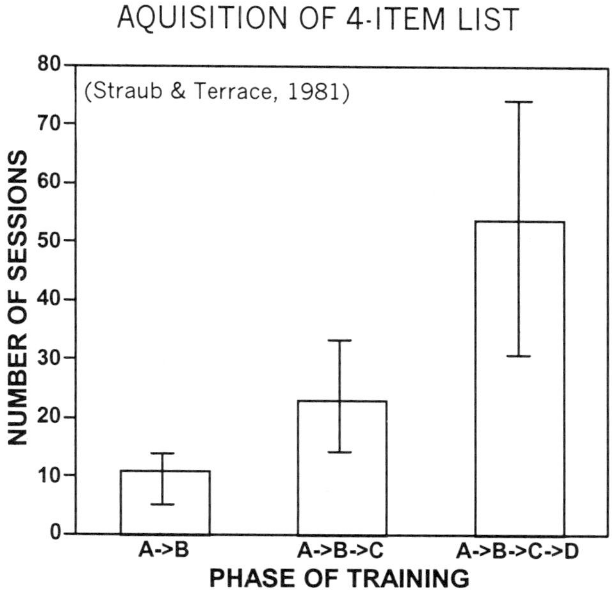

Figure 1. Average number of sessions needed by pigeons to satisfy accuracy criterion during successive phases of training on a 4-item list. See Straub and Terrace (1981) for further details.

lengthened. The relevant data are shown in Figure 1. Straub & Terrace then administered a subset test that consisted of the six 2-item pairs that could be derived from a 4-item list (AB, AC, AD, BC, BD & CD). For 5 of the 6 subsets, accuracy of responding to the subsets exceeded the level predicted by chance and was uniformly high across all subsets. The one exception was subset BC. The relevant data are shown in Figure 2.

Accurate performance to the subsets BC and BD is puzzling for different reasons. Chance performance on the subset BC is puzzling because subjects completed the required sequence (A→B→C→D) correctly on 75% of the trials during a criterial session that was administered immediately prior to the subset test. Indeed, the conditional probability of responding to C (given a response to B), was greater than 0.9. The contrast between the chance level of the subset test. Indeed, the conditional probability of responding to C (given a response to B), was greater than 0.9. responding to BC during the subset test and the high transitional probability of responding correctly to BC during the 4-item phase of training, suggests that the response to C was conditional upon the sequence A→B. The high level of accuracy to subset BD is equally puzzling as it provided neither the opportunity to start the sequence at A nor the opportunity to respond to C before D.

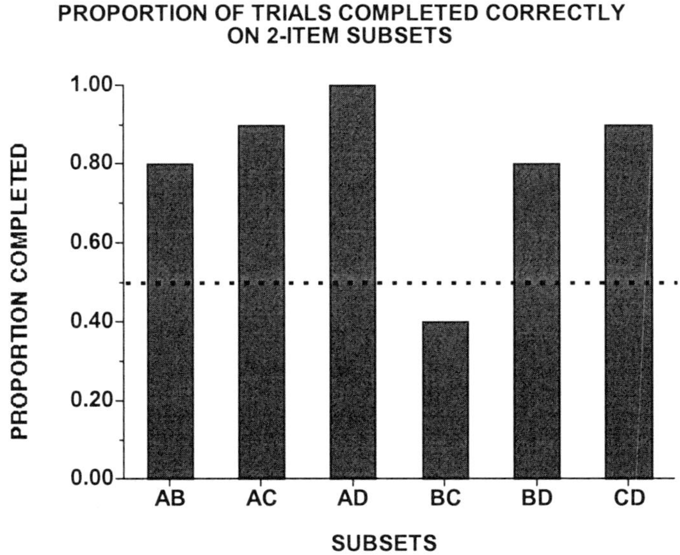

Figure 2. Proportion of trials completed correctly on each 2-item subset that can be derived from the list A→B→C→D. See Straub and Terrace (1981) for further details.

The latencies of the responses to the first and second item of each subset provide important clues as to the manner in which subjects represent the sequence. The latency of the first response to subsets beginning with A was approximately half a second faster than the latency of the first response to subsets beginning with either B or C. The relevant data are shown in Figure 3. For all subsets, the latency to the second item was shorter than the latency to the first item. These data suggest that pigeons adopted the following rules when responding on the subset test:

i. respond first to item A.
ii. respond last to item D.
iii. respond to any other item by default.

Rules *i-iii* predict the latencies shown in Figure 3. The short latencies to the first item of subsets that began with A (AB, AC, and AD) follow from the subjects' extensive histories of responding to A. It should take less time to apply rule *i* to subsets beginning with A than to apply rule *ii* to subsets that don't begin with A. Application of the default rule (*iii*) predicts shorter latencies to the second item of each subset than to the first item. Rules *i-iii* predict accurate performance on all subsets that contain an end item (AB, AC, AD, BD and CD). By contrast, they provide no basis for selecting B or C.

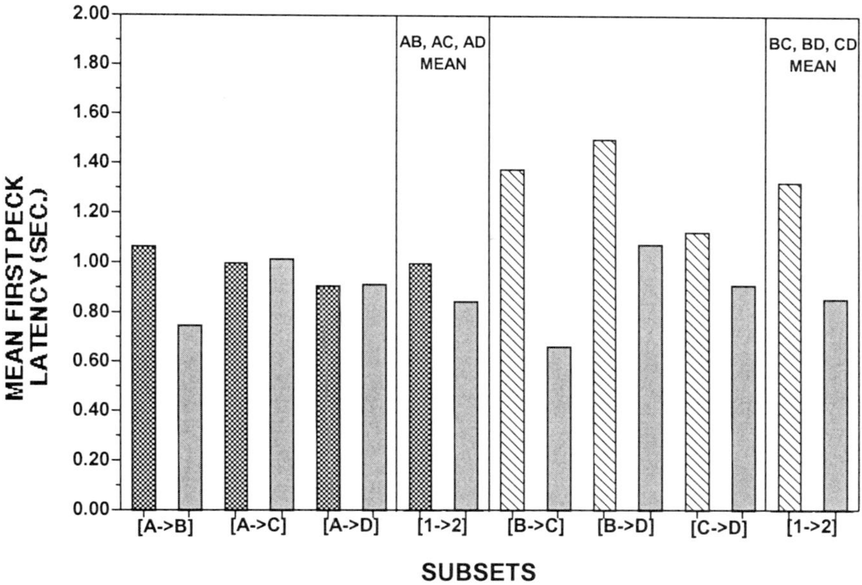

Figure 3. Average latency of responding to each item during 2-item subset tests. See Straub and Terrace (1981) for further details.

CHUNKING

The extensive training that pigeons need to master a 4-item list (more than 3 months) suggests that 4 items may approach the limit of their memory span. For human subjects the classic remedy for overcoming limitations of memory span is to reorganize unrelated list items into chunks (Miller, 1956). The efficacy of that approach was evaluated with pigeons that were trained to learn 5-item lists composed of colors and achromatic geometric forms (Terrace, 1991; Terrace, 1987). To differentiate the types of items used on each list, colors are represented by unprimed letters; achromatic geometric forms, by primed letters. Two lists provided a basis for chunking similar items: A→B→C→D'→E' and A→B→C→D→E'. Control groups learned lists in which colors and forms were interspersed: A→B'→C→D'→E and A→B→C'→D→E, or which consisted of 5 colors: A→B→C→D→E.

Lists that provided a basis for chunking were learned twice as rapidly as those that could not be chunked. The relevant data are shown in Figure 4. These results are consistent with the hypothesis that pigeons could organize the segregated lists into two chunks: [A→B→C] & [D'→E'] and [A→B→C→D] & [E'].

The number of sessions needed to master successive phases of training provides another basis for differentiating the chunking and the

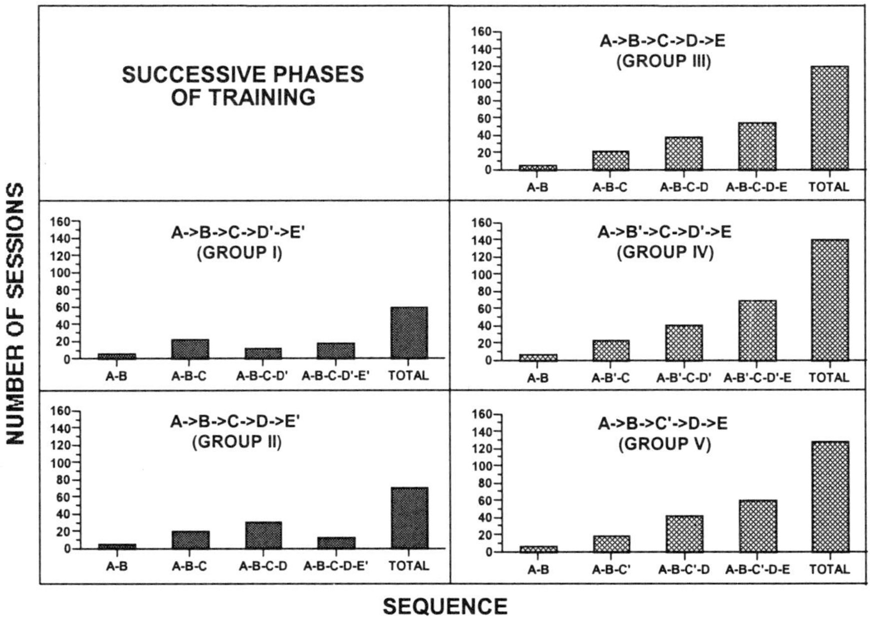

Figure 4. Average number of sessions needed to satisfy accuracy criterion during each phase of training and during all sessions for chunking (left-hand panels) and non-chunking groups (right-hand panels). See Terrace (1987) for additional details.

control groups. The relevant data is shown in Figure 4. The control groups needed progressively more time to master a particular phase each time a new item was added. That pattern of acquisition is consistent with the results of the Straub and Terrace (1981) study in which pigeons were trained to execute 4-item lists (cf. Figure 1). A different pattern was observed in the case of the two chunking groups. The phase of training in which an achromatic item was added to the colored items was completed more rapidly than the previous phase (in which the subject was required to produce a list containing one *fewer* item). The advantage of the two chunking groups cannot be attributed to release from proactive inhibition (Keppel & Underwood, 1962). List A→B→C'→D→E, which provides an opportunity for the release of proactive inhibition at the start of training on the A→B→C'→D phase, took as much time to master as lists A→B'→C→D'→E and A→B→C→D→E.

Additional evidence that pigeons chunk similar items on clustered lists was provided by analyses of the time it took each group to execute the 5-item list and by their performance on subset tests (Terrace, 1991; H. Terrace & S. Chen, 1991). As shown in Figure 5, the two "chunking" groups executed their lists more rapidly than any of the control groups (on

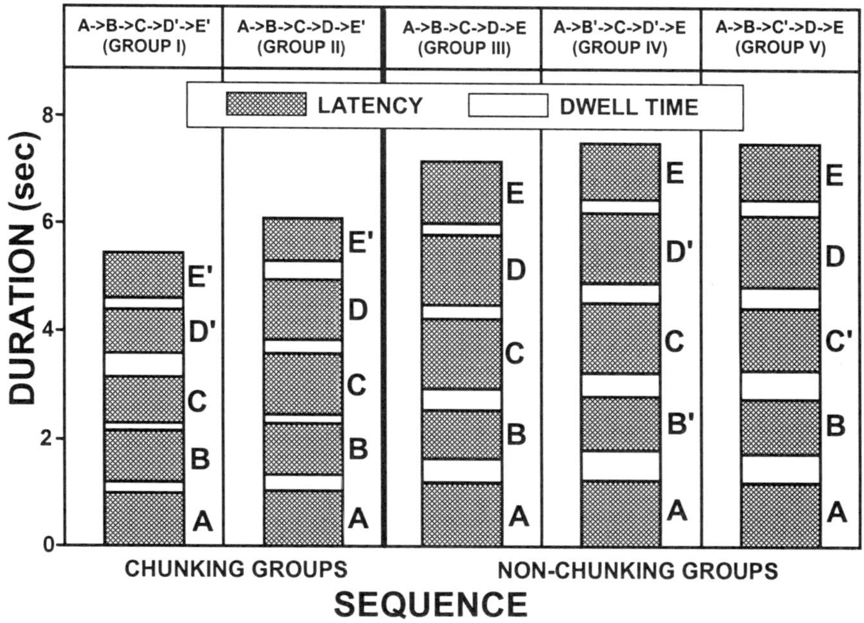

Figure 5. Average duration of each correctly completed sequence for each of the chunking and non-chunking groups. See Terrace and Chen (1991) for additional details.

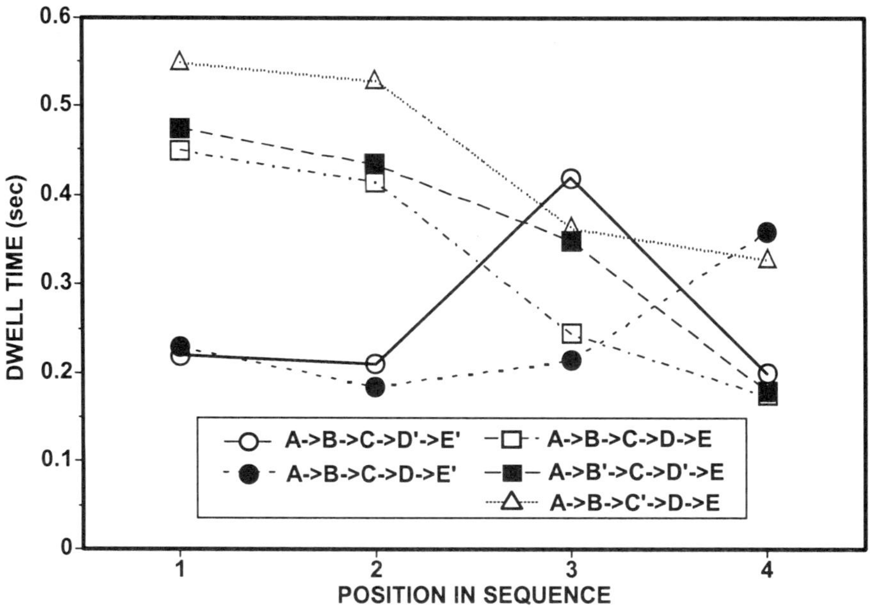

Figure 6. Average dwell time for each of the chunking and non-chunking groups. See Terrace and Chen (1991) for additional details.

average, 5.8 vs. 7.2 sec.). Curiously, these data appear to be the only data in the animal *and* human literatures on serial learning which show that chunked sequences are executed more rapidly than unchunked sequences (Terrace & Chen, 1991a).

Figure 5 shows two components of the time needed to execute a simultaneous chain: latency and dwell times. Latency is the time that precedes the initial response to each item. Dwell time is the interval between the first and the last response to each item. If pigeons responded only once to each item, dwell time would be zero. Since pigeons tend to make multiple responses to each item, dwell time is typically longer than the latency of the first response to that item. The data presented in Figure 5 show that dwell time varied considerably within lists and between groups.

Figure 6 shows that dwell time increased at the item that preceded purported chunk boundaries. For the non-chunking groups, dwell time decreased gradually as the pigeon worked its way through the sequence and showed no abrupt increases. Taken together, the temporal data shown in Figures 5 and 6 supports the hypothesis that the pigeons chunk list items that are segregated into qualitatively different segments. The latencies of the first response to successive items were shorter for the two chunking groups than they were for the two non-chunking groups. That factor resulted in faster times for executing the entire list on the part of the two chunking groups. Short latencies may reflect a relatively rapid search time for locating qualitatively similar items, in this instance (chromatic stimuli). The dramatically longer dwell times at the last item of purported chunks suggest that the pigeons used that time to locate the remaining (achromatic) items.

After subjects satisfied the accuracy criterion on a 5-item list, they were given a two-item subset test. Rules similar to those derived for 4-item lists (*i-iii*) provided a basis for predicting performance on each of the 10 types of subset that could be derived from a 5-item list. Subjects from all groups would be expected to respond accurately to the 7 subsets that contain either a start or an end item (A, E and E'). However, different predictions follow for the chunking and the control groups in the case of the 3 subsets that were composed of interior items. If the list A→B→C→D'→E' was parsed as two chunks, [A→B→C] & [D'→E'], and if those chunks were functionally equivalent to 3- and 2-item lists, subjects should respond at a greater than chance level of accuracy to all of the 3 "internal" subsets that can be generated from the original list (BC, BD' & CD'). Similarly, if the list A→B→C→D→E' was executed as the chunks [A→B→C→D] & [E'], subjects should respond at a greater than chance level of accuracy to the subsets BD & CD', but not to the subset BC. The 3 control groups would be expected to respond to subsets composed of interior items at chance levels of accuracy (subsets BC, CD & BD on list A→B→C→D→E, subsets B'C,

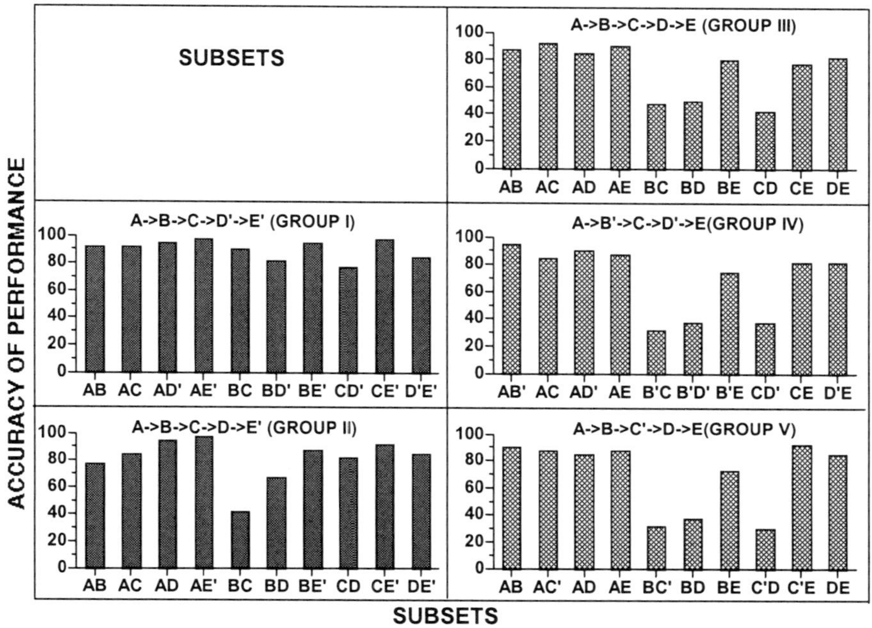

Figure 7. Proportion of trials completed correctly to each of ten 2-item subsets that can be derived from 5-item lists. Chunking groups are on the left; non-chunking groups on the right. See Terrace and Chen (1991) for additional details.

CD' & B'D' on list A→B'→C→D'→E, and subsets BC', C'D & BD on list A→B→C'→D→E).

These predictions were confirmed for each of the 50 subsets that were tested after subjects mastered the accuracy criterion (5 lists x 10 subsets for each list.) All groups responded at high levels of accuracy to subsets that contained the first or the last items (A, E or E'). The chunking groups responded at similar levels of accuracy to subsets that contained purported chunk boundaries. By contrast, accuracy on internal subsets that lacked a chunk boundary did not exceed the level predicted by chance. The relevant data are shown in Figure 7.

WHAT IS A CHUNK?

The facilitory effects of clustering similar items on a 5-item list appear to be prima facia evidence of chunking by pigeons. However, that evidence does not stand up to scrutiny when evaluated as a means of enhancing STM, the defining characteristic of a chunk proposed by Miller. The basic function of a chunk is to enhance STM. Yet studies of animal chunking have relied exclusively on tasks that require long-term rather that short-term memory. This is true not only of the experiments on simultaneous chaining described in the previous section, but also of experiments in which animals learned rules concerning the spatial organization of different reinforcers (Dallal & Meck, 1990), the monotonicity of changes in the relative magnitude of reinforcers (Capaldi et al., 1990; Hulse, 1978), and the temporal and spatial patterns of reinforcers (Fountain et al., 1984). In each instance, the same items were repeated in the same sequence on each trial. That would rule out the limited capacity of short-term memory as an explanation of the facilitory effects of grouping particular sets of stimuli during training. Instead, these effects appear to result from organizational processes that occur during the retrieval of familiar information from LTM.

To distinguish between the organizational principles used to encode new information in STM and to retrieve familiar information from LTM, I will refer to the former as *input* chunking, and to the latter as *output* chunking. Postulating a second type of chunking does, of course, raise the same definitional questions that apply to the general concept of chunking. In the case of output chunks, however, some recent experiments on the execution of simultaneous chains by monkeys and college students suggest that the temporal organization of a sequence can be used to define output chunks. These experiments and their background are reviewed in the next section.

LIST LEARNING BY MONKEYS

D'Amato & Colombo (1988) used the simultaneous chaining paradigm to train monkeys to produce arbitrary 5-item lists. Of minor interest was their finding that monkeys acquired 5-item lists more rapidly than pigeons. Of greater significance, were the results of a 2-item subset test. Unlike pigeons, monkeys responded accurately to all 10 of the subsets that can be derived from a 5-item list of heterogeneous items. Of particular significance is their ability to respond accurately to subsets composed exclusively of items from the middle of a list (BC, CD and BD). As shown earlier (in Figure 7), pigeons responded at chance levels of accuracy to subsets drawn from lists on which items weren't clustered. The accuracy of each species on 2-item subsets is shown in Figure 8.

Monkeys and pigeons also differed with respect to the latencies of their responses to the first and second items of each subset. The top portion of Figure 9 shows the latency of responding to the first item of a two-item

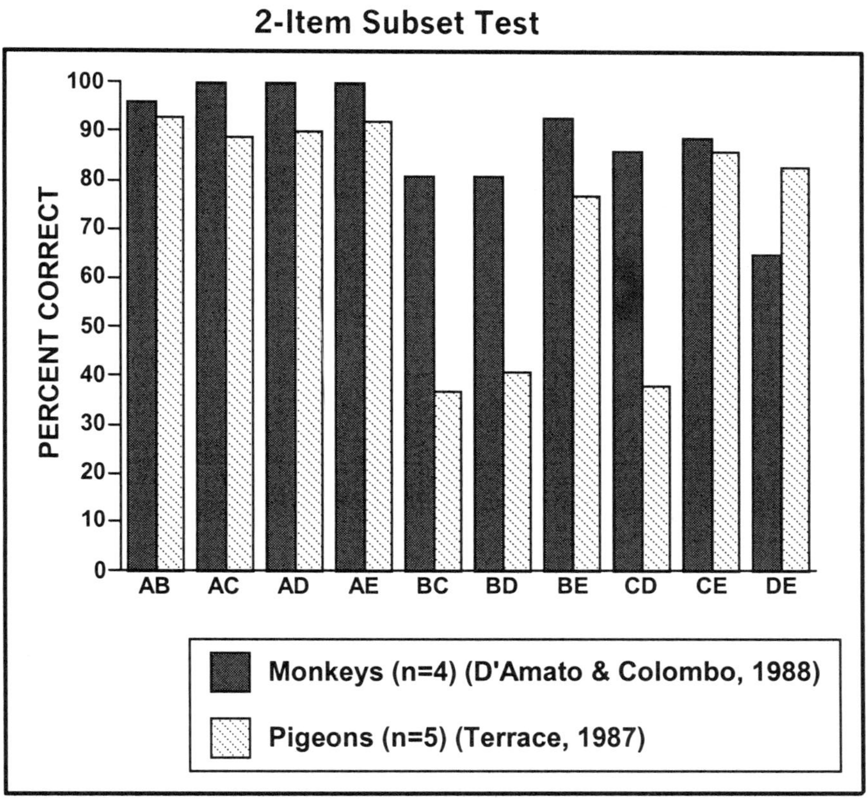

Figure 8. Average percentage of correct responses to each of the ten 2-item subsets that can be derived from 5-item lists by monkeys (D'Amato and Colombo, 1988) and pigeons (Terrace, 1987).

subset. For monkeys, the latency of responding to the first item increased monotonically with the position of that item on the original list. For pigeons, the position of the first item on the original list had no effect on latency. As can be seen in the bottom portion of Figure 9, the latency of the monkeys' responses to the second item also increased monotonically as a function of the number of items on the original list that intervened between subset items. For pigeons, the size of that interval had no effect. These data show that, unlike pigeons, monkeys form a linear representation of a list. Functions similar to those shown in Figure 9 have also been obtained from rhesus monkeys, who were trained to produce 4- and 6-item lists (Ohshiba, 1997; Swartz et al., 1991b), and from 4-year-old children, who were trained to produce a 5-item list (McGonigle & Chalmers, 1996)

Another important difference between the serial skills of monkeys and pigeons was the ease of acquiring new lists. Pigeons showed no signs of improvement on successive 3- or 4-item lists, each composed of novel

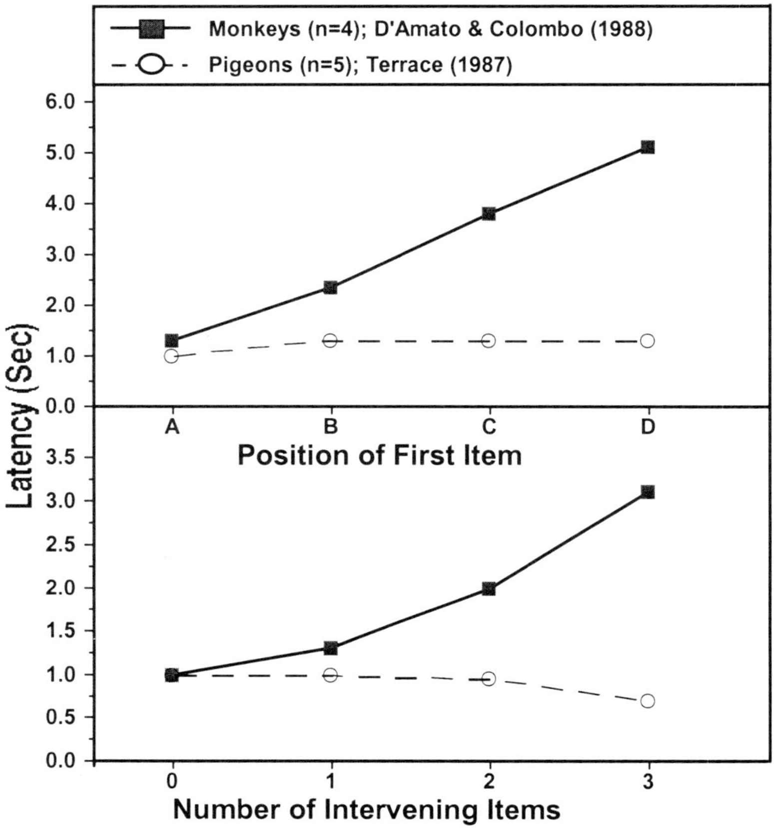

Figure 9. Latency of response to first and second items of subsets. The top panel shows the mean latency of responding (in seconds) to the first item of two-item test pairs as a function of that item's position on the original list (A→B→C→D→E). The bottom panel shows the mean latency of responding to the second item of two-item subset pairs as a function of the number of items that intervened between those items on the original list. Filled squares show data from monkeys (D'Amato and Colombo, 1988); open triangles, data from pigeons (Terrace, 1987).

items (digitized color photographs of natural scenes). Monkeys trained to learn successive 4- and 6-item lists of different photographs became progressively more efficient at mastering each list (Chen et al., 1991; Swartz et al., 1991a). Indeed, after mastering approximately a dozen 4-item lists by the successive phase method, monkeys were able to learn new 4-item lists on which *all* items were displayed from the start of training (Chen et al., 2000).

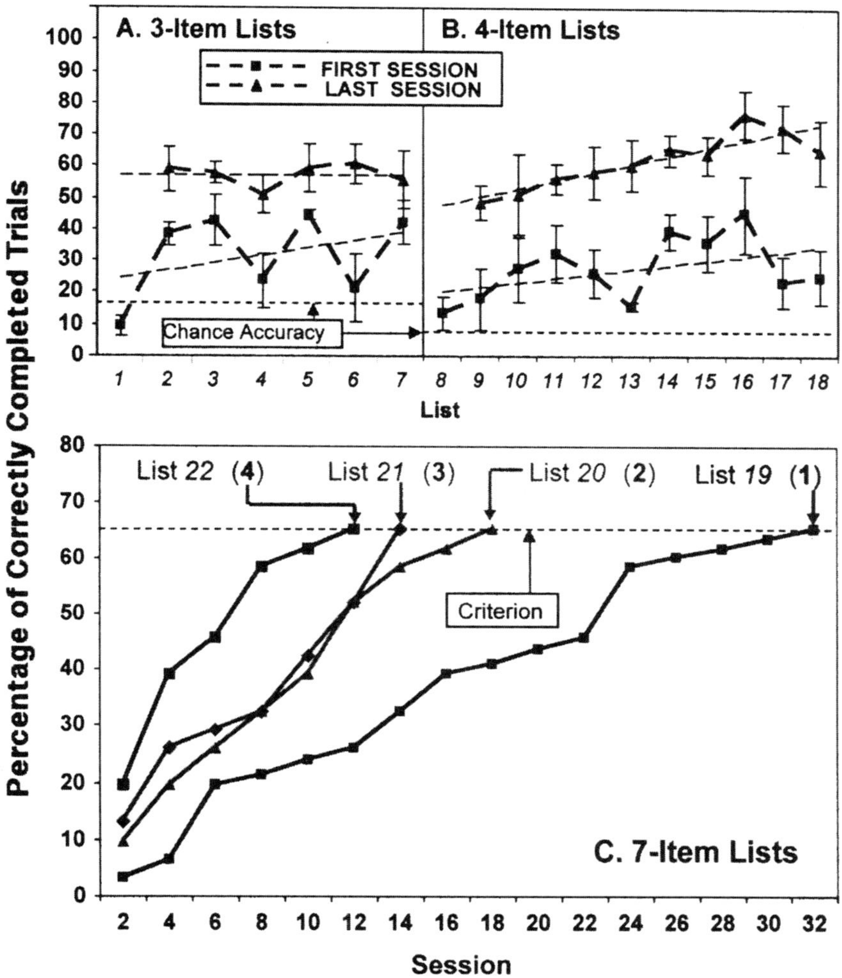

Figure 10. Learning curves for 3-, 4- and 7-item lists trained in experiment by Terrace (2001). Panels A & B. Each panel shows the percentage of correctly completed trials during the first and the last (lower and upper functions, respectively) sessions of training on 3-item lists (A) and 4-item lists (B). The dashed lines show the best-fit linear models. Panel C. Each function shows the mean accuracy of responding on each 7-item list during even-numbered sessions. Note that the abscissa in Panel C is Session (not List).

A recent experiment by Terrace (2001) showed that monkeys could learn 7-item lists on which all items were introduced at the start of training. Experimentally naïve monkeys were first trained on 3- and 4-item lists on which all items were presented from the start of training. The monkeys were then trained in the same manner on four 7-item lists. As shown in Figure 10, the monkeys not only mastered each list but they did so with progressively fewer trials on each new list. To place this achievement in perspective, the reader should note that the probability of guessing correctly the ordinal position of each item at the start of training on a 7-item list is 1/7! = .0002 (assuming no backward errors). Thus, monkeys are not only capable of learning arbitrary lists as long as phone numbers, but they are also became progressively more adept at devising trial and error strategies for determining the ordinal positions of each item during the course of mastering successive lists.

KNOWLEDGE OF ORDINAL POSITION

The availability of list-sophisticated monkeys provided an opportunity to evaluate their knowledge of the ordinal position of list items with a "derived list" paradigm used previously with human subjects (Ebbinghaus, 1964; Ebenholtz, 1963). In Ebenholtz's experiment, two groups of college students learned two 10-item lists of nonsense syllables. All of the items of List 1 were novel. Half of the items of List 2 were drawn from List 1. The remaining items were new. Items derived from List 1 occupied every other position on List 2. For Group I, the original ordinal positions of the derived items were maintained on List 2. For Group II, they were changed. This arrangement insured that the subjects of each group had to learn the same number of new item-item associations while mastering their derived lists.

If a subject's knowledge of the original list was limited to item-item associations, both derived lists should be equally difficult. This was not the case. Group I mastered its derived list more rapidly than Group II. Indeed, Group II required as many trials to learn its derived list as a control group needed to learn a single list. The positive transfer shown by Group I provides compelling evidence that subjects acquired knowledge of the ordinal position of list items while learning List 1.

Ebenholtz's test of ordinal knowledge was adopted for two monkeys (Franklin and Rutherford) who learned to produce 4-item lists on which all items were present from the start of training. Four derived lists, each containing 4 items, were composed of items drawn from four previously learned 4-item lists (Chen et al., 1997). The composition of the original and the derived lists is shown in Figure 11. Each item's original ordinal position was maintained on two of the derived lists. The original ordinal position of the items was changed on the other two derived lists. All items on the derived lists were equally familiar since each of the original lists was trained to the same accuracy criterion. Also, because each list

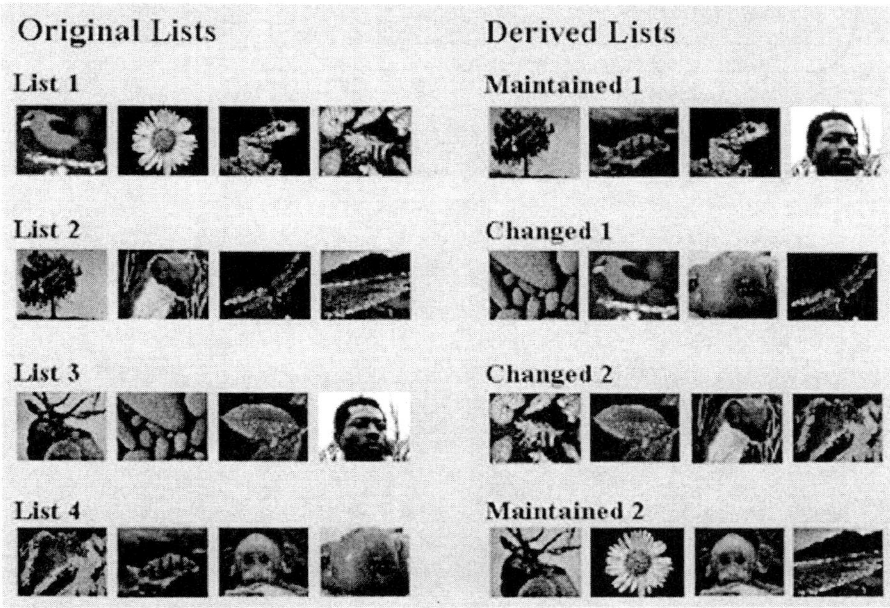

Figure 11. Composition of original and derived lists used in experiment by Chen, et al. (1997). See text for additional details.

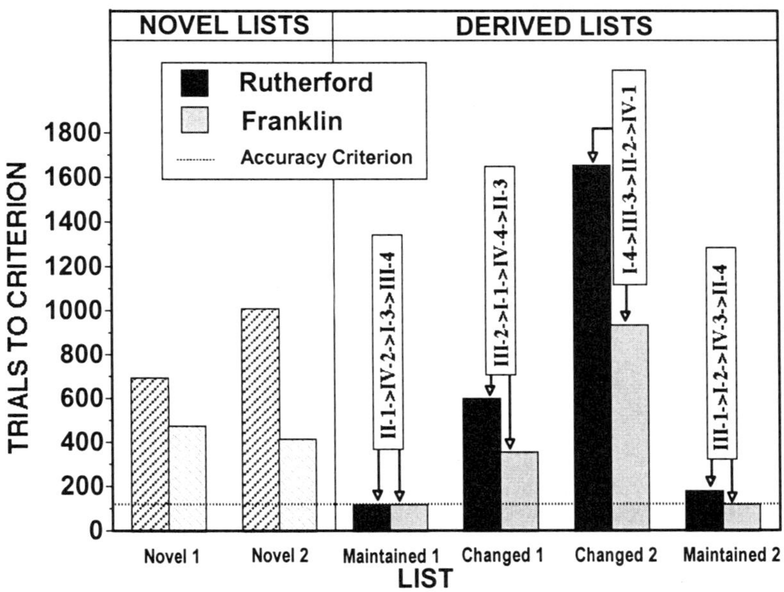

Figure 12. Number of trials needed to satisfy the accuracy criterion on four derived lists and on two novel lists. The novel lists were the last lists learned by Rutherford and Franklin prior to this experiment (Swartz et al., 1991). See Figure 11 for the order in which the derived lists were trained.

contained only one item from each of the previously learned lists, all previously acquired item-item associations were irrelevant on both the maintained and changed lists.

The maintained lists could be executed correctly from the start of training by using each item's original ordinal position as a basis for ordering the newly juxtaposed items. On changed lists, the correct sequence could only be determined by trial and error. To the extent that the monkeys acquired knowledge of each item's original ordinal position, the maintained lists should be easier to acquire than the changed lists.

As shown in Figure 12, the lists on which each item's ordinal position was maintained were acquired rapidly and with virtually no errors. The derived lists on which each item's ordinal position was changed were as difficult to learn as novel lists. The only explanation of the dramatic difference between the amount of training needed to learn maintained and changed lists is that the subjects were able to retrieve the ordinal positions of items from previously learned lists while learning the derived lists on which the ordinal positions of list items were maintained.

TEMPORAL ORGANIZATION OF SIMULTANEOUS CHAINS

Thus far, our description of performance on simultaneous chains has focused on accuracy of responding during list learning, subset tests and tests of ordinal knowledge using derived lists. Another important aspect of simultaneous chains is their temporal organization. Reaction times (RTs) to the first item of a list and interresponse times (IRTs) between subsequent items can provide information about how a subject plans a sequence.

Consider the following two strategies for planning a sequence. The first is to search for $item_1$, respond to it, search for $item_2$, respond to it, and so on, until the sequence is completed. Another strategy is to scan all of the items (or some subset of the sequence), and to then devise a plan for executing the entire sequence (or some subset thereof) before making the first response. The application of these strategies would result in different temporal patterns of responding to successive items of the list. If subjects adopted a "select-one-item-at-a-time" strategy, they would need progressively less time to select each item. With a "plan-the-sequence-first" strategy, the RT of the first response should be long and the IRTs between the remaining responses should be relatively short. Initial analyses of mean RT and IRTs of two monkeys trained on four 6-item lists (Bugs and Garbo), supported the "plan-the-sequence-first" strategy. As shown in Figure 13, the mean RT to $item_1$ was long and the mean IRTs to the subsequent 5 items were uniformly short.

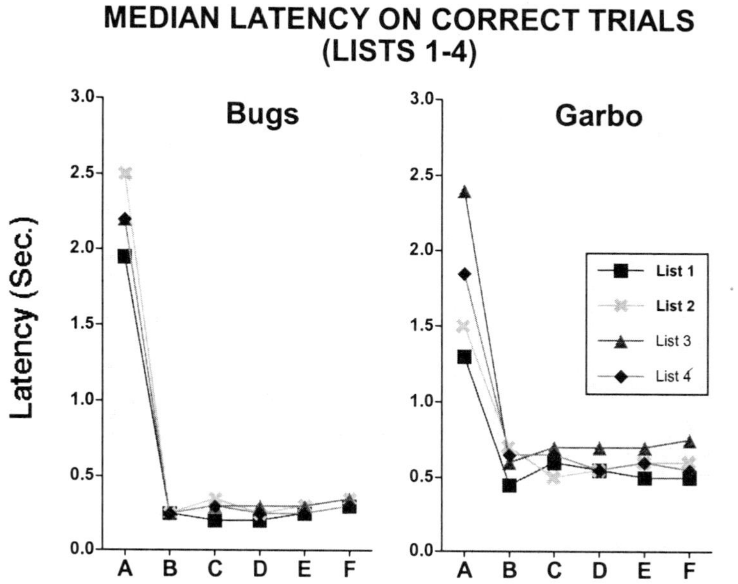

Figure 13. Average latency of responding by monkeys to each item on four 6-item lists. See text for additional details.

EVIDENCE OF TEMPORAL CHUNKING

A replication of the multiple-list experiment with human subjects led to the unexpected discovery that the uniformly short IRT functions obtained from Bugs and Garbo were artifacts of averaging and that pauses occurred on most trials, albeit at different positions. The reliability with which pauses occurred, both on correct and incorrect trials, suggests that they could be used to define the boundaries of output chunks (Terrace et al., 1996).

The procedure and the apparatus used to train human subjects was similar to that used to train monkeys to produce 6-item lists. Human subjects (N = 40) learned 4 eight-item lists composed of achromatic nonsense geometric shapes (Terrace et al., 1996). One of those lists and two of the hundreds of different configurations of the list items on which subjects were trained are shown in Figure 14.

Following a 3-item practice list, subjects were told to determine, by trial and error, the correct order in which to respond to 8 items displayed on the monitor. As expected, human subjects learned their lists much more rapidly than monkeys. Details of the list-acquisition process for each species can be found in (Swartz et al., 1991) and (Jaswal, 1995). The mean latency functions obtained from human subjects are shown in Figure 15. As was true of monkeys, the mean latency of responding to the first item (2-3 sec) was longer than the uniformly shorter mean IRTs between responses to subsequent items (0.75-1.5 sec). For both species, the long latency of the

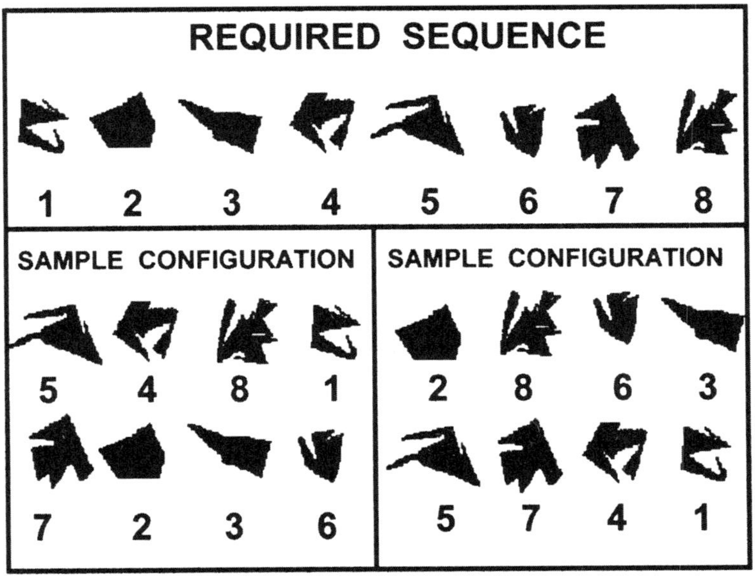

Figure 14. Stimuli used to construct 8-item lists for human subjects trained by the simultaneous chaining paradigm. See text for additional details.

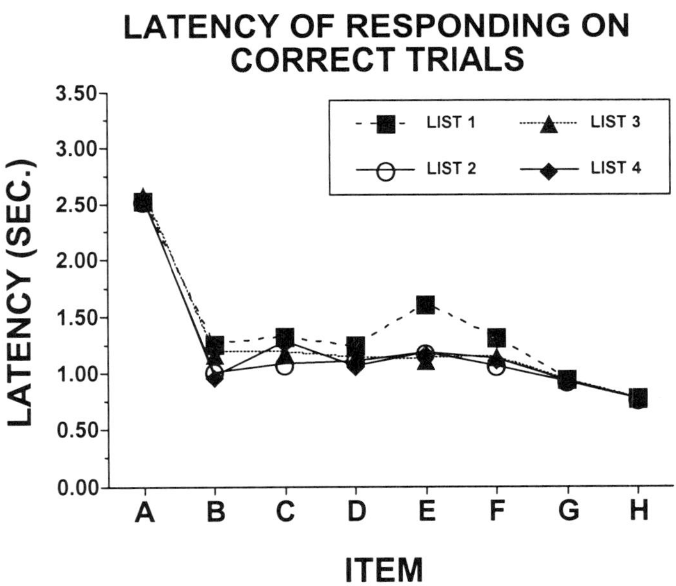

Figure 15. Average latency of responding by human subjects to each item on four 8-item lists. See text for additional details.

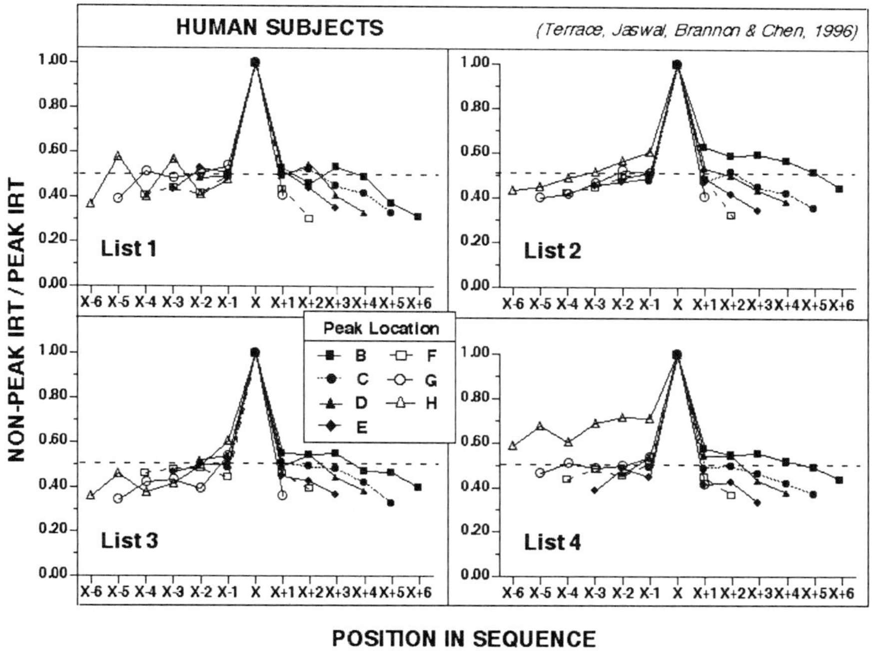

Figure 16. Relative magnitude the longest IRT of human subjects at each position of four 8-item lists. For each trial, the IRT's at each position were divided by the longest latency on that trial (excluding the latency to A). The longest latency is represented as X, and the other items correspond to their ordinal distance from X (X-6 to X+6). The range of the number of correct trials used to calculate each of the data points for positions B-G was 20-84; for position H, 6-17. See text for additional details.

response to the first item appears to reflect the time needed to orient to the array of list items and to search for the initial items (Sternberg et al., 1982).

A molecular analysis of these data revealed that the uniformly short IRTs shown in Figure 15 were artifacts of averaging IRTs, across *subjects* and *trials*, and that, on most trials, one of the IRTs was significantly longer than the others. The longer IRT could not be detected in the average functions shown in Figure 14 because the location of the pause varied from trial to trial.

An analysis of each subject's IRTs on *each* correct trial showed that pauses occurred on virtually every trial, typically after one of the first few responses. Pauses were approximately twice as long as other IRTs. Figure 16 shows relativized data for each of the four lists on which human subjects were trained. For example, when the response to C had the longest latency, the location marked X-1 refers to the latency of the response to item B, while X refers to the latency of the response to C. X+1 refers to

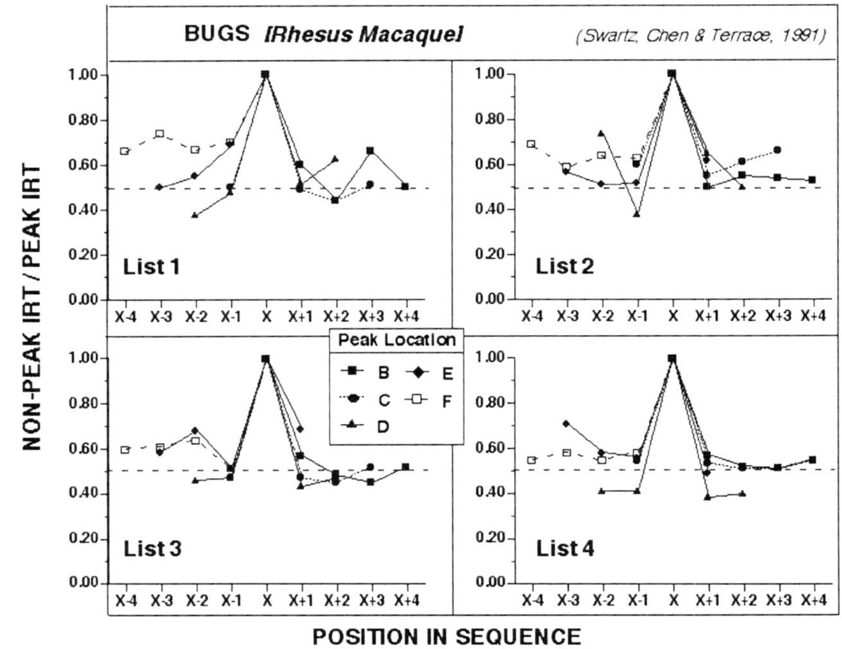

Figure 17. Relative magnitude of the longest IRT of Bugs (*rhesus macaque*) at each position on four 6-item lists. For each trial, the IRT's at each position were divided by the longest latency on that trial (excluding the latency to A). The longest latency is represented as X, and the other items correspond to their ordinal distance from X (X-4 to X+4). The range of the number of correct trials used to calculate each of the data points for positions B-E was 6-30; for position F, 2-10. See text for additional details.

the latency of the response to item D, X+2 to the latency of the response to item E, and so on. On trials on which the latency of the response to D was longest, (solid triangles), X-2 refers to the latency of the response to B, X-1 refers to the latency of the response to C, X to the latency of the response to D, X+1 to the latency of the response to E, and so on. The values of each function were determined by locating the longest IRT on each trial and then calculating the relative magnitude of the IRT's at other positions. By definition, the maximum value of each function is 1.0. Analogous functions were obtained from a molecular analysis of Bugs' and Garbo's IRT data. These are shown in Figures 17 (Bugs) and 18 (Garbo).

The procedure used to train monkeys and human subjects to execute simultaneous chains lacked any contingencies that favored short or long IRTs at any point of the required sequence. Subjects had ample time

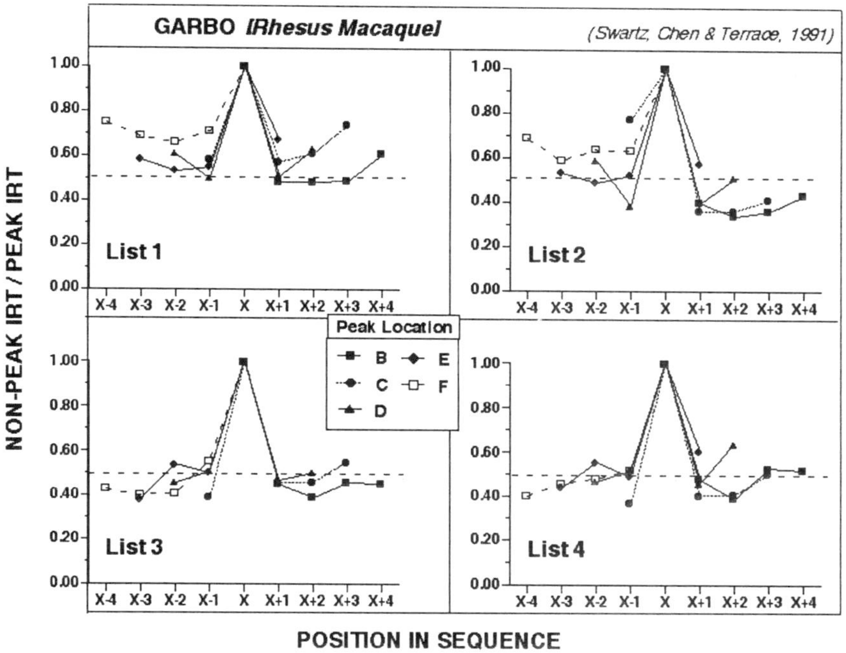

Figure 18. Relative magnitude of the longest IRT of Garbo (*rhesus macaque*) at each position on four 6-item lists. For each trial, the IRT's at each position were divided by the longest latency on that trial (excluding the latency to A). The longest latency is represented as X, and the other items correspond to their ordinal distance from X (X-4 to X+4). The range of the number of correct trials used to calculate each of the data points for positions B-F was 6-34. See text for additional details.

to complete each trial. Indeed, less than 0.5% of all trials were ended because a subject failed to respond to the list items in the time allotted for each trial (20 sec. for human subjects; 15 sec. for monkeys). The only requirement for reinforcement was responding to list items in the correct order, at any pace.

OTHER EVIDENCE OF PAUSING

Studies of human list learning seldom analyze IRTs. The few that have also reported pauses during the execution of sequences. For example, Thorpe and Rowland (1965) described an experiment in which subjects learned to produce 9-item lists of numbers. Subjects paused spontaneously, typically after every 3rd response, as they executed those lists. Similar results were reported by Ryan (1969), Wilkes (1975), Wilkes & Kennedy, 1969, and by

Wilkes, et al., 1972). As in the experiments described above, there were no constraints on the temporal pattern of responding on each trial.

Multiple pauses were also observed in an experiment in which human subjects were trained to reproduce 12-item lists on which the items were presented successively (Brannon, 1996). Following the presentation of the 12th item, all items were displayed simultaneously. The subject's task was to respond to the items in the order in which they were presented. Thirteen of the 14 subjects paused at least twice on correctly completed trials, typically after sequences of 3 or 4 items.

Pauses play an important role in Johnson's (1972) model of chunking which characterizes chunks as "opaque containers". Johnson's model predicts long IRTs and high transitional error probabilities (TEPs) between chunks but short IRTs and low TEPs within chunks. Johnson confirmed these predictions in a task in which subjects were asked to recall lists composed of clusters of letters (Johnson, 1970).

The paradigm that Johnson used differed in two respects from those used to train simultaneous chains. The lists on which Johnson's subjects were trained were segregated into temporally defined chunks throughout training. Accordingly, the pauses Johnson observed could reflect the structure provided by the experimenter. By contrast, no temporal structure was provided for either monkey or human subjects in the experiments on simultaneous chaining. The 2nd difference is that TEPs cannot be calculated for performance on simultaneous chains because trials were terminated after any error. The pauses shown in Figures 16-18 are nevertheless consistent with Johnson's conceptual analysis of chunks. As suggested by Johnson, pauses occur because subjects need more time to retrieve a chunk of list items from LTM than they need to execute a response to a particular item from a chunk that has been downloaded into working memory. Each chunk is held in working memory until the subject responds to the relevant items in the correct order. Subjects then retrieve another chunk, respond to the items it contains, and so on, until they complete the list.

THE OPTIMAL SIZE OF A CHUNK

Estes' (1972) model of chunking shows how a chunk size of 3 requires fewer associative and inhibitory connections between list items than do larger chunks. Using a different set of premises, McGonigle and Chalmers (1996) show how the "exponential explosion of combinatorial possibilities" with increasing chunk size favors an output chunk size of 3-4 items. Wickelgren (1964, 1957) and Broadbent (1975) have also provided empirical evidence that the optimal size of input chunks is 3-4 items. An extensive review of experiments on the capacity of STM concluded that both empirical and theoretical analyses of chunking yielded an optimal chunk size of 4 items (Cowan, 2001).

SIGNIFICANCE OF PAUSES

The spontaneous occurrence of pauses during the execution of a simultaneous chain is significant because they reflect a *self*-imposed organization of list items. In contrast to experiments in which the temporal structure of a sequence is constrained by the subject's verbal history (e.g., Bousefield, 1953) or by the experimenter (e.g., Bower and Winzenz, 1972; Johnson (1972), the spontaneous chaining paradigm provides none. Pauses serve a useful function in that they reduce the load on working memory during the execution of a list. Less attention is needed to prepare a relatively short sequence that, on average consists of 3-4 items, than 6-8 item.

Another important feature of pauses is the variability of their location from trial to trial. Although pauses occurred mainly after the response to the 3rd or 4th items, the distribution of pause locations was broader. This variability of the location of output chunks is to be contrasted with the lack of variability in the structure of input chunks. In experiments on STM, list items vary from trial to trial and subjects encode items in a fixed manner as they memorize them. During training with the simultaneous chaining paradigm, the same items are presented on each trial and the required sequence doesn't vary.

INPUT VS. OUTPUT CHUNKS

Input chunks reflect the limitations of working memory during the encoding of new information, how new information is stored in long-term memory, and how it is retrieved during recall. Output chunks reflect the organization of over-learned motor programs that are generated "on-line" in working memory.

The literature on the organization of sequential behavior rarely distinguishes between input and output chunks [see Estes (1972) & Schneider and Detweiler (1987) for exceptions]. Miller (1956) made the case for the importance of organizing new material into input chunks to compensate for the limitations of working memory. Earlier, Lashley (1951) noted the importance of hierarchically organized motor programs in his analysis of the inability of linear models to account for the speed with which skilled sequences were executed. Chomsky (1957) made an analogous argument in his influential analyses of grammatical structure (Chomsky,

1957, 1965). The experiments on simultaneous chaining described in this chapter indicate that language is not crucial for generating hierarchical motor programs spontaneously

Unfortunately, the influence of Lashley's and Chomsky's views have created the widespread impression that the problem of serial organized behavior has been solved. Instead of asking about the origins of serially organized behavior, cognitive scientists (mainly theoretical linguists and investigators of artificial intelligence) take for granted the availability of ordered sets, recursive strings and other formal primitives in their theories of seriation. The result is confusion between theories of the internal representation of serial knowledge in STM and LTM ("competence") and theories of the execution of serially organized behavior ("performance"). That confusion is avoided by the distinction between input and output chunks.

CONCLUSIONS

The concept of chunking has had a powerful influence on modern studies of cognition, in particular, analyses of serially organized behavior. However, the uncritical application of this concept has led to the neglect of operational definitions of a chunk. The few definitions of chunks that have been proposed assume that chunks are organized verbally [e.g., (Simon, 1974) and (Shiffrin & Nosofsky, 1994)]. Such definitions are of no value in analyses of chunking in animals.

The problem of chunking can be clarified by distinguishing between input and output chunks. Input chunks, which are of fixed length and which have zero variability, reflect the limited capacity of working memory to encode new information. Output chunks, whose length can vary, reflect attentional limitations of working memory during the generation of rotely learned sequences that are composed of familiar items. The results of recent experiments on list learning by monkeys and humans confirm the validity of this distinction.

ACKNOWLEDGMENTS

This research was supported by grants from the National Institute of Mental Health (MH40462), the Whitehall Foundation and NATO. Correspondence concerning this article should be addressed to H. S. Terrace, Department of Psychology, 418 Schermerhorn Hall, Columbia University, New York, NY 10027; email: terrace@columbia.edu.

REFERENCES

Anderson, J. R., & Bower, G. H. (1974). *Human associative memory.* Washington, DC: Hemisphere Publishing.

Atkinson, R. C., & Shiffrin, R. M. (1971). The Control of Short-term memory. *Scientific American,* **225**, 82-90.

Baddeley, A. (1981). The Concept of Working Memory: A View of its Current State and Probable Future Development. *Cognition,* **10**, 17-23.

Baddeley, A. (1986). *Working Memory.* Oxford, England: Clarendon Press.

Baddeley, A. (1992). Is Working Memory Working? The Fifteenth Bartlett Lecture. *Quarterly Journal of Experimental Psychology,* **44A**(1), 1-31.

Bousfield, A. K., & Bousfield, W. A. (1966). Measurement of clustering and of sequential constancies in repeated free recall. *Psychological Report,* **19**, 935-942.

Bousfield, W. A. (1953). The occurrence of clustering in the recall of randomly arranged associates. *Journal of General Psychology,* **49**, 229-240.

Bousfield, W. A., Puff, C. R., & Cowan, T. M. (1964). The development of constancies in sequential organization during repeated free recall. *Journal of Verbal Learning and Verbal Behavior,* **3**, 489-495.

Bower, G. (1972a). Perceptual Groups as coding units in immediate memory. *Proceedings of the Psychonomic Society,* **27**(4), 217-219.

Bower, G. (1972b). A selective review of organizational factors in memory. In E. Tulving & W. Donaldson (Eds.), *Organization of Memory* (pp. 93-137). New York, NY: Academic Press.

Bower, G. H., & Winzenz, D. (1969). Group structure coding and memory for digit series. *Journal of Experimental Psychology Monographs,* **80**(2).

Brannon, E. (1996). *Measures of chunking.* Unpublished MA, Columbia University, New York.

Broadbent, D. E. (1975). The Magic Number Seven After Fifteen Years. In A. W. Kennedy, Alan (Ed.), *Studies in Long-Term Memory* (pp. 3-18). London: John Wiley & Sons.

Capaldi, E. J., Miller, D. J., Alptekin, S., & Barry, K. (1990). Organized responding in instrumental learning: Chunks and superchunks. *Learning and Motivation,* **21**, 415-433.

Capaldi, E. J., Nawrocki, T. M., Miller, D. J., & Verry, D. R. (1986). Grouping chunking memory and learning. *The Quarterly Journal of Experimental Psychology,* **38B**, 53-80.

Chen, S., Swartz, K., & Terrace, H. S. (1997). Knowledge of the Ordinal Position of List Items in Rhesus Monkeys. *Psychological Science,* **8**, 80-86.

Chen, S., Swartz, K., & Terrace, H. S. (2000). Serial Learning by Rhesus Monkeys: II. Learning 4-item Lists by Trial & Error. *Journal of Experimental Psychology: Animal Behavioral Processes,* **26**, 274-285.

Chen, S., Swartz, K. B., & Terrace, H. S. (1991). *Preliminary evidence for the development of learning set for 4-item lists by rhesus monkeys.* Paper presented at the Paper presented at annual meeting of the Eastern Psychological Association.

Chomsky, N. (1957). *Syntactic structures.* The Hague: Mouton Publishers.

Chomsky, N. (1965). *Aspects of the Theory of Syntax.* Cambridge, MA: MIT Press.

Cowan, N. (2001). The magical number 4 in short-term memory: A reconsideration of mental storage capacity. *Behavioral and Brain Sciences,* **24**, 87-185.

Craik, F. I. M., & Tulving, E. (1975). Depth processing and the retention of words in episodic memory. *Journal of Experimental Psychology,* **104**, 268-294.

Crowder, R. G. (1976). *Principles of learning and memory.* Hillsdale, NJ: Lawrence Erlbaum Associates.

Dallal, N. L., & Meck, W. H. (1990). Hierarchical structures: Chunking by food type facilitates spatial memory. *Journal of Experimental Psychology: Animal Behavior Processes,* **16**(1), 69-84.

Ebbinghaus, H. (1964). *Memory: A contribution to experimental psychology* (Originally published 1885; translated 1913. ed.). New York, NY: Dover.

Ebenholtz, S. M. (1963). Serial learning: position learning and sequential associations. *Journal of Experimental Psychology,* **66**, 353-362.

Estes, W. K. (1972). An associative basis for coding and organization in memory. In A. W. Melton & E. Martin (Eds.), *Coding processes in human memory* (pp. 161-190). Washington, DC: Winston.

Fountain, S. B., & Annau, Z. (1984). Chunking sorting and rule learning from serial patterns of brain-stimulation reward by rats. *Animal Learning & Behavior,* **12**, 265-274.

Fountain, S. B., Henne, D. R., & Hulse, S. H. (1984). Phrasing cues and hierarchical organization in serial learning in rats. *Journal of Experimental Psychology: Animal Behavior Processes,* **10**, 30-39.

Gardner, H. (1985). *The Minds New Science: A History of the Cognitive Revolution.* New York, NY: Basic Books Inc.

Hull, C. L. (1932). The goal gradient hypothesis and maze learning. *Psychological Review,* **39**, 25-43.

Hull, C. L. (1952). *A behavior system: An introduction to behavior theory concerning the individual organism*. New Haven, CT: Yale Univ. Press.

Hulse, S. H. (1978). Cognitive structure and serial pattern learning by animals. In S. Hulse & H. Fowler & W. K. Honig (Eds.), *Cognitive processes in animal behavior* (pp. 311-340). Hillsdale, NJ: Erlbaum.

Jaswal, V. (1995). *Acquisition of 8-item simultaneous chain by human subjects*. Unpublished manuscript, New York.

Johnson, N. F. (1970). Chunking and organization in the process of recall. In G. H. Bower (Ed.), *The Psychology of Learning and Motivation* (Vol. IV). New York, NY: Academic Press.

Johnson, N. F. (1972). Organization and the concept of a memory code. In A. W. Melton & E. Martin (Eds.), *Coding processes in human memory* (pp. 125-159). Washington, DC: Winston.

Keppel, G., & Underwood, B. J. (1962). Proactive inhibition in short-term retention of single items. *Journal of Verbal Learning and Verbal Behavior,* **1**, 153-161.

Lashley, K. S. (1951). The problem of serial order in behavior. In L. A. Jeffries (Ed.), *Cerebral mechanisms in behavior* (pp. 112-136). New York, NY: John Wiley & Sons.

Mandler, G., & Dean, P. J. (1969). Seriation: Development of serial order in free recall. *Journal of Experimental Psychology,* **81**(2), 207-215.

McGonigle, B., & Chalmers, M. (1996). The ontology of order. In L. Smith (Ed.), *Critical Readings on Piaget* (pp. 279-311). New York: Routledge.

Miller, G. A. (1956). The magical number seven plus or minus two: Some limits on our capacity for processing information. *Psychological Review,* **63**(2), 81-96.

Mishkin, M., & Petri, H. L. (1984). Memories & habits: Some implications for the analysis of learning and retention. In L. R. Squire & N. Butters (Eds.), *Neuropsychology of Memory.* (pp. 287-296). New York, NY: Guilford.

Murdock, B. B. (1993). TODAM2: A model for the storage and retrieval of item, associative, and serial-order information. *Psychological Review,* **100**(2), 183-203.

Ohshiba, N. (1997). Memorization of serial items by Japanese monkeys, a chimpanzee, and humans. *Japanese Psychological Research,* **39**, 236-252.

Osgood, C. E. (1953). *Method and theory in experimental psychology*. New York, NY: Oxford University Press.

Premack, D. (1976). *Intelligence in Ape and Man*. Hillsdale, NJ: Lawrence Erlbaum Associates.

Rumbaugh, D. M. (1977). *Language learning by a chimpanzee: The Lana project*. New York, NY: Academic Press.

Ryan, J. (1969). Grouping and short-term memory: different means and patterns of grouping. *Quarterly Journal of Experimental Psychology,* **21**, 137-147.

Schneider, W., & Detweiler, M. (1987). A connectionist/control architecture for working memory. *The psychology of Learning motivations, 21*.

Shiffrin, R. M., & Nosofsky, R. M. (1994). Seven Plus or Minus Two: A Commentary on Capacity Limitations. *Psychological Review,* **101**(2), 357-361.

Simon, H. A. (1974). How Big Is a Chunk? *Science,* **183**, 482-488.

Skinner, B. F. (1938). *The behavior of organisms*. New York, NY: Appleton-Century-Crofts.

Small, W. S. (1900). An experimental study of the mental processes of the rat. *American Journal of Psychology,* **11**, 80-100.

Squire, L. R. (1986). Mechanisms of memory. *Science,* **232**, 1612-1619.

Sternberg, S., Knoll, R. L., & Wright, C. E. (1982). *Control of Rapid Action Sequences in Speech and Typewriting*. Murray Hill, NJ: Bell Laboratories.

Straub, R. O., Seidenberg, M. S., Bever, T. G., & Terrace, H. S. (1979). Serial learning in the pigeon. *Journal of the Experimental Analysis of Behavior,* **32**, 137-148.

Swartz, K. B., Chen, S., & Terrace, H. S. (1991). *Acquisition of 6-item lists by rhesus monkeys.* Paper presented at the Eastern Psychological Association annual meeting.

Swartz, K. B., Chen, S., & Terrace, H. S. (1991a). Serial Learning by Rhesus Monkeys. I: Acquisition and Retention of Multiple Four-item lists. *Journal of Experimental Psychology: Animal Behavior Processes,* **17**, 396-410.

Swartz, K. B., Chen, S., & Terrace, H. S. (1991b). *Acquisition of 6-item lists by rhesus monkeys.* Paper presented at the Eastern Psychological Association annual meeting.

Terrace, H. (1991). Chunking during serial learning by a pigeon: I. Basic evidence. *Journal of Experimental Psychology: Animal Behavior Processes,* **17**(1), 81-93.

Terrace, H., & Chen, S. (1991). Chunking during serial learning by a pigeon: II. Integrity of a chunk on a new list. *Journal of Experimental Psychology: Animal Behavior Processes,* **17**(1), 94-106.

Terrace, H., & Chen, S. (1991). Chunking during serial learning by a pigeon: III. What are the necessary conditions for establishing a chunk?

Journal of Experimental Psychology: Animal Behavior Processes, **17**(1), 107-118.

Terrace, H. S. (1979). Is problem solving language? A review of Premack's Intelligence in Apes and Man. *Journal of the Experimental Analysis of Behavior,* **31**, 161-175.

Terrace, H. S. (1984). Simultaneous chaining: The problem it poses for traditional chaining theory. In M. L. Commons & R. J. Herrnstein & A. R. Wagner (Eds.), *Quantitative Analyses of Behavior: Discrimination Processes* (pp. 115-138). Cambridge, MA: Ballinger Publishing Co.

Terrace, H. S. (1987). Chunking by a pigeon in a serial learning task. *Nature,* **325**, 149-151.

Terrace, H. S. (2001). Serial Expertise and the Evolution of Language. In A. Wray & J. R. Hurford & F. Newmeyer (Eds.), *The Transition to Language*. Oxford: Oxford University Press.

Terrace, H. S., Jaswal, V., Brannon, E., & Chen, S. (1996). What is a chunk? Ask a monkey. *Abstracts of Psychonomic Society,* **1**, 35.

Terrace, H. S., Son, L., & Brannon, E. (2000). The development of serial expertise by rhesus macaques. *Nature*, (under review).

Thorpe, C. E. a. R., George E. (1965). The Effect of "Natural" Grouping of Numerals on Short Term Memory. *Human Factors,* **7**, 38-44.

Underwood, B. J. (1957). Interference and forgetting. *Psychological Review,* **64**, 49-60.

Weiskrantz, L. (1970). A Long-Term View of Short-Term Memory in Psychology. In G. Horn & R. A. Hinde (Eds.), *Short-term Changes in Neural Activity and Behaviour.*

Wickelgren, W. A. (1964). Size of rehearsal group and short-term memory. *Journal of Experimental Psychology,* **68**(4), 413-419.

Wickelgren, W. A. (1967). Rehearsal grouping and the hierarchical organization of serial position cues in short-term memory. *Quarterly Journal of Experimental Psychology,* **19**, 97-102.

Wilkes, A. L. (1975). Encoding Processes and Pausing Behaviour. In A. a. W. Kennedy, Alan (Ed.), *Studies in Long-Term Memory* (pp. 19-42): John Wiley & Sons.

Wilkes, A. L., & Kennedy, R. A. (1969). Relationship Between Pausing and Retrieval Latency in Sentences of Varying Grammatical Form. *Journal of Experimental Psychology,* **79**(2), 241-245.

Wilkes, A. L., Lloyd, P., & Simpson, I. (1972). Pause measures during reading and recall in serial list learning. *Quarterly Journal of Experimental Psychology,* **24**, 48-54.

3

SERIAL LEARNING BY RHESUS MONKEYS

Strategies for List Learning and Execution

Karyl B. Swartz and Sharon A. Himmanen
Lehman College of The City University of New York

Like most interesting questions in comparative psychology, the question of serial organization of behavior is not new. Hamilton (1911) demonstrated species differences in strategies for solving a rule-based four-choice problem, with monkeys and humans more likely to make choices based on relevant information obtained from previous trials than dogs, cats, or a horse. Although Hamilton's task was not a serial task, an efficient solution did require that the subject base its behavior on the series of responses executed during the current and immediately previous trial. Memory was implied, as was the rational use of information.

Tinklepaugh (1932) found that chimpanzees and macaque monkeys were successful in remembering the location of food hidden in one of a pair of containers when individual pairs were located in a series of different rooms, or when multiple pairs were arranged in a circle around the subject in one room. The two monkeys scored below the two chimpanzees in the 10-room condition. In the single-room multiple pairs condition, chimpanzees performed at high accuracy levels with as many as 16 pairs. Monkeys performed well with four pairs but performance declined with six and eight pairs.

Although not as proficient as chimpanzees in Tinklepaugh's studies, macaque monkeys have demonstrated the acquisition of sophisticated strategies. Learning set ("learning to learn"; Harlow, 1949) involves the use of information obtained on the first trial of a novel problem to identify the correct stimulus in a novel two-choice discrimination task. Once monkeys have acquired learning set, they perform at above-chance levels on the second trial of each novel problem, independent of first-trial success. Furthermore, Rumbaugh and Washburn (1996) describe the successful acquisition by rhesus monkeys of a set of complex cognitive tasks, some using serial order information to guide performance.

Lashley's (1951) pivotal discussion of the basis of serial behavior prompted experimental psychologists to address this problem from a cognitive perspective (see, e. g., Terrace, 1984). Over the past several years, we have been investigating how monkeys organize serial order information. Rhesus monkeys (*Macaca mulatta*) have been trained to learn arbitrary lists of photographic items presented on a touch-sensitive video screen. The results of a series of studies suggest that monkeys quickly develop a strategy for acquiring lists that is used in solving subsequent problems. The present chapter will examine serial learning in rhesus monkeys by looking at executive function demonstrated by (1) representation of serial order in memory, and (2) use of strategies to solve a problem of serial order. Two strategies will be described here, strategies for acquiring a list and strategies for executing a list on a particular trial.

THE PARADIGM

We have produced a group of list-sophisticated rhesus monkeys and are studying the response strategies that develop as they learn lists. Monkeys are trained to produce lists with the simultaneous chaining procedure developed for pigeons by Terrace (Straub, Seidenberg, Bever, & Terrace, 1979) and extended to monkeys (Swartz, Chen, & Terrace, 1991). Photographic stimuli are presented on a touch-sensitive color video monitor. The monkey's task is to touch the items in a predetermined order. The positions of the stimuli change on each trial to prevent learning of a motor sequence. Successful execution of the list produces a small food reward; errors end the trial and produce a time-out period during which the lights in the stimulus chamber are extinguished. Initially, monkeys were trained in a phase-wise progression, starting with the first item and adding items to the end of the list as the monkey mastered each phase. However, once the animals became experienced with the task, we were able to eliminate early phases of training, and we now have a group of monkeys who can acquire novel lists with all four items present from the first day of training (Swartz, Chen, & Terrace, 2000). Dependent measures of major interest are trials to criterion for each phase or each list, error patterns, and response times to each item.

WHAT DO WE KNOW ABOUT LIST-LEARNING BY RHESUS MONKEYS?

In a previous report (Swartz, et al., 1991) we demonstrated that monkeys could acquire several four-item lists, that they remembered these lists over long time periods, and that they could produce multiple lists in a single session with no decrease in accuracy or increase in response time. Our data suggest that monkeys develop a strategy for acquiring lists which they learn during acquisition of the first list. Following acquisition of the first list, the number of trials to criterion on subsequent lists decreases, especially during the early phases. Monkeys who learned multiple four-item lists showed dramatic decreases in the number of trials to reach criterion in the 2- and 3-item phases, although trials to criterion for the 4-item phase remained high and variable.

The data suggest that during the acquisition of the first list, monkeys develop an understanding of the sequential nature of the task, i. e., they understand that the list is defined by the sequence of items. Further, the almost-perfect performance in Phase 2 suggests that the first item is defined by the one-day (60 trials) experience with that item. Error patterns show that the point of difficulty for lists of three or more items is primarily at the N-1 position and that the most common errors are forward-skip errors, usually one-skip errors (see Swartz et al., 1991). The suggestion from these data is that the strategy developed by these animals for acquiring 4-item lists presented in the successive phase manner is to integrate the first N-1 items into the list and then respond to the last item by default.

The finding that these animals can execute four lists within a single session with no decrease in accuracy suggests that they represent multiple lists in reference memory (Swartz et al., 1991). Further, the data from subset tests suggest that the representation is linear. Following acquisition of their first four lists, two monkeys reacquired each list and were given a subset test following the reacquisition of each list. All possible pairs of items from a four-item list were presented during these subset tests. Accuracy on all subsets, including those containing only interior items (B and C), and those not containing adjacent items (e., g., A→C) was high. Response times from subsets showed some evidence of linear representation, and clear knowledge of the relative ordinal position of individual items. The suggestion here is that execution of a particular list was not solely a function of simple pair-wise associations between adjacent items. The response time to the first item of a two-item pair was in general longer when that item occurred later in the list than earlier in the list. Further, the response time to the second item of a pair increased as the number of items intervening between first and second item increased.

STRATEGIES FOR ACQUIRING LISTS

From the previous data, it was clear that monkeys understand the sequential nature of the task, that they use this understanding in acquiring novel lists, and that they also acquire information about relative ordinal position of specific list items as they acquire a new list. In order to determine further how effectively they could use this knowledge we conducted two studies. In one, we successively eliminated the early phases of training to determine whether monkeys could use their knowledge about the structure of a list to acquire a novel list beginning with four items (Swartz, et al., 2000). In the second study, we asked whether monkeys could use their knowledge of ordinal position to acquire novel lists composed of familiar items (Chen, Swartz, & Terrace, 1997).

Successive Elimination of Phases

Animals who have learned several lists using the successive phase method can acquire lists in which early phases have been eliminated. Three monkeys who had learned several four-item lists were trained to begin novel lists with two, then three, then four items present on the first day of training. They were able to acquire this task, and now begin all novel lists with all four items present from the beginning (Swartz et al., 2000). Subjects were able to begin lists with four items present with no increase in overall trials to criterion.

It appears that the strategy used here is: establish A, determine the relative positions of B and C; then respond to D by default. There is evidence of a trial-and-error strategy in establishing the first item. At the beginning of each novel list all four items are tried, but A is rather quickly established within the first 100-200 completed trials. There is more difficulty with interior items. The middle items must be disambiguated, which is more difficult here than in the successive phase method. Once A is established, error patterns suggest the next step is to disambiguate B from C and to determine their relative ordinal positions. Responses to D are most likely by default, as the stimulus display remains available for one more response following the successful completion of the first three items, and the only item remaining is D.

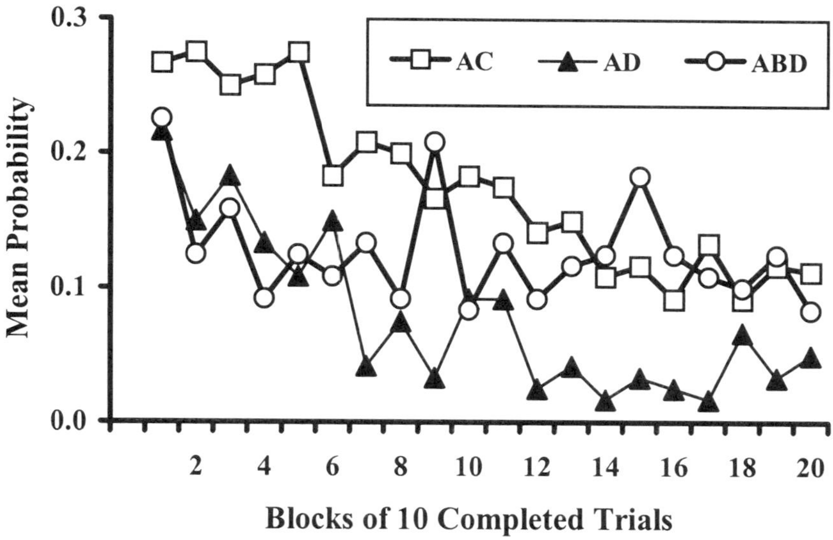

Figure 1. Distribution of forward errors across acquisition for lists beginning with four items present and position of all items variable, Franklin and Rutherford combined.

While A→B is being established, B has not been completely disambiguated from C. That is, most errors are one-skip errors such as A→C, or A→B→D errors, suggesting confusion of B and C. To address this hypothesis, we performed an analysis of error patterns on six lists learned by two experienced monkeys (Franklin and Rutherford; see Swartz et al., 2000, for additional analyses of these acquisition data). Figure 1 shows the proportion of the three possible forward errors, A→C, A→B→D, and A→D, for Franklin and Rutherford, over the first 200 completed trials that began with a correct initial response to A. Backward errors were not included because of their low incidence. A two-way (Error Type X Block) repeated measures ANOVA yielded significant main effects of Error Type [$F(2,10)=5.01$, $p<.031$] and of Block [$F(19,95)=6.32$, $p<.000$], but no significant interaction. Post-hoc analyses using Tukey's HSD showed that overall errors decreased as a function of blocks (acquisition). The incidence of A→D errors was significantly lower than A→C errors ($p<.05$) and the incidence of A→B→D errors was intermediate and not different from either. The low incidence of A→D errors combined with the higher incidence of A→C and A→B→D errors support the interpretation that once A is established, B and C are disambiguated and a correct response to D occurs by default.

Table 1
Composition and Sequence of Derived Lists

Original Lists	Item order	Items
List 1	A1→B1→C1→D1	bird→flower→frog→shells
List 2	A2→B2→C2→D2	tree→weasel→dragonfly→water
List 3	A3→B3→C3→D3	elk→rocks→leaves→person
List 4	A4→B4→C4→D4	mountain→fish→monkey→tomato

Derived Lists	Item order	Items
Maintained 1	A2→B4→C1→D3	tree→fish→frog→person
Changed 1	B3→A1→D4→C2	rocks→bird→tomato→dragonfly
Changed 2	D1→C2→B3→A4	shell→leaves→weasel→mountain
Maintained 2	A3→B1→C4→D2	elk→flower→monkey→water

Order of Training Derived Lists	
Franklin:	Maintained 2→Changed 2→Changed 1→Maintained 1
Rutherford:	Maintained 1→Changed 1→Changed 2→Maintained 2

Knowledge of Ordinal Position of List Items

The results of the subset tests reported by Swartz et al. (1991) suggested that lists are represented linearly and that the monkeys have some knowledge of relative position of individual items; however, it was not clear to what extent they represented the absolute ordinal position of individual items. To investigate knowledge of ordinal position, we presented the animals with reconstituted lists, that is, novel lists composed of familiar items from previously learned lists (Chen et al., 1997). In two lists the ordinal position of each item was maintained relative to its original position, and in the other two the original ordinal position of each item was violated. The original lists were well-learned lists, having gone through acquisition and two occurrences of reacquisition, with the second reacquisition beginning at the four-item phase.

These lists were rearranged to produce four derived lists. One item from each of the previous lists was included in each derived list. In two of the derived lists, the ordinal position of each item was maintained relative to its position in the original list (Maintained). In the remaining two derived lists, the ordinal position of the items differed from that in the original lists (Changed). Table 1 shows the rearrangement of lists.

All lists were acquired beginning with four items. If the animal has knowledge of the ordinal positions of particular items, the prediction would be rapid (and practically errorless) acquisition of lists in which the ordinal position was maintained, and slower acquisition of lists in which the ordinal position of items was violated. That, in fact, was what obtained. The two Maintained lists were learned rapidly and with few errors whereas the two

Changed lists took longer to acquire. With respect to the pattern of errors, we should be able to predict what kinds of errors are made based on the original ordinal positions of the items. For example, on Changed List 1, with the order, $B_3 \rightarrow A_1 \rightarrow D_4 \rightarrow C_2$, the prediction was that the largest number of errors at the beginning would be forward errors at the first item position, and that they would be responses to the current item B (previous item A). Similarly, for Changed List 2, with the order $D_1 \rightarrow C_3 \rightarrow B_2 \rightarrow A_4$, the prediction was that the largest number of errors again would be at the first item and that they would be responses to current item D (previous A). These predictions were confirmed.

The distribution of responses showed that knowledge of ordinal position is the only thing that can explain such high levels of accuracy on maintained lists, especially for interior items (B and C) which present the most difficulty during acquisition of novel lists. Despite their efficient use of ordinal position, the animals were sensitive to contingencies and changed responses somewhat on their second Maintained list. That list followed two Changed lists in which ordinal position did not accurately predict the correct sequence. These results suggest, then, that the animal represents the ordinal position of an item as well as its identity in reference memory and uses that information to respond to new configurations of familiar items. However, when use of that knowledge does not work, the animal can shift item-related strategies and use his understanding of the sequential nature of the task to acquire new sequences of familiar items.

Strategies for Executing Lists

Response time patterns during execution of lists show a long response time at the beginning of the list, followed by shorter, equivalent response times at all other positions (Swartz et al., 1991). That long response time to the first item could be a function of several factors: (1) "Start-up" time, i. e, time taken to orient to the screen, (2) Searching for the first item only, (3) Searching for and locating the first *N* items. This latter explanation would imply a plan for the execution of the list, generated at the beginning of a trial. It is unlikely that start-up time is a sufficient explanation for this long initial response time, as our observations of the subjects showed that they were extremely attentive to the screen during the ITI. Further, there was a warning signal—two seconds before the onset of a trial, the lights in the chamber were dimmed—which should reduce the start-up effect. The second possibility, searching for the first item only, is not a sufficient explanation in light of shorter response times to the intermediate items, and the failure to find a negative correlation between the response times to the first and second items (see Swartz et al., 1991).

To tease apart these explanations, we presented novel lists to two list-sophisticated monkeys (Franklin and Rutherford), with the position of the items systematically presented in fixed and variable positions (Himmanen & Swartz, in preparation). Note that one aspect of the simultaneous chaining procedure was that the position of each item varied on each trial, in order to prevent the possibility that the monkeys could learn the list as a motor sequence. In this study, we presented six lists with all items presented in all possible positions (All variable), six lists with the first item always appearing in the same position on each trial (A fixed), and four lists with all items appearing in the same position on each trial (All fixed). In a final condition, we presented two lists whose order was determined solely by item position. That is, four identical versions of a single item were presented, and the list was defined by the spatial location of the four items (Single item, all fixed).

In the All variable condition, the position of each item was unknown to the animal at the beginning of a trial. In the A fixed condition, the position of the first item was always known to the animal but the position of each of the remaining three items was unknown. In the All fixed condition, the position of all items was known to the subject. In this condition, a list could be acquired on the basis of item order or as a motor sequence alone. In the Single item, all fixed condition, the motor sequence was the only way of acquiring the list.

The first prediction was that providing information about the position of list items should make a list easier to acquire. However, trials to criterion decreased only in the condition in which all positions were fixed and list items differed from one another. The fourth condition, in which the four items were the same and the list order was defined solely by position, was very difficult for the monkeys. It took more sessions for the animals to learn these lists, as they failed to respond during the initial sessions of these lists. Once the animals began responding to the stimulus display, the number of trials to criterion did not differ from other conditions.

With respect to response time, if the long response time to A is a function of "start-up" time, it should be unaffected by knowledge of the position of the first item (i. e., it should not vary across conditions). If the long response time to A is a function of locating that item only, the response time to A should decrease when its position is known, and the response time to B should increase. If the long response time to A involves locating the first N items, the response time should be shorter when the position of A is known and should decrease even more when the position of all items is known.

An analysis of response times across acquisition for the four types of list showed that the response time to A decreased when the location of any or all items was fixed, hence start-up time is not the sole explanation for the long initial response time. Response time to A decreased when the position of that item was fixed, but response time to B did not increase, eliminating the hypothesis that the long initial response time involved a search for the first item only. When the position of all items was fixed (All fixed and Single item, all fixed conditions), the response times to the first three items decreased across training. The interesting aspect of this finding was that the middle response times (B and C) decreased to the level of the last item (D) in these conditions.

The implication of these data is that the long response time to the first item may involve the location of the first N items, and we think in this case that N=3. The hypothesis that N=3 is suggested by the short response time to the last item when some or all items are in variable positions. As suggested previously (Swartz et al., 1991), acquisition data support the interpretation that the last item is responded to by default, a suggestion that is consistent with these data. The obtained decrease in response times to the internal items during the two conditions in which the sequence was defined by location suggests to us that the subjects shifted strategy from that which they used in learning earlier lists to one based on location.

This suggestion is supported by the data from the fourth condition. In this condition, the four items of the list were identical, and the order of the list was dictated by item positions. Any solution based on a fixed motor sequence or item spatial location had been eliminated in all previous training (Chen et al., 1997; Swartz, et al. 1991, 2000) by randomly varying the positions of all items across trials. When the correct sequence was dictated by spatial position, and a fixed motor sequence was the most efficient solution, the animals found this type of list extremely difficult to learn, shown by their failure to respond across the first several sessions of each of this list type. It was so difficult to maintain their performance in this condition that only two lists of this type were presented. This difficulty, however, stands in contrast to an obtained change in response times across acquisition. By the end of training of each list, response times to all four positions had significantly decreased, suggesting that the animals had adopted a response strategy based only on location.

The above change in response times suggests flexibility in the use of spatial position information, and is consistent with the obtained flexibility in the use of ordinal position information reported earlier (Chen, et al., 1997). However, the rapidity of change differed. In the case of ordinal position, new information about the ordinal position of a familiar item was rather rapidly incorporated into the animals' response patterns and was present from the beginning of the novel lists following the first Changed list. In the case in which a change in the entire strategy for solution was required, the animals were not so quick to change, and showed difficulty with the second list defined only by spatial location. The change in response times across acquisition suggests that they did, within each list, modify response patterns. Superordinate strategies, developed at the task level, appear to change with some difficulty, whereas strategies at the specific list item level are more easily modified when they no longer work.

OVERALL SUMMARY

To summarize, our data suggest that monkeys acquire an overall strategy for list-learning that they use in acquiring all lists following the first. That strategy includes knowledge of the sequential nature of the task, and use of information regarding relative and absolute ordinal positions of items. Acquisition of lists involves establishing three components of a list: beginning, middle, and end, in that order. Monkeys are able to represent multiple lists in reference memory, and are able to use their representation of the list to execute it on individual trials. That representation is linear, is anchored by the first item, and includes ordinal position information. In executing a particular list, the monkey uses the representation of the list, and plans the execution of the list at the beginning of a trial. The last item is executed by default, suggesting that the animal not only plans the execution of the initial items in the list, but also keeps track of where he is in the list as he executes it. Strategies related to the task of list learning are modified only with difficulty whereas strategies using specific item information are subject to change when that information no longer works.

ACKNOWLEDGMENTS

This research was supported by grants from NIMH (MH40462 and MH49747), NIH (GM08225), and the Whitehall Foundation. We thank Janis Eisner for her help in manuscript preparation.

REFERENCES

Chen, S., Swartz, K. B., & Terrace, H. S. (1997). Knowledge of the ordinal position of list items in rhesus monkeys. *Psychological Science*, **8**, 80-86.

Hamilton, G. V. (1911). A study of trial and error reactions in mammals. *Journal of Animal Behavior*, **1**, 33-66.

Harlow, H. F. (1949). The formation of learning sets. *Psychological Review*, **56**, 51-65.

Himmanen, S. A., & Swartz, K. B. (in preparation). Response time patterns during acquisition of four-item lists by rhesus monkeys.

Lashley, K. S. (1951). The problem of serial order in behavior. In L. A. Jeffress (Ed.), *Cerebral mechanisms in behavior: The Hixon Symposium* (pp. 112-146). New York: Wiley.

Rumbaugh, D. M., & Washburn, D. A. (1996). Attention and memory in relation to learning: A comparative adaptation perspective. In G. R. Lyon & N. A. Krasnegor (Eds.), *Attention, memory, and executive function* (pp. 199-219). Baltimore, MD: Paul H. Brooks Publishing Co.

Straub, R. O., Seidenberg, M. S., Bever, T. G., & Terrace, H. S. (1979). Serial learning in the pigeon. *Journal of the Experimental Analysis of Behavior*, **32**, 137-148.

Swartz, K. B., Chen, S., & Terrace, H. S. (1991). Serial learning by rhesus monkeys: I. Acquisition and retention of multiple four-item lists. *Journal of Experimental Psychology: Animal Behavior Processes*, **17**, 396-410.

Swartz, K. B., Chen, S., & Terrace, H. S. (2000). Serial learning by rhesus monkeys: II. Learning four-item lists by trial and error. *Journal of Experimental Psychology: Animal Behavior Processes*, **26**, 274-285.

Terrace, H. S. (1984). Animal cognition. In H. L. Roitblat, T. G. Bever, & H. S. Terrace (Eds.), *Animal Cognition* (pp. 7-28). New Jersey: Lawrence Erlbaum.

Tinklepaugh, O. L. (1932). Multiple delayed reaction with chimpanzees and monkeys. *Journal of Comparative Psychology*, **13**, 207-243.

4

MONKEY VISUAL AND AUDITORY MEMORY

Anthony A. Wright
University of Texas Medical School at Houston

One of the earliest studies of memory (Ebbinghaus, 1895) showed that list memory, when plotted according to the list's serial position, was U-shaped. Memory was best at the beginning (primacy effect) and the end (recency effect) of the list, and poorest in the middle. In the late '50s, researchers showed that the recency effect, thought to be the short-term memory (STM) component, could be selectively eliminated from the primacy effect and middle portion, thought to be the long-term memory (LTM) component, by delaying recall and filling the delay with distracter activity (e.g., Brown, 1958; Peterson and Peterson, 1959). This manipulation was the beginning of the modern study of short-term memory and helped bring about the so-called cognitive revolution. There has been considerable emphasis on the discovery of dissociable memory processes and development of theories to explain them. Some memory processes (e.g., language based) may be unique to humans, but general theories of memory should also apply, at least in part, to animals as well as humans.

Poor performance by animals in list-memory tasks has been a major impediment to exploring whether animals have similar memory processes (e.g., STM, LTM) as humans. Before we began our experiments, monkeys had performed less than 70 percent correct with three visual items in a list (e.g., Davis & Fitts, 1976; Devine & Jones, 1975; Eddy, 1973; Gaffan, 1977). More recently, procedures have been developed which produce better animal list-memory performance, and have opened the possibility of comparing animal and human memory processes (e.g., Bolhuis and van Kampen, 1988; Buchanan, Gill, and Braggio, 1981; Castro and Larsen, 1992; Kesner and Novak, 1982; Reed, Chih-ta, Aggleton, and Rawlins, 1991, Harper, McLean, and Dalrymple-Alford, 1993; Roberts and Kraemer, 1981; Sands and Wright, 1980a,b; Thompson and Herman, 1977; Wright, Santiago, Sands, Kendrick, and Cook 1985).

In one of our studies, large numbers of colorful "travel-slide" pictures were used to test monkey visual list memory (Sands & Wright, 1980a). These procedures contrasted with the small number (e.g., 6) of abstract shapes and colors used in previous investigations of nonhuman primate memory (e.g., Gaffan, 1977). Our original rationale had been that monkeys might remember pictures of real-world objects better than abstract shapes. As it turned out, the larger number of stimuli used was more critical than their familiarity (Wright, 1990; Wright, Urcuioli, & Sands, 1986). Training began with pairs of pictures. The monkeys moved a lever (right/left) to indicate whether or not the pair of pictures were the "same" (right movement) or "different" (left movement). After several months of training with pairs of pictures, simultaneously presented, they learned the same/different concept. The simultaneous same/different task (both pictures present at the same time) was changed to a delayed same/different task, and then the first item was expanded into a list.

Shown in Figure 1 is performance of a monkey, Oscar, with lists of 10 and 20 items. Each item is presented for 1s with a 1s (ISI) separation. The serial position functions show the benchmark "U" shape. Performance is good at the beginning and the end of the list and poorer in the middle. As with humans, the good performance at the end of the list (recency effect) could be taken to represent short-term memory (STM) and the good performance at the beginning of the list (primacy effect) could be taken to represent long-term memory (LTM). This was the first demonstration of a primacy effect for a nonhuman animal.

The reason for why our monkey's performance was better than that from previous studies, is shown in an experiment of ours where a small number of memory items were repeated. We used the same pool size (6) and the same list length (3) as was used in a study by Gaffan (1977). With these conditions, items had to be repeated many times across trials. At best, list item repetitions had to begin by the third trial. We compared results from this condition, to those from a trial-unique condition where items were seen only one time each daily session. As shown in Figure 2, performance was 93% in the trial-unique condition and a paltry 70% in the highly-repeating 6-item pool condition. This was virtually the same performance level as found by Gaffan (1977) and others testing monkeys with 3-item lists (e.g., Eddy, 1973; Devine & Jones, 1975).

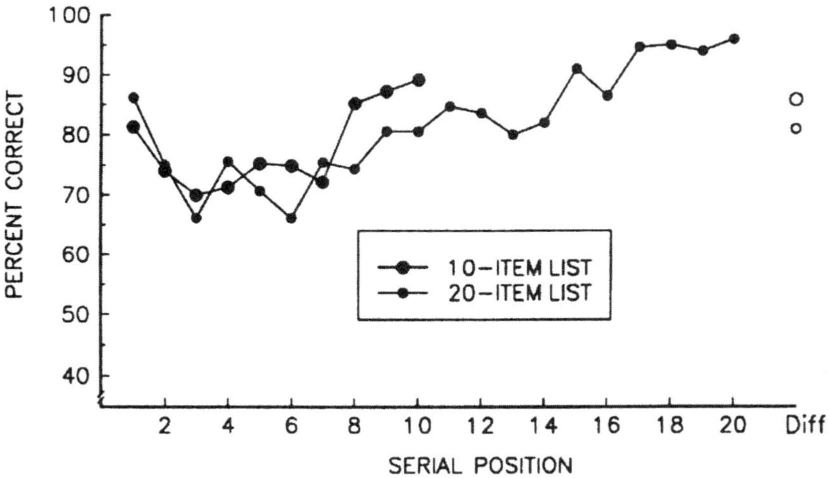

Figure 1. Memory performance by a monkey, Oscar, with 10 and 20 list items. Performance on "different" trials (Diff) is shown to the right of the functions for trials where the probe or test item matched no list item.

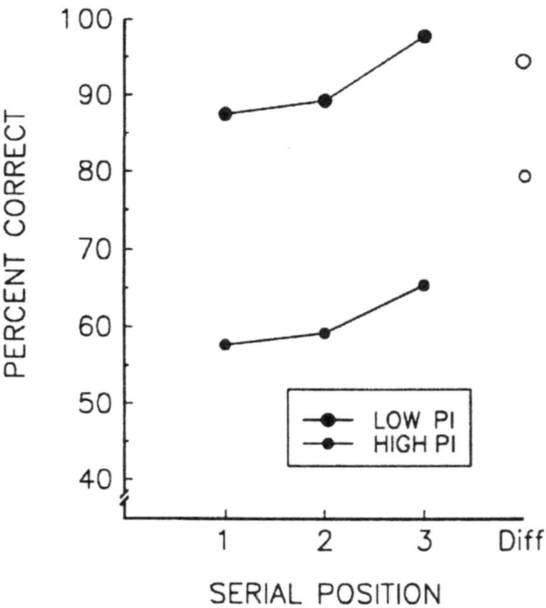

Figure 2. Memory performance by a monkey, Oscar, with 3-item lists where the picture stimuli were selected from either a 211-item pool where proactive interference was low (LOW PI) or were selected from a 6-item pool where proactive interference was high (HIGH PI).

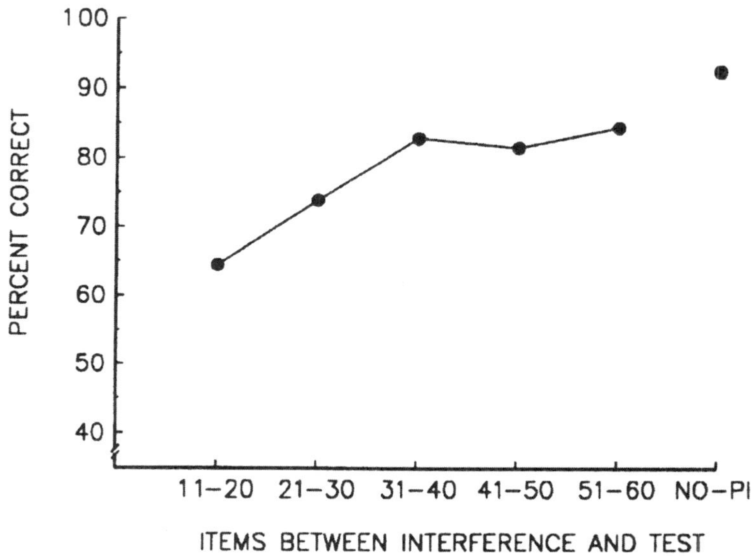

Figure 3. Memory performance by a monkey, Oscar, showing the effect of proactive interference on 10-item list memory as a function of interpolated items between prior exposure to an item and its appearance as a probe item on a "different" trial.

Repeating list items creates proactive interference (PI), interference from previous items. In order to explain how PI affects memory performance, consider the situation where a subject is in the middle of a session and has completed 20-30 trials. Let us say that a probe item currently being evaluated actually does not match any list item (a Different trial). Because there are only six total items, the three list items of the current trial will have been seen many times previously. Having seen the probe item in a previous list, maybe just in the immediately preceding list, confuses the subject. There is a tendency to think that it was in the list being tested, and consequently a tendency to make a mistake. Thus, the locus of confusion of the (repeated-item) PI effect is the probe item on Different trials.

The far-reaching effects of proactive interference are shown in Figure 3. In this experiment, interfering items were placed in previous lists. These interfering items would later appear as probe items on Different trials. The number of intervening 10-item lists or total number of intervening items between the interfering item and the probe test was the variable of primary interest in this experiment.

Figure 3 shows that when the interfering item occurred in the previous list (10-20 items intervening) performance was only 64% correct. This compares to 93% correct for the no-interference condition which is a

29% performance difference. Notice that even when 51-60 items, or 6, 10-item lists, intervened between the interfering items and the probe test item, there was still a 10% interference effect relative to the no-interference condition.

The relatively accurate performance resulting from minimizing proactive interference allowed the retention of memory to be tested. In this experiment, we tested two (different) rhesus monkeys with 4-item lists and varied the retention interval across sessions (in a blocked design). The procedures were the same as those previously described except that the monkeys were free roaming in the test cage and reached between cage bars to manipulate the lever.

The serial position functions in Figure 4 show changes in memory as a function of retention interval. As the retention interval or delay was lengthened, the recency effect dissipated. Additionally, on the immediate test there was no primacy effect. Primacy appeared only after a delay. These results show that primacy and recency can be separated and dissociated by retention interval. Once considered the 'benchmark' of list memory, the U-shaped serial position function is actually a transitional function, between an all-recency function and an all-primacy function. These results raised the question of whether such serial position function changes with retention interval were characteristic of memory systems generally, or were particular to visual memory. In order to answer this question, similar tests were conducted on the auditory memory of rhesus monkeys.

MONKEY AUDITORY MEMORY

Two rhesus monkeys (*Macaca mulatta*) were tested for their auditory memory. There were not many indications from the literature as to what might be the outcome. Previous attempts to train monkeys in auditory list-memory tasks had not been successful (e.g., Thompson, 1980; Mishkin, personal communication). Even comparatively simple auditory discrimination tasks have not been easy for monkeys to learn (e.g., Cowey, 1968; D'Amato, 1973; D'Amato & Salmon, 1982; Dewson & Cowey, 1969; Dewson, Wertheim, and Lynch, 1968; Thompson, 1980; Wegener, 1964). The monkeys apparent inability to learn abstract concepts with auditory stimuli prompted speculation that this was a measure of the monkey's inferior intellect (D'Amato, Salmon, and Colombo, 1985). For auditory memory research, this is a particular problem because concept learning provides the foundation for successful list-memory performance. These difficulties or inabilities to learn with auditory stimuli have puzzled researchers because monkeys are responsive to vocalizations in wild and

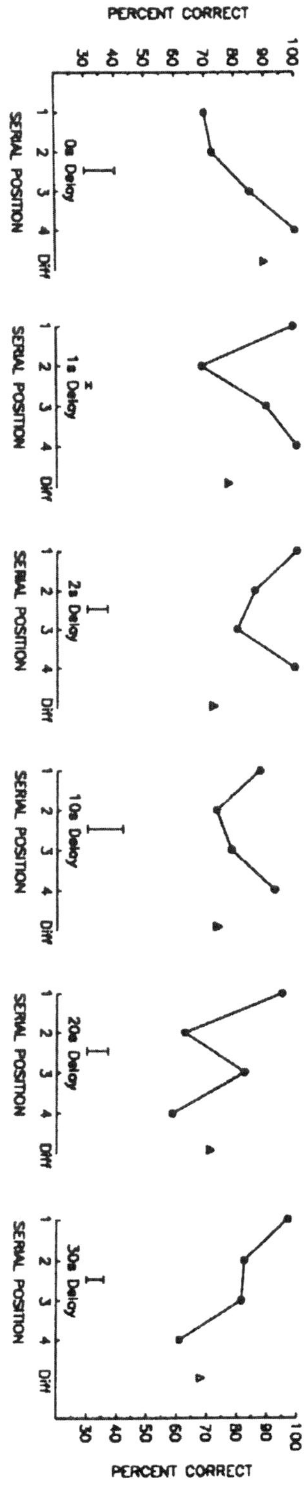

Figure 4. Average 4-item visual list memory for two rhesus monkeys under different retention delays between the last list item (4) and the probe item.

laboratory (Beecher, Petersen, Zoloth, Moody, & Stebbins, 1979; Green & Marler, 1979; Seyfarth, Cheney, & Marler, 1980; Waser, 1975, 1977; Waser and Waser, 1977; Zoloth, Petersen, Beecher, Green, Marler, Moody, & Stebbins, 1979), and their hearing is sensitive (Stebbins, 1970, 1971, 1973).

The human memory literature also provides few clues as to the possible outcomes of these experiments. Human auditory memory experiments routinely use longer lists, verbal stimuli (words), recall procedures, and little or no manipulation of retention interval.

Our first attempts to train monkeys in the auditory-memory task were also unsuccessful. Our rationale had been to use a large number of auditory stimuli to eliminate repeated-item PI. Sounds were presented from upper and lower pairs of stereo speakers, and a monkey moved a lever to indicate same or different for pairs of natural/environmental sounds. Even after 1 1/2 years of training, still there was no learning. We even tried multi-media training (visual plus auditory stimuli) followed by visual-stimulus fading with a well-trained visual list-memory subject, Oscar, the subject of the experiments presented in Figures 1-3. Still no learning. We then changed the basic design of the task and, among other things, took advantage of the more rapid learning resulting from response contact close to the source of the sound (e.g., Harrison, Downey, Iverson & Pratt, 1977).

Two experimentally naive monkeys were trained to touch the sound sources—the speakers. A test cage was modified so that monkeys could touch copper screens (connected via high-impudence circuit to a computer) in front of 3 speakers located on 3 cage walls (Wright, Shyan, & Jitsumori, 1990), as depicted in Figure 5. Monkeys were trained to touch the center speaker during a single list sound and then received a test sound "played" simultaneously from both side speakers. The test matched the center-speaker sound on half the trials, and on these trials monkeys were required to touch the right-side speaker to obtain a 3.5cc reward of Tang orange drink. On the other half of the trials, the test was different from all list sounds, and the monkey was required to touch the left side speaker to obtain the juice reward. Juice rewards were dispensed adjacent to each side speaker. In preliminary training, the probe sounds came only from the correct side to help learning. Then intensity from the incorrect side speaker was gradually increased over several weeks of training. The monkeys learned the abstract concept necessary for successful list-memory performance.

In the auditory list memory experiment, a list of 4 different sounds (2-s duration, 1 s separation of sounds) was played from the center speaker. Following the list, there was a retention delay. The delay was fixed for a 32-trial session, but varied in randomized blocks across session. The same 6 retention delays tested in the visual list memory experiment were tested in this experiment. After the delay, a single test sound was "played"

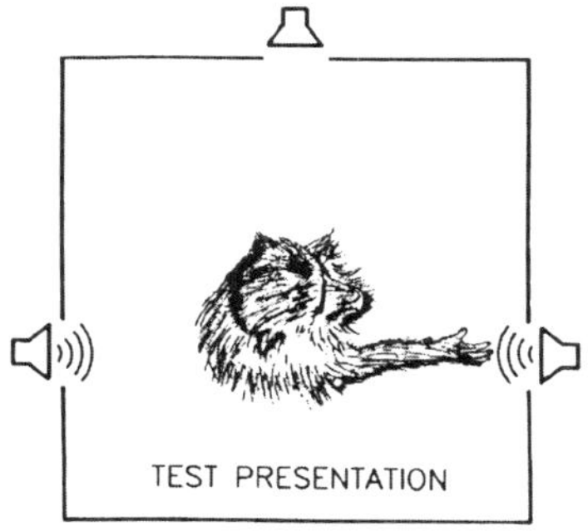

Figure 5. Diagram of auditory memory procedure. Monkey subjects touched speakers (copper mesh) from which sounds were "played".

simultaneously from the two side speakers for 6s. Choice responses were accepted after 2s of test sound. The test matched one list sound on half the trials, and the monkey was required to touch the right-side speaker to obtain a 3.5cc reward of Tang orange drink. On the other half of the trials, the test sound was different from all list sounds (unfilled points), and the monkey was required to touch the left side speaker to obtain a juice reward. Juice rewards were dispensed adjacent to each side speaker. Incorrect choices were not rewarded and were followed by a 30s time-out. A pseudorandom intertrial interval of 12 s to 28 s in duration followed correct choices or time-outs. There were 520 environmental sounds used (e.g., fog horn, adding machine, Morse code, stagecoach passing, frog croaks, shortwave tuning, door knocking, seagulls, French siren, crow caws, sonar pings, music box, angry cat, boat creaking, police whistle, tea kettle, crickets chirping, cow mooing, dentist drill, safe dial, steamboat whistle, cable car, rooster crows, etc.) Sounds for lists and trial sequences varied daily.

The average (mean) results for the subjects are shown in Figure 6. Also shown are the average results from the visual list memory experiment (Figure 4). Both experiments were similar: 2 rhesus monkeys, 4 memory

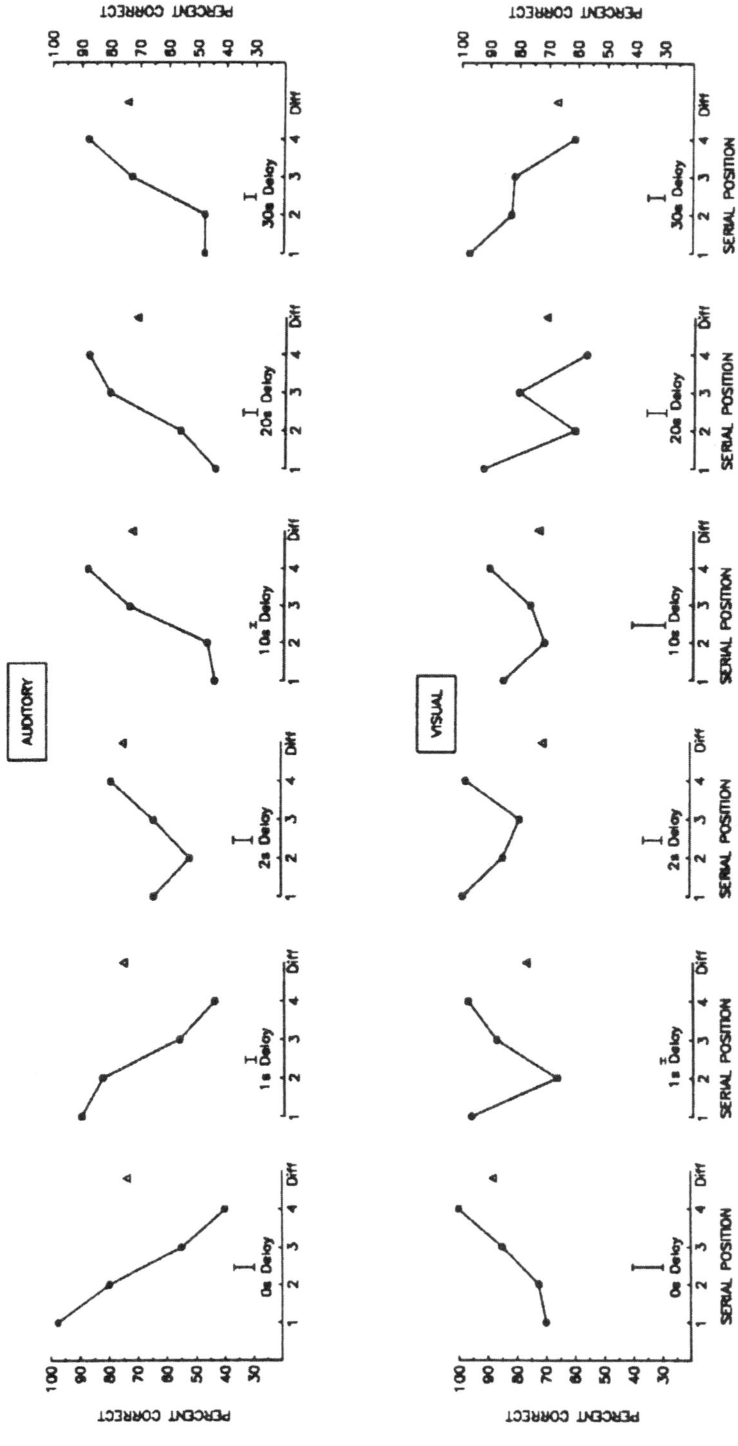

Figure 6. Average 4-item auditory list memory for two rhesus monkeys under different retention delays between the last list item (4) and the probe item compared to visual list memory from Figure 4.

list items, single-item tests at the same retention intervals in the same blocked design, and a large item pool of memory items presented trial unique. The most striking difference is the opposite shape for auditory and visual serial positions functions. On the immediate test (0s), auditory memory shows a primacy effect but no recency effect, whereas visual memory shows a recency effect but no primacy effect. At a 2s delay, functions from both modalities are U-shaped; if these had been the only delays tested auditory and visual list memory would have seemed rather similar. At long delays, functions from the different modalities diverge once again; auditory memory shows a strong recency effect but no primacy effect, whereas visual memory shows a strong primacy effect but no recency effect.

The serial position functions for each modality change dynamically with retention interval. Primacy and recency effects are dissociated by the parameter of retention interval. This dissociation indicates that primacy and recency are a function of different underlying memory process or mechanisms. Because the changes in primacy and recency are opposite for visual and auditory memory, the underlying memory process or mechanisms are going to be different. One possibility is that they share at least some common properties, but their primacy/recency roles are reversed. Different shaped serial position functions for auditory and visual memory may constrain theoretical accounts, and help converge on the true underlying processes. In order to put these results in perspective, they are discussed in the next sections in relation of some memory theories past and present.

THEORETICAL IMPLICATIONS

Decay Models

An aspect of the results that poses special problems for some memory theories is the improvement in memory shown for a portion of visual and auditory memory. Improvement of memory is contrary to the most common notion that all memory fades with time. Notice in Figure 6 that visual primacy increases 27% (70-97%) and auditory recency increases 50% (40-90%). These increases in memory are counter to theories that memory fades with time due to unlearning (e.g., Melton & Irwin, 1940), decay (e.g., Brown, 1958; Wingfield & Byrnes, 1972), displacement (e.g., Waugh & Norman, 1965), acid bath (Posner & Konick, 1966), or ratios of forgetting intervals (e.g., Glenberg, Bradley, Kraus, & Renzaglia, 1983; Glenberg, Bradley, Stevenson, Draus, Tkachuk, Gretz, Fish, & Turpin, 1980).

Modal Model

Perhaps the most influential memory model has been the dual-store or modal model (Atkinson & Shiffrin, 1968; Waugh & Norman, 1965). This model uses decay as a theoretical construct, but only for short-term memory (STM). By contrast, long-term memory (LTM) is thought to be comparatively immune to decay. The two memory stores are related by rehearsal. Rehearsal moves items from STM (before they decay) to LTM. Evidence for this limited capacity STM "buffer" was provided by results showing that the recency effect dissipated with retention interval when the retention interval was filled with distracter activity (e.g., Gardiner, Thompson, & Maskarinec, 1974; Glanzer & Cunitz, 1966; Glanzer, 1972; Postman & Phillips, 1965; Roediger & Crowder, 1975). Distractor activity prevented rehearsal so items in the STM buffer (represented by the recency effect) could not be maintained. Thus, STM or primary memory was thought to be reflected by the recency effect. LTM was thought to be reflected by the primacy effect. Evidence for LTM and the primacy effect being created by rehearsal was provided by the frequency of overt rehearsals matching the primacy effect of the serial position function (Rundus, 1971). This was correlational evidence which does not identify the cause of the primacy effect. Nevertheless, rehearsal, STM-LTM, and the modal model continue their influence (e.g., Modigliani & Hedges, 1987).

Over and above the obvious problem of monkeys (actively) rehearsing, the dual-store memory model has trouble accounting for certain aspects of the results presented in Figure 6. The initial absence and later appearance of the visual primacy effect cannot be the result of rehearsal, as one might predict from the modal model. In order to retrieve items at short delays so that they could be rehearsed, they would have to be remembered. But they were not remembered at the short delays. Auditory memory results pose additional problems for the Modal model. Auditory memory initially shows a strong primacy effect (LTM), and no recency effect (STM). But how can there be LTM with no STM, when, according to the theory, rehearsal of STM is supposed to produce LTM?

Associative Memory Models

Memory models, particularly associative or network memory models, have been highly fashionable. Some of the more well known ones go by acronyms of TODAM, SAM, CHARM, MINERVA2, standing for Theory of Distributed Associative Memory, Search of Associative Memory, etc. These associative-network theories of memory typically do not deal with memory changes over time (e.g., MINERVA2—Hintzman, 1988; CHARM—Metcalfe-Eich, 1982; Matrix Model—Humphreys, Pike, Bain &

Tehan, 1989; SAM—Gillund & Shiffrin, 1984; REM—Shiffrin & Steyvers, 1997). One model, TODAM (Murdock, 1982; Murdock & Kahana, 1993), does have a parameter, alpha, for decay of memory with time. But this is monotonic decay, a decreasing slope of the recency effect with retention. This model in its present form probably would not be able to account for the increasing memory of the visual primacy effect or the auditory recency effect. Another model only deals with paired-associate (single-item) memory, but it was developed to account for the memory increase of the spontaneous-recovery effect (Mensink & Raaijmakers, 1988). It remains to be seen whether a similar model could adequately account for the increases in serial position effects shown here.

Other Connectionist Models

Back-propagation connectionist models (e.g., Rumelhart, Hinton, & Williams, 1986) have been used to model learning and memory processes. Typically, forgetting in these models is a matter of over-writing and unlearning, a kind of catastrophic unlearning (e.g., McCloskey & Cohen, 1989). While some models may be less susceptible to catastrophic unlearning (e.g., Sloman & Rumelhart, 1992) they all have great difficulty in dealing with improvements in memory with time (spontaneous recovery), where something forgotten (or extinguished in the case of learning) recovers over time. In the auditory memory case, the improvement in recency memory with time is an absolute rise in performance (not just a relative change to recency) and cannot be explained by unlearning of the primacy items. In the visual case, primacy memory improves with time, also a rise in absolute (not just relative) performance. Postman and Underwood (e.g., Postman, Stark & Fraser, 1968; Underwood, 1948b) in their explanation of paired-associate memory with interference theory used the term spontaneous recovery to refer to similar memory increases with time.

Interference Theory

The serial position function is not a static U-shaped function, as once thought. Any explanation of the monkey's visual and auditory memory results would seem to depend upon processes that themselves are dynamically changing. Before the cognitive revolution, dynamically changing interference processes were the focus of memory research. Proactive interference (PI), interference from preceding items on the memory item, was shown to be weak immediately after learning and grew slowly with time (e.g., Keppel & Underwood, 1962; Postman and Phillips, 1965; Postman, Stark, & Fraser, 1968). Retroactive interference (RI), interference from following items on the memory item, was shown to be strong immediately after learning but dissipated rapidly (e.g., Briggs, 1954; Ceraso & Henderson, 1966; Forrester, 1970; Koppenaal, 1963; Postman, Stark & Fraser, 1968; Postman, Stark, & Henschel, 1969; Shulman & Martin, 1970; Silverstein, 1967; Slamecka, 1966; Underwood, 1948a, 1948b). Applying these facts to the interference within list items in visual memory, an initial high RI would mean that the last items would interfere with memory of the first items and thereby prevent a primacy effect. Dissipation of RI would allow the primacy effect to appear, and produce a U-shaped serial position function. Since PI was low immediately after learning, the first list items would not interfere with memory of the last list items. The slow growth of PI would mean that eventually the first list items would interfere with memory of the last items and eventually eliminate the recency effect. If analogous interference processes were operating in auditory memory, then the roles of PI and RI would have to be reversed (see Wright, 1999).

Human Modality Effect

Human auditory memory has not been as thoroughly investigated as human visual memory. Serial position functions like those shown in Figure 7 for auditory memory have not been shown for human subjects. There are probably several reasons for why these investigations have not been conducted. Human memory experiments use much longer lists than the four items used in the monkey memory experiments. The practical advantage of longer lists is that ceiling effects can be avoided. The disadvantage is that as the list becomes longer, it is more difficult to identify changes in primacy and recency, particularly those occurring at short delays. For longer lists, some of changes shown in this article at short delays will have run their course well before there is an opportunity to test them. Another reason for why similar investigations have not been conducted with humans is that free-recall is used (almost exclusively) with humans instead of a serial-probe-recognition (SPR) procedure like the one was used here. The SPR procedure controls time between item presentation and test, but only one item per list is tested. The free-recall procedure allows time between presentation and test to vary, in part, according to the subject's recall order, but all items are tested. Because recall order is left to the subject, they can first recall the more easy to remember items (e.g., visual recency items) and thereby build a little delay. This delay might then allow (previously) more difficult to remember items (e.g., visual primacy items) to be recalled.

Despite these procedural differences, there is a well documented modality effect in human memory. This is a recency advantage for auditory over visual memory (e.g., Crowder, 1986; Engle & Mobley, 1976; Gathercole & Gardiner, 1982). Despite difficulties of comparing human and monkey memory results, longer lists and free-recall used in human memory would tend to be more comparable to the longer as opposed to shorter delay conditions of the monkey memory investigations. At longer retention intervals with monkeys, visual recency has dissipated and auditory recency has strengthened. The visual serial position functions tend toward all-primacy functions, whereas auditory serial position functions tend toward all-recency functions. The modality effect may result from similar trends and different shapes of the auditory vs. visual serial position functions. If they do, then there might be expected to be a primacy advantage for visual memory over auditory memory. And, indeed, there is some evidence for a visual primacy advantage (e.g., Engle & Mobley, 1976).

A variety of theories have been advanced to explain the modality effect. They include **PAS** a sensory or echoic storage of the auditory item that was thought to provide the auditory recency advantage (Crowder & Morton, 1969). But a sensory store doesn't make sense from the standpoint of the auditory serial position functions of Figure 6 because the recency effect is absent just after study, when one would expect the sensory store to be strongest, and grows with time when one would expect a sensory store to fade with time. Other explanations of the modality effect focus on language related processing (Morton, Marcus & Ottley, 1981; Richardson, 1979; Shand & Klima, 1981; Surprenant, Pitt & Crowder, 1993). But it cannot be language because monkeys do not have language. Perhaps, the human modality effect will be shown to depend upon similar differences in shapes of the serial position functions with retention interval.

CONCLUDING REMARKS

The contrasting shaped functions for auditory and visual memory raise questions regarding their possible practical value and purpose. Practical advantages could result from coding to-be-remembered items both visually and auditorily. From the opposite shaped serial position functions shown in Figure 6, it can be seen that if visual memory is poor, for example the first items on immediate testing, then auditory memory will be good. Likewise, if auditory memory is poor, for example the first items after a long retention delay, then visual memory will be good. There is considerable evidence that humans cross-modality code information. We frequently pronounce or silently-mouth visual material (e.g., Crowder, 1983). We often visualize auditory material such as environmental sounds or words (e.g., Campbell and Dodd, 1980). Thus, if humans were to show opposite shaped serial position functions like the monkeys, then this would explain the practical value of cross-modality coding. While there is no evidence (that I am aware) about human auditory serial position functions from similar experiments, there is evidence that human visual serial position functions change in similar ways with retention delay (Korsnes, 1995; Korsnes and Gilinsky, 1993; Neath, 1993; Neath and Knoedler, 1994; Wright et al., 1985).

From the standpoint of monkeys, cross-modality coding is unlikely. Monkeys do not readily code items (Neiworth & Wright, 1994), and behaviors dependent upon coding (e.g., rehearsal) are generally nonexistent (Cook, Wright, & Sands, 1991). The purpose of different shaped serial position functions for animals might be related specific associations in learning. Animals tend to associate visual stimuli with food better than auditory stimuli, and to associate auditory stimuli with danger better than visual stimuli (e.g., Shapiro, Jacobs, & LoLordo, 1980). With this in mind, consider the different shaped serial position functions of Figure 7. If an animal has been having good luck foraging in a patch, for example, it will need to remember to continue feeding there (visual recency, short delay). On the other hand, if an animal feeds in a depleting patch (e.g., berries that are ripe in the morning but are quickly depleted), then the animal will need to remember to go to this patch first thing in the morning after an overnight delay (visual primacy, long delay). Now, consider auditory memory. If an animal hears a danger sound, it will need to know where the sound came from (auditory primacy, short delay) to determine if the sound is coming toward it and which way to escape. On the other hand, if an animal hears a danger sound but the sound abruptly stops, then the animal will need to remember where the sound was last heard (auditory recency) to avoid a possible waiting predator (auditory recency, long delay).

ACKNOWLEDGMENTS

The research and preparation of this manuscript was supported by NIH grants MH 32302 and DA 10715 to the author. The author thanks Jacquelyne J. Rivera for her careful conduct of many of the experiments reported here.

REFERENCES

Atkinson, R. C. & Shiffrin, R. M. (1968). Human memory: A proposed system and its control processes. In K. W. Spence & J. T. Spence (Eds.), *The psychology of learning and motivation* (Vol. 2, pp. 89-105). New York: Academic Press.

Beecher, M., Petersen, M., Zoloth, S., Moody, D., & Stebbins, W. (1979). Perception of conspecific vocalizations by Japanese monkeys (*Macaca fuscata*). *Brain, Behavior and Evolution*, **16**, 443-460.

Bolhuis, J. J., & van Kampen, H. S. (1988). Serial position curves in spatial memory of rats: primacy and recency effects. *The Quarterly Journal of Experimental Psychology*, **40**, 135-149.

Briggs, G. E. (1954). Acquisition, extinction, and recovery functions in retroactive inhibition. *Journal of Experimental Psychology*, **47**, 285-293.

Buchanan, J. P., Gill, T. V., & Braggio, J. T. (1981). Serial Position and clustering effects in chimpanzee's "free recall." *Memory and Cognition*, **9**, 651-660.

Brown, J. (1958). Some tests of the decay theory of immediate memory. *Quarterly Journal of Experimental Psychology*, **10**, 12- 21.

Campbell, R. & Dodd, B. (1980). Hearing by eye. *Quarterly Journal of Experimental Psychology*, **32**, 85-99.

Castro, C. A., & Larsen, T. (1992). Primacy and recency effects in nonhuman primates. *Journal of Experimental Psychology: Animal Behavior Processes*, **18**, 335-340.

Ceraso, J., & Henderson, A. (1966). Unavailability and associative loss in RI and PI: Second try. *Journal of Experimental Psychology*, **72**, 314-316.

Cook, R. G., Wright, A. A., & Sands, S. F. (1991). Interstimulus interval and viewing time effects in monkey list memory. *Animal Learning & Behavior*, **19**, 153-163.

Cowey, A. (1968). Discrimination. In L Weiskrantz (Ed.) *Analysis of behavioral change* (pp. 189-238). NY: Harper & Row.

Crowder, R. G. (1983). The purity of auditory memory. *Philosophical Transactions of the Royal Society, Section B*, **302B**, 251-265.

Crowder, R. G. (1986). Auditory and temporal factors in the modality effect. *Journal of Experimental Psychology: Learning, Memory, and Cognition*, **2**, 268-278.

Crowder, R. G. & Morton J. (1969). Precategorical acoustic storage (PAS). *Perception & Psychophysics*, **5**, 365-378.

D'Amato, M. R. (1973). Delayed matching and short-term memory in monkeys. In G. H. Bower (Ed.) *The Psychology of Learning and Motivation: Advances in Theory and Research*, **7**, New York: Academic Press, pp. 227-269.

D'Amato, M. R. & Salmon, D. P. (1982). Tune discrimination in monkeys (Cebus apella) and rats. *Animal Learning & Behavior*, **10**, 126-134.

D'Amato, M. R., Salmon, D. P., & Colombo, M. (1985). Extent and limits of the matching concept in monkeys (*Cebus apella*). *Journal of Experimental Psychology: Animal Behavior Processes*, **11**, 35-51.

Davis, R. T., & Fitts, S. S. (1976). Memory and coding processes in discrimination learning, In D. L. Medin, W. A. Roberts, & R. T. Davis (Eds.), *Processes of Animal Memory*. Hillsdale, N. J.: Erlbaum.

Devine, J. V., & Jones, L. C. (1975). Matching-to-successive samples: A multiple-unit memory task with rhesus monkeys. *Behavior Research Methods and Instrumentation*, **7**, 438-440.

Dewson, J. H., III & Cowey, A. (1969). Discrimination of auditory sequences by monkeys. *Nature*, **222,** 695-697.

Dewson, J. H., III, Wertheim, G. A., & Lynch, J. C. (1968). Acquisition of successive auditory discrimination in monkeys. *Journal of the Acoustical Society of America*, **43**, 162-163

Ebbinghaus, H. E. (1902). *Grundzuge der Psychologie*. Leipzig: Von Viet, 1902.

Eddy, D. R. (1973). *Memory processing in Macaca speciosa: Mental processes revealed by reaction time experiments*. Unpublished doctoral dissertation, Carnegie-Mellon University.

Engle, R. W. & Mobley, L. A. (1976). The modality effect: What happens in long-term memory? *Journal of Verbal Learning and Verbal Behavior*, **15**, 519-527.

Forrester, W. E. (1970). Retroactive inhibition and spontaneous recovery in the A-B, D-C paradigm. *Journal of Verbal Learning and Verbal Behavior*, **9**, 525-528.

Gaffan, D. (1977). Recognition memory after short retention intervals in fornix-transected monkeys. *Quarterly Journal of Experimental Psychology*, **29**, 577-588.

Gardiner, J. M., Thompson, C. P. & Maskarinec, A. S. (1974). Negative recency in initial free recall. *Journal of Experimental Psychology*, **103**, 71-78.

Gathercole, S. E., Gardiner, J. M., & Gregg, V. H. (1982). Modality and phonological similarity effects in serial recall: Does one's own voice play a role? *Memory & Cognition*, **10**, 176-180.

Glanzer, M. (1972). Storage mechanisms in recall. In G. H. Bower and J. T. Spence (Eds.) *The Psychology of Learning and Motivation*. (Vol. 5, pp. 129-193) NY: Academic Press.

Glanzer, M., & Cunitz, A. R. (1966). Two storage mechanisms in free recall. *Journal of Verbal Learning and Verbal Behavior*, **5**, 351-360.

Glenberg, A. M., Bradley, M. M., Kraus, T.A., & Renzaglia, G. J. (1983). Studies of the long-term recency effect: Support for a contextually guided retrieval hypothesis. *Journal of Experimental Psychology: Learning, Memory, and Cognition*, **9**, 231-255.

Glenberg, A. M., Bradley, M. M., Stevenson, J. A., Kraus, T. A., Tkachuk, M. J., Gretz, A. L., Fish, J. H., & Turpin, B. M. (1980). A Two-process account of long-term serial position effects. *Journal of Experimental Psychology: Learning, Memory, and Cognition*, **6**, 355-369.

Green, S. & Marler, P. (1979). The analysis of animal communication. In Marler, P. & Vandenberg, J. G. (Eds.) *Social Behavior and Communication*, Vol. 3, New York: Plenum.

Gillund, G. & Shiffrin, R. M. (1984). A retrieval model for both recognition and recall. *Psychological Review*, **19**, 1-65.

Harper, D. N., McLean, A. P., & Dalrymple-Alford, J. C. (1993). *Journal of Experimental Psychology: Animal Behavior Processes*, **19**, 307.

Harrison, J. M., Downey, P., Iverson, S. D. & Pratt, S. R. (1977). Control of responding by location of auditory stimuli: Adjacency of sound and response. *Journal of the Experimental Analysis of Behavior*, **28**, 243-251.

Hintzman, D. L. (1988). Judgments of frequency and recognition memory in a multiple-trace model. *Psychological Review*, **95**, 528-551.

Humphreys, M. S., Pike, R., Bain, J. D., & Tehan, G. (1989). Global matching: A comparison of the SAM, Minerva II, Matrix, and TODAM models. *Journal of Mathematical Psychology*, **33**, 36- 67.

Keppel, G. & Underwood, B. J. (1962). Proactive inhibition in short-term retention of single items. *Journal of Verbal Learning and Verbal Behavior*, **1**, 153-161.

Kesner, R. P. & Novak, J. M. (1982). Serial position curve in rats: role of the dorsal hippocampus. *Science*, **218**, 173-175.

Koppenaal, R. J. (1963). Time changes in the strengths of A-B, A-C lists: Spontaneous recovery. *Journal of Verbal Learning and Verbal Behavior*, **2**, 310-319.

Korsnes, M. S. (1995). Retention intervals and serial list memory. *Perceptual and Motor Skills*, **80**, 723-731.

Korsnes, M. S. & Gilinsky, S. A. (1993). Aging and serial list picture memory. *Perceptual and Motor Skills*, **76**, 1011-1014.

McCloskey, M. & Cohen N. J. (1989). Catastrophic interference in connectionist networks: The sequential learning problem. In G.H. Bower (Ed.) *The Psychology of Learning and Motivation* (Vol. 24, pp. 109-165). San Diego, CA: Academic Press.

Melton, A. W. & Irwin, J. M. (1940). The influence of degree of interpolated learning on retroactive inhibition and the overt transfer of specific responses. *American Journal of Psychology*, **53**, 173-203.

Mensink, G. & Raaijmakers, J. G. W. (1988). A model for interference and forgetting. *Psychological Review*, **95**, 434- 455.

Metcalfe-Eich, J. (1982). A composite holographic associative recall model. *Psychological Review*, **89**, 627-661.

Modigliani, V. & Hedges, D. G. (1987). Distributed rehearsals and the primacy effect in single-trial free recall. *Journal of Experimental Psychology: Learning, Memory & Cognition*, **13**, 426-436.

Morton, J., Marcus, S. M. & Ottley P. (1981). The acoustic correlates of "speechlike": A use of the suffix effect. *Journal of Experimental Psychology: General*, **110**, 568-593.

Murdock, B. B. (1982). A theory for the storage and retrieval of item and associative information. *Psychological Review*, **89**, 609-626.

Murdock, B. B., Jr. & Kahana M. J. (1993). List-strength and list-length effects: Reply to Shiffrin, Ratcliff, Murnane and Nobel (1993). *Journal of Experimental Psychology: Learning, Memory, and Cognition*, **19**, 1450-1453.

Neath, I. (1993). Distinctiveness and serial position effects in recognition. *Memory & Cognition*, **21**, 689-698.

Neath, I., & Knoedler, A. J. (1994). Distinctiveness and serial position effects in recognition and sentence processing. *Journal of Memory and Language*, **33**, 776-795.

Neiworth, J. J. & Wright, A. A. (1994). Monkeys (*Macaca mulatta*) learn category matching in a nonidentical same/different task. *Journal of Experimental Psychology: Animal Behavior Processes*, **20**, 429-435.

Peterson, L. R. & Peterson, M. J. (1959). Short-term retention of individual verbal items. *Journal of Experimental Psychology*, **58**, 193-198.

Posner, M. I. & Konick, A. W. (1966). On the role of interference in short-term retention. *Journal of Experimental Psychology*, **72**, 221-231.

Postman, L., & Phillips, L. (1965). Short-term temporal changes in free-recall. *Quarterly Journal of Experimental Psychology*, **17**, 132-138.

Postman, L., Stark, K. & Fraser, J. (1968). Temporal changes in interference. *Journal of Verbal Learning and Verbal Behavior*, **7**, 672-694.

Postman, L., Stark, K. & Henschel, D. (1969). Conditions of recovery after unlearning. *Journal of Experimental Psychology*, **82**, (1, Pt. 2).

Reed, P., Chih-Ta, T., Aggleton, J. P. & Rawlins, J. N. P. (1991). *Journal of Experimental Psychology: Animal Behavior Processes*, **17**, 36.

Richardson, J. T. E. (1979). Precategorical acoustic storage and postcategorical lexical storage. *Cognitive Psychology*, **11**, 265-286.

Roberts, W. A. & Kraemer, P. J. (1981). Recognition memory for lists of visual stimuli in monkeys and humans. *Animal Learning and Behavior*, **9**, 587-594.

Roediger, H. L. III & Crowder, R. G. (1975). Spacing of lists in free recall. *Journal of Verbal Learning and Verbal Behavior*, **14**, 590-602.

Rumelhart, D. E., Hinton, G. E. & Williams R. J. (1986). Learning internal representations by error propagation. In D. E. Rumelhart, J. L.

McClelland & the PDP Research Group (Eds.) *Parallel Distributed Processing: Explorations in the Microstructure of Cognition: Vol.1. Foundations*. (pp. 318- 362). Cambridge, MA: Bradford.

Rundus, D. (1971). Analysis of rehearsal processes in free recall. *Journal of Experimental Psychology*, **89**, 63-77.

Sands, S. F. & Wright, A. A. (1980a). Primate memory: retention of serial list items by a rhesus monkey. *Science*, **209**,938- 940.

Sands, S. F. & Wright (1980b). Serial probe recognition performance by a rhesus monkey and a human with 10- and 20- item lists. *Journal of Experimental Psychology: Animal Behavior Processes*, **6**, 386-396.

Seyfarth, R. M., Cheney, D. & Marler, P. (1980). Monkey responses to different alarm calls: evidence of predator classification and semantic communication. *Science*, **210**, 801-803.

Shand, M. A & Klima, E. S. (1981). Nonauditory suffix effects in congenitally deaf signers of American Sign Language. *Journal of Experimental Psychology: Human Learning and Memory*, **7**, 464- 474.

Shapiro, K. L., Jacobs, W. J., & LoLordo, V. M. (1980). Stimulus-reinforcer interactions in Pavlovian conditioning of pigeons: Implications for selective associations. *Animal Learning & Behavior*, **8**, 586-594.

Shiffrin, R. M., & Steyvers, M. (1997). A model for recognition memory: REM—retrieving effectively from memory. *Psychonomic Bulletin & Review*, **4**, 145-166.

Shulman, H. G. & Martin, E. (1970). Effects of response set similarity on unlearning and spontaneous recovery. *Journal of Experimental Psychology*, **86**, 230-235.

Silverstein, A. (1967). Unlearning, spontaneous recovery, and the partial-reinforcement effect in paired-associate learning. *Journal of Experimental Psychology*, **73**, 15-21.

Slamecka, N. J. (1966). A search for spontaneous recovery of verbal associations. *Journal of Verbal Learning and Verbal Behavior*, **5**, 205-207.

Sloman, S. A. & Rumelhart, D. E. (1992). Reducing interference in distributed memories through episodic gating, In A. F. Healy, S. M. Kosslyn & R. M. Shiffrin (Eds.). *From learning theory to connectionist theory: Essays in honor of William K. Estes* (Vol. 1. pp. 227-248). Hillsdale, NJ: Erlbaum.

Stebbins, W. C. (1970). Studies of hearing and hearing loss in the monkey. In W. C. Stebbins (Ed.) *Animal Psychophysics* (pp. 41-66). NY: Appleton-Century-Crofts.

Stebbins, W. C. (1971). Hearing. In A. M. Schrier & F. Stollnitz (Eds.) *Behavior of nonhuman primates* (Vol. 3, pp. 159-192). NY: Academic Press.

Stebbins, W. C. (1973). Hearing of old world monkeys. *American Journal of Physical Anthropology*, **38**, 357-364.

Surprenant, A. M., Pitt, M. A. & Crowder, R. G. (1993). Auditory recency in immediate memory. *Quarterly Journal of Experimental Psychology*, **46**, 193-223.

Thompson, R. K. R. (1980). Auditory cued reversal and matching-to-sample learning by rhesus monkeys. *Antropologia Comtemporanea*, **3** 284-292.

Thompson, R. K. R. & Herman, L. M. (1977). *Science*, **195**, 501.

Underwood, B. J. (1948a). Retroactive and proactive inhibition after five and forty-eight hours. *Journal of Experimental Psychology*, **38**, 29-38.

Underwood, B. J. (1948b). "Spontaneous recovery" of verbal associations. *Journal of Experimental Psychology*, **38**, 429-439.

Vaughan, W. Jr. and Greene, S. L. (1984). Acquisition of absolute discriminations in pigeons. In: M. L. Commons & R. J. Herrnstein (Eds.) *Quantitative Analysis of Behavior* Vol. IV pp. 231-238.

Waser, P. M. (1975). Experimental playbacks show vocal mediation if intergroup avoidance in forest monkeys. *Nature*, **255**, 56-58.

Waser, P. M. (1977). Sound localization by monkeys; A field experiment. *Behavioral Ecology and Sociobiology*, **2**, 427-431.

Waser, P. M. & Waser, M. S. (1977). Experimental studies of primate vocalization: Specializations for long-distance propagation. *Zietzchrift fur Tierpsychology*, **43**, 239-263.

Waugh, N. C. & Norman, D. A. (1965). Primacy memory. *Psychological Review*, **72**, 89-104.

Wegener, J. G. (1964). Auditory discrimination behavior of normal monkeys. *Journal of Auditory Research*, **4**, 81-106.

Wingfield, A. & Byrnes, D. L. (1972). Decay of information in short-term memory. *Science*, **176**, 690-692.

Wright, A. A. (1990). Memory processing by pigeons, monkeys, and people. In G. H. Bower (Ed.), *The Psychology of Learning and Motivation* (Vol. 23, pp. 25-70). New York: Academic press.

Wright, A. A. (1999). Auditory list memory and interference processes in monkeys: *Journal of Experimental Psychology: Animal Behavior Processes*, **25**, 284-296.

Wright, A. A., Santiago, H. C., Sands, S. F., Kendrick, D. F. & Cook, R. G. (1985). Memory Processing of serial lists by pigeons, monkeys, and people. *Science*, **229**: 287-289.

Wright, A. A., Shyan M. R. & Jitsumori, M. (1990). Auditory *same/different* concept learning by monkeys. *Animal Learning & Behavior*, **18**, 287-294.

Wright, A. A., Urcuioli, P. J., & Sands, S. F. (1986). Proactive interference in animal memory research. In: D. F. Kendrick, M. Rilling & R. Denny (Eds.) *Theories of Animal Memory*, Lawrence Erlbaum, Englewood Cliffs, N.J., pp. 101-125.

Zoloth, S., Petersen, M., Beecher, M., Green, S., Marler, P., Moody, D. & Stebbins, W. (1979). Species-specific perceptual processing of vocal sounds by old world monkeys. *Science*, **204**, 870-873.

5

THE DISCRIMINATIVE STIMULUS AND RESPONSE ENHANCING PROPERTIES OF REWARD PRODUCED MEMORIES

E. J. Capaldi
Purdue University

INTRODUCTION

In this chapter I present a theory which is capable of explaining a wide variety of learning phenomena. Two such phenomena will be emphasized here: serial learning and reward magnitude effects. I shall focus exclusively on a particular type of serial learning, reward outcome serial learning. In that type of serial learning, the stimuli which constitute the series are different qualities or quantities of food items presented successively in a regular order which must be learned. In other types of serial learning, not of immediate concern here, the stimuli might be geometric forms, color, etc. which are presented simultaneously rather than successively (e.g., D'Amato, 1991; Terrace, 1991).

The present approach, unlike others, suggests that any investigation that provides organisms with two or more different magnitudes or qualities of reward on different trials is a variety of reward outcome serial learning and this includes irregular as well as regular schedules of reward (see e.g., Capaldi, 1994). Reward outcomes possess both a discriminative stimulus property and a response enhancing property. For example, larger rewards are discriminably different from smaller rewards and larger rewards enhance behavior more strongly than smaller rewards. Both of these properties must be recognized by any investigation that varies reward magnitudes whether it is called a serial learning investigation or a reward magnitude investigation. But, in general, and as I will emphasize later, serial learning investigators have tended to ignore the response enhancing effects of reward magnitudes. Too, reward magnitude investigators have ignored the sort of discriminative stimulus properties of reward which could explain serial learning. My assumption is that the two areas are inextricably bound together and each can advance only by attending to the other.

The model described here is an extension and an elaboration of my sequential view which posits that distinctive memories are associated with distinctive reward events. These reward related memories, like other stimuli, have two functions, a discriminative or directive function and a response enhancing function. I present here a new type of evidence both for the discriminative stimulus function of reward produced memories and for the response enhancing function of reward produced memories.

The theory presented here is different in one respect or another from various alternative views suggested by others. The new data reported here cannot be explained by these alternative views and in some cases contradicts them. This does not mean, necessarily, that I reject entirely these alternative views. Indeed, in one place or another, I have suggested that each of the alternatives may be useful for explaining some sorts of phenomena, however limited (e.g., Capaldi, 1967, 1994). In any event, the alternatives fall into two classes, theories which attempt to explain the effects of reward and reward magnitude on learning and performance and theories of serial learning. My plan in this chapter is as follows. After briefly describing these alternative views, I supply a short introduction to my own view. I then supply evidence consistent with my view. Having examined that evidence I then go on to indicate what its general implications may be for my view and the alternative views.

ALTERNATIVE THEORIES

In this section I briefly describe two alternative views of how reward affects performance and learning and two alternative theories of serial learning.

(a) ***Reward events affect association formation***. According to this view suggested initially by Hull in 1943 and subsequently accepted by others (e.g., Capaldi, 1967; Rescorla & Wagner, 1972), reward events are necessary for association formation to occur, the bigger the reward the stronger the association. Stronger associations promote more vigorous responding. According to this view, organisms have no knowledge or representation of reward events.

(b) *Reward events affect expectancy*. According to this view, organisms do have knowledge of reward events. There are many versions of the expectancy view (see e.g., Amsel, 1992; Capaldi, 1994; Domjan, 1993; Hull, 1931; Rescorla & Solomon, 1967; Spence, 1956; Trapold & Overmier, 1972). According to the most popular version, expectancies are established via the Pavlovian conditioning of the unconditioned consummatory response to various cues. The conditioned form of eating is called r_g. Bigger rewards produce bigger r_gs and bigger r_gs produce more vigorous behavior (see e.g., Spence, 1956).

(c) *Serial learning and position-item association*. According to this view, series are learned because animals form associations between items and cues associated with the position of the item in the series. The position-item view has been applied to reward outcome serial learning (e.g., Roitblat, Polage, & Scopatz, 1983) to other forms of serial learning mentioned earlier (e.g., Terrace, 1991) and to all forms of serial learning (D'Amato, 1991).

(d) *Serial learning and rules*. According to this view, animals are capable of isolating the rule or rules a series may contain. The simpler the rule structure of a series the more rapidly the series will be learned (e.g., Fountain, 1990; Hulse, & Dorsky, 1977).

The Sequential View and Memory

In this section I provide a brief introduction to the sequential view which will be expanded later. According to the sequential view, the order in which events occur (e.g., phases of an experiment, various magnitudes of reward) has a profound effect on learning and thus performance (see e.g., Capaldi, 1992). The reason for this is that animals have memory. Consider, for example, a sequence of two trials, a nonrewarded or 0 pellet trial followed by a rewarded trial of 10 0.45 g pellets. On the 0 pellet, trial the memory of nonreward will be stored. The memory of nonreward may be retrieved on the 10 pellet trial where it may become a signal for 10 pellets. Thus, what is learned under the sequence of 0-10 differs radically from what is learning under the sequence 10-0 (the memory of reward becomes a signal for nonreward).

Memories of reward events, as indicated, both enhance and direct behavior. The enhancing and discriminative functions for reward produced memories may be illustrated by considering some older data collected in my laboratory. If rats in a runway are presented with a single alternating series of rewarded and nonreward trials, they may come with training to run more rapidly on the rewarded than on the nonrewarded trials, i.e., they may come to respond discriminatively. To explain discriminative responding of this sort we need to identify the stimuli which mediate responding on the rewarded trials and the nonrewarded trials. I have suggested that on

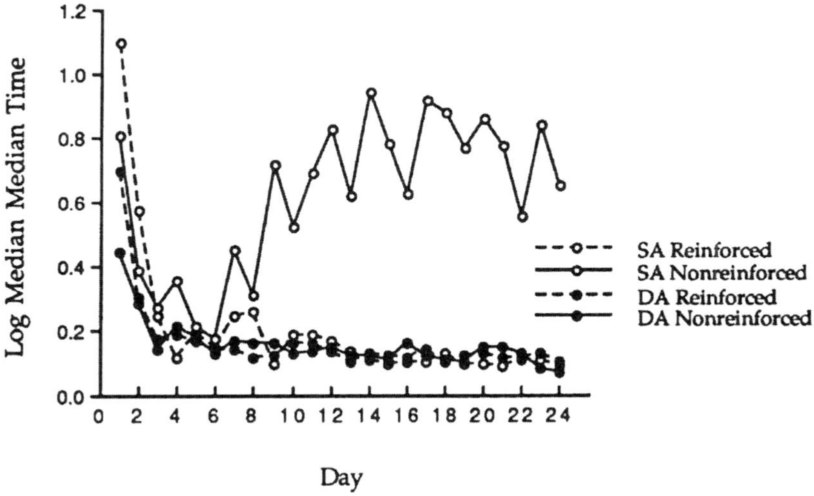

Figure 1. Running time on rewarded and on nonrewarded trials on each day of training for rats given a schedule of rewarded and nonrewarded trials either according to a single alternation schedule or a double alternation schedule.

rewarded trials the rat remembers being nonrewarded on the previous trial and that on nonrewarded trials the rat remembers being rewarded on the previous trial. The memory of nonreward becomes a signal for reward (an S+ cue) and the memory of reward becomes a signal for nonreward (an S- cue). Thus, single alternation responding illustrates the discriminative or directive function of reward produced memories.

Figure 1 shows data Bloom and I reported some time ago (1961). It shows running time of rats in a runway on rewarded and on nonrewarded trials on each of 24 days of training at 12 trials each day. There were two groups of rats. One group received a single alternation schedule of rewarded and nonrewarded trials, the other group a double alternation schedule. As can be seen with training, the single alternation group but not the double alternation group showed discriminative responding, faster running on rewarded than on nonrewarded trials. Such discriminative responding, as indicted, illustrates the directive or discriminative function of reward produced memories.

Attend now to the responding of the single alternation group on Days 1 and 2 of training. On these days, rats in the single alternation group ran faster on nonrewarded trials, i.e., on trials following reward (TFR) than on reward trials i.e., on trials following nonreward (TFN). This TFR-TFN effect illustrates the response enhancing function of reward produced

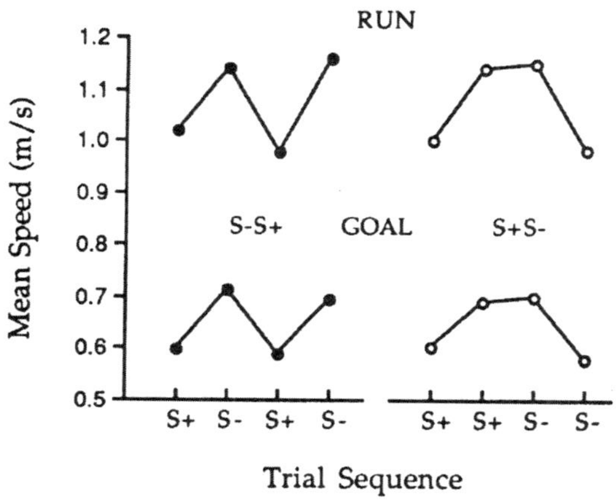

Figure 2. Running speed on Day 4 of training for rats rewarded (S+) in a runway of one brightness and nonrewarded (S-) in a runway of another brightness. The X-axis shows the schedule of S+ and S- trials received by two different groups on Day 4, Group S+S-S+S- (left panel) and Group S+S+S-S- (right panel). The top portion of Figure 2 shows speeds in the run section of the apparatus, the bottom portion shows goal speeds.

memories. That is, all else equal, rats respond more vigorously in the presence of the memory of reward (on TFR) than in the presence of the memory of nonreward (on TFN).

All subsequent figures presented in this chapter report running speeds rather than running times as does Figure 1.

Figure 2 illustrates the enhancing function of reward produced memories in a different learning situation (Capaldi, Berg, & Morris, 1975). In that situation, two groups of rats were rewarded in a runway of one brightness e.g., black, and they were nonrewarded in a runway of a different brightness e.g., white. The X-axis shows the sequence of S+ and S- trials received by each of the groups on Day 4 of acquisition. The top portion of Figure 2 shows speeds in the run section of the apparatus, the bottom portion shows speeds in the goal section of the apparatus. As can be seen in the left portion of Figure 2, the group receiving the sequence S+, S-, S+ , S- on Day 4 ran faster in the S- alternative, i.e., on TFR, than in the S+ alternative, i.e., on TFN. This is called paradoxical discrimination, a phenomenon also shown by the single alternation group of Bloom and Capaldi early in training. As the right side of Figure 2 shows, the group trained S+, S+, S-, S- also manifested the TFR-TFN effect. In an unpublished study from my laboratory, the TFR-TFN effect (on running speed) was obtained in a choice situation. The TFR-TFN effect is very general.

The data shown in Figure 2 illustrate the considerable power of the TFR-TFN effect in a dramatic way as indicated in a subsequent shift phase (extinction). In the shift phase employed immediately subsequent to Day 4 it became clear that the S+ brightness cue had a stronger capacity to elicit responding than the S- brightness cue since in extinction rats ran faster in S+ than in S- (see Capaldi et al., 1975). However, the TFR-TFN effect completely overcame those tendencies, resulting in the animals running more slowly in the acquisition phase in more excitatory S+ alternative on TFN than in the less excitatory S- alternative on TFR.

The TFR-TFN effect shown by the single alternation group in Figure 1 was not evident on Day 3. This does not mean that the enhancing effect on behavior of reward produced memories disappears following some small number of training trials. Rather, I suggest, it reflects the general fact that under any schedule the effect of the memories on behavior will with training ultimately become more appropriate to the reward schedule associated with those memories. For example, under the single alternation schedule, the memory of nonreward is consistently rewarded resulting eventually in fast running while the memory of reward is consistently nonrewarded, resulting eventually in slow running. In other words when rats manifest appropriate pattern running under the single alternation schedule of reward, the signaling function of the reward produced memories is in opposition to the response enhancing function of the reward produced memories, a matter given considerable attention in Experiment 1B.

Brief Description of Procedure Employed Here

In the three as yet unpublished studies reported in this chapter, rats were trained in a runway, running speed being the dependent variable. In all three of the investigations reward produced memories served as the cues for reward (S+ cue) and for nonreward (S- cue).

Each group received trials in pairs each day. Each group received two different pairs. For example, a group might receive a reward of two pellets followed 30 s later by a reward of 10 pellets, a 2-10 pair. The other pair received by this group would be 10-0, 0 meaning nonreward. The 2-10 pair and the 10-0 pair each occurred twice each day in irregular order. Pairs were separated by about a 10 min time interval. Presenting the pairs irregularly precluded the rat from correctly determining whether it would receive 2 pellets or 10 pellets on the first trial of the pair. However, the second trial of the pair could be correctly anticipated providing:

(a) The rat remembered the reward outcome on the first trial of the pair. Thus, in the above example, the memory of 2 pellets, S^2 could signal 10 pellets and the memory of 10 pellets, S^{10}, could signal 0 pellets. In conventional terminology, S^2 which signals reward is an S+ cue, while S^{10} which signals nonreward is an S- cue. Aside from the reward produced memories there is no other basis for responding discriminatively on Trial 2 of the pairs.

(b) Correct responding on Trial 2 of the pairs, fast running to 10 pellets, slow running to 0 pellets, requires more than remembering the reward outcome on Trial 1. The rat must also be able to discriminate between the memory of 10 pellets, S^{10}, and the memory of 2 pellets, S^2. While rats are quite capable of discriminating between S^{10} and S^2, they might have a hard time discriminating between, say, the memory of 7 pellets, S^7, and the memory of 5 pellets, S^5.

In Experiment 1A, the stimuli to be discriminated were the same in both groups, S^{10} and S^2. In one group, S^{10} was the S+ cue, S^2 the S- cue. In the other group, S^2 was the S+ cue, S^{10} the S- cue. An analogous manipulation in a conventional discrimination learning investigation might involve the following. In a black vs. white discrimination, black would be S+ for one group, white for the other. Experiment 1 was concerned with the response enhancing effects of reward magnitudes.

In Experiment 2, the memories to be discriminated were more similar in two of the groups (S^0 vs. S^2) than in two other groups (S^0 vs. S^{10}). In two of the groups, S^0 occurred on larger reward trials and in two other groups on small reward trials. Analogously, a conventional discrimination learning investigation might vary the similarity of the S+ and S- cues (e.g., white vs. black or gray) and magnitude of reward in S+ (e.g., 2 vs. 10 pellets). Experiment 2 was concerned with the discriminative properties of reward magnitudes.

In Experiment 3, as in Experiment 2, similarity of the S+ and S- cues (memories) was varied (S^5- vs. either S^7+ or S^1+) but the hypothesis of major concern in Experiment 3 was somewhat different from that tested in Experiment 2. Experiment 3 was concerned with evaluating the rule learning view of serial learning. All three experiments were relevant to the position-item view of serial learning.

Experimental Data

The Response Enhancing Effects of Reward Produced Memories

As evidence for the response enhancing effects of reward produced memories, the TFR-TFN data so far presented, are deficient in a number of respects. For one, the effect may not reflect response enhancement but extinction of responding on TFN and the reacquisition or additional

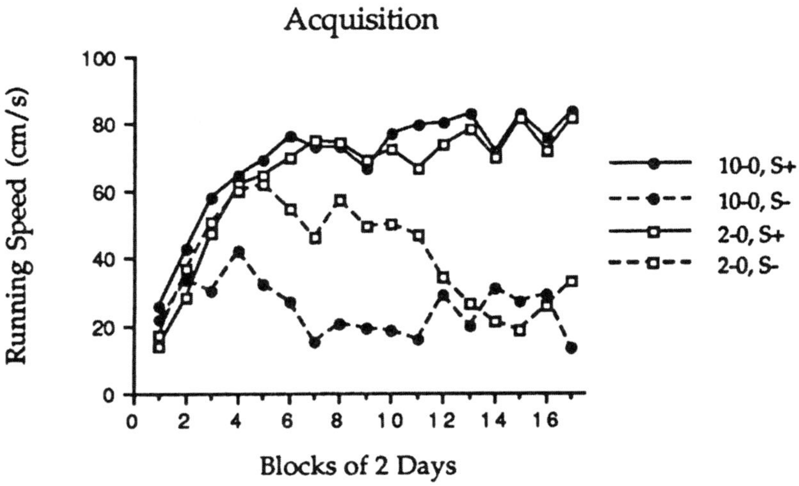

Figure 3. Running speed in blocks of two days on S+ and S- trials for Group 2-0 and Group 10-0 for Trial 2 only.

acquisition of responding on TFR. Too, response enhancing effects have not been demonstrated for reward outcomes other than reward and nonreward. The two experiments reported in this section remedy those deficiencies and firmly establish, in my view, that all reward produced memories have a response enhancing effect on behavior.

Experiment 1A

There were two groups of rats. One group, Group 2-0, received the pairs 10-10 and 2-0. The other group, Group 10-0, received the pairs 2-10 and 10-0. Major interest, as indicated, was in running speed on Trial 2 of the pairs to 10 pellets (run fast) and to 0 pellets (run slow). Group 2-0 could use S^{10} to signal 10 pellets and S^2 to signal 0 pellets. Group 10-0 could use S^2 to signal 10 pellets and S^{10} to signal 0 pellets.

From a discrimination standpoint, the two groups have a comparable problem on Trial 2, i.e., to discriminate between S^2 and S^{10}. If S^2 and S^{10} have no response enhancing function we could expect Groups 2-0 and 10-0 not to differ on Trial 2. But if S^{10} enhances behavior more strongly than S^2, we may expect Group 2-0 to run more slowly on 0 pellet trials (2-0 pair) than Group 10-0 (10-0 pair). Figure 3 shows the running speeds of Groups 2-0 and 10-0 on their S+ and S- trials of Trial 2. Trial 1 data are not shown. The groups did not differ on Trial 1 ($F < 1$).

As may be seen in Figure 3, running speed on S+ trials was greater following 10 pellets than 2 pellets as it was on S- trials, but clearly the effect was bigger on S- trials, i.e., Group 2-0 acquired the tendency to run slowly on S- trials much sooner than Group 10-0. This difference was highly significant. Eventually, Group 10-0 came to run as slowly on S- trials as Group 2-0 but this required considerable training. By Block 12, for example, each group has experienced its S- pair, 2-0 or 10-0, 44 times. Clearly , it was much more difficult to acquire slow running on S- trials when remembering S^{10} (Group 10-0) than when remembering S^2 (Group 2-0). The fact that Group 10-0 eventually came to run as slowly as Group 2-0 on S- trials may be explained as follows. With training, S^{10} came to elicit the expectancy of 0 pellets. The expectancy of 0 pellets elicited slow running, obscuring the tendency of S^{10} to elicit fast running. This explanation implies the following. Early in training greater vigor of responding due to greater reward magnitudes is due primarily to the energizing effects of reward produced memories. As training progresses, expectancies increasingly enter the picture. The expectancies are elicited by memories, among other cues (see Capaldi, 1994; Capaldi, Birmingham, & Alptekin, 1995). Vigor of responding will be some resultant of (a) the particular reward produced memory eliciting responding and (b) the particular expectancy which comes to be elicited. The memory and the expectancy may be congruent - i.e., both may promote the same vigor of responding or they may be incongruent, i.e., one may promote vigorous responding the other not.

As may be seen in Figure 3, the tendency run was more slowly on 0 pellet trials following 2 pellets (2-0) than following 10 pellets (10-0) did not appear on the first block of trials as might be expected. Perhaps this occurred because the rats had not been given enough pretraining to overcome their fear of the apparatus. Once fear diminished, the tendency for rats to ran faster after 10 pellets than after 2 pellets manifested itself.

Experiment 1B

That memories and expectancies may produce effects later in training which are either congruent or incongruent is indicated by an experiment reported in 1983 (Capaldi, Nawrocki, & Verry). While the implications of the findings obtained in Experiment 1B are very clear, they were not fully appreciated until after obtaining the data of Experiment 1A. In Experiment 1B rats were trained pretty much as in Experiment 1A except rather than being trained with pairs of trials they were trained with triplets, i.e., series of three trials. Each of two groups received two different triplets, each triplet occurring twice each day in irregular order. Trials of a triplet were separated by about 30 s, triplets by about 20 min. One group, Group 20-0-20|0-0-0, received the triplets 20-0-20 and 0-0-0 (20 indicating 20 pellets).

The other group, Group 0-0-20|20-0-0, received the triplets, 0-0-20 and 20-0-0. While the running speeds of Groups 20-0-20|0-0-0 and 0-0-20|20-0-0 on Trials 1 and 2 of the triplets is meaningful in certain respects (see Capaldi et al., 1983), responding on Trial 3 of the triplets speaks most clearly to the response enhancing effects of S^{20} and S^{0} on responding and is exclusively emphasized here.

Discrimination responding on Trial 3 of the triplets cannot be mediated by the reward outcome on Trial 2 because it was 0 pellets in both groups in both series. To determine whether or not Trial 3 would terminate in either 0 or 20 pellets, the rats had to remember the Trial 1 reward outcome. This was either 20 pellets or 0 pellets in both groups. Thus, both groups in Experiment 1B, as in Experiment 1A, have a discrimination problem of comparable difficulty. That difference between the groups is this: the enhancing effects of the reward produced memories are compatible with appropriate discrimination responding in Group 20-0-20|0-0-0 and incompatible with appropriate discrimination responding in Group 0-0-20|20-0-0. Thus, in Group 20-0-20|0-0-0, 20 pellets on Trial 1 signaled 20 pellets on Trial 3 (enhancing fast running) and 0 pellets on Trial 1 signed 0 pellets on Trial 3 (enhancing slow running). But in Group 0-0-20|20-0-0, 20 pellets on Trial 1 signaled 0 pellets on Trial 3 (enhancing fast running), while 0 pellets on Trial 1 signaled 20 pellets on Trials 3 (enhancing slow running).

Figure 4 shows the Trial 3 running speeds of Group 20-0-20|0-0-0 and Group 0-0-20|20-0-0 on 0 pellet trials and on 20 pellets trials in blocks of two days. As may be seen in Figure 4, Group 20-0-20|0-0-0 ran faster than Group 0-0-20|20-0-0 on 20 pellet trials and slower than Group 0-0-20|20-0-0 on 0 pellet trials. Note especially that the response enhancing effects of Trial 1 reward on running speeds in Group 20-0-20|0-0-0 were still manifest on Day 40 of training and showed no sign of abating. Too, it is clear that the response enhancing function of rewards extend back to more trials than the immediately preceding trial.

The two experiments reported in this section support the following additional conclusions concerning the response enhancing function of reward produced memories. First, the enhancing effects of reward produced memories are manifested in more or less pure form early in training. It seems likely that these enhancing effects are unmodified by extensive training. However, these enhancing effects can be obscured by training. This occurs because the memories are rewarded according to some reward schedule and so memories come to elicit expectancies appropriate to that reward schedule. As one example, under the single alternation schedule the memory of 0 pellets (run slow) comes to elicit the expectancy of reward (run fast, see Capaldi, 1994). In this instance the response enhancing effect of the memory of nonrward (run slow) and the response enhancing effect of the

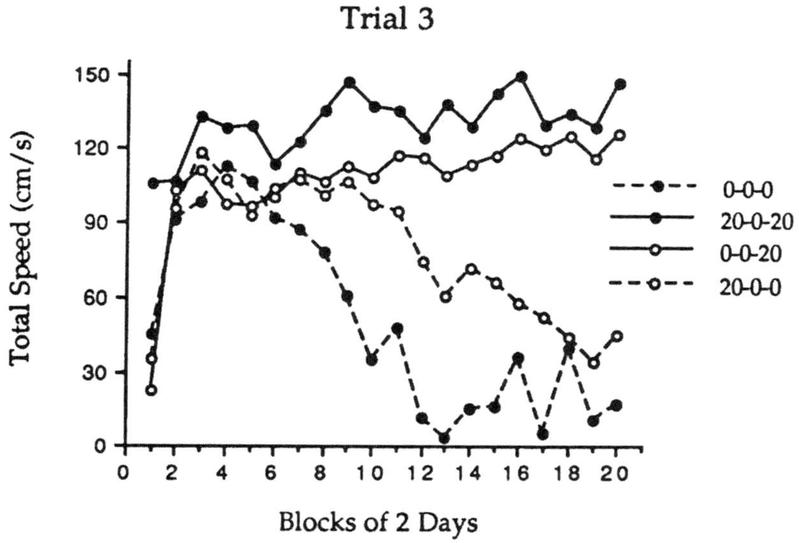

Figure 4. Running speed on Trial 3 on blocks of two days for two groups of rats. One group received the two series 20-0-20 and 0-0-0, the other group received the series 0-0-20 and 20-0-0.

expectancy of reward (run fast) are in opposition and vigor of responding will be some resultant of the two tendencies. In a recent paper, Capaldi et al. (1995) presented evidence of a different kind for the proposition that memories of reward events come to elicit expectancies of reward events. This view was called the tandem hypothesis (see also Capaldi, 1994). According to the tandem hypothesis, memories and expectancies work in tandem such that expectancies of reward are elicited by memories of reward.

Another conclusion supported by Experiments 1A and 1B is that greater reward magnitudes e.g., S^{20}, S^{10}, enhanced behavior more greatly than memories associated with smaller reward magnitudes, e.g., S^2 or with nonreward, S^0. Finally, the enhancing effects of reward produced memories are not due to extinction on TFN and reacqusition on TFR. The extinction reacquisition hypothesis could not possibly explain the results obtained in the two experiments reported in this section. According to the extinction view, the groups employed in Experiments 1A and 1B should not have differed.

Let me underline clearly a major implication of the data reported in this section. It is common to suggest that animals respond more vigorously to larger reward than to smaller rewards because of expectancy. I accept that view but suggest that it is only half the story. The other half is this. Rats respond more vigorously when they remember larger rewards than when they remember smaller rewards. The two tendencies, memory and expectancy, depending upon the reward schedule, can operate either in the same direction, i.e., both enhance responding in a similar manner or in different directions, i.e., one enhances behavior more than the other.

The Discriminative Function of Reward Produced Memories

The sequential hypothesis suggests that distinctive reward events produce distinctive memories and that discriminating between these memories explains a variety of conventional learning phenomena. To illustrate this proposition, let S^{10}, S^2 and S^0 be associated with, respectively, a large magnitude of consistent reward (CRF), a small magnitude of CRF, and nonreward (extinction). I suggest that it is easier to discriminate between S^{10} and S^0 than between S^2 and S^0. This assumption explains, among other things, why extinction is more rapid following a large magnitude of CRF than a small magnitude of CRF (see e.g., Wagner, 1960). This is because less generalized excitatory tendency is supplied to S^0 by the dissimilar stimulus S^{10} than by the similar stimulus S^2. The directive function of memories can explain other phenomena. Consider that under CRF, excitation associated with S^{10} will generalize to S^2. This generalized excitation will be less than that directly acquired by S^2 because the memory S^{10} differs substantially from the memory S^2 (see e.g., Capaldi, 1967). This explains, among other things, that a shift from large reward CRF to small reward CRF produces a successive negative contrast effect, i.e., the speeds of the shifted group fall below those of the small reward control group because of generalization decrement.

There are two problems with explaining conventional learning phenomena as above. First, it may not be entirely obvious to some that shifts to, e.g., extinction involve discrimination learning. This is especially the case, I would guess, where another theory of extinction is accepted, e.g., that extinction involves unlearning (e.g., Rescorla & Wagner, 1972) or that it involved the learning of incompatible responses (e.g., Amsel, 1958, 1992). Second, there are many instances in which the sequential explanation is only one among other plausible explanations. For example, various versions of expectancy theory have been used to explain why extinction is more rapid the larger the magnitude of CRF (e.g. Amsel, 1992; Wagner, 1961).

The two experiments reported below have the following noteworthy characteristics. They employ reward produced memories as the S+ and S- cues, and so they are explicit, albeit unconventional, forms of discrimination learning. If in this situation the memories have the characteristics we have applied to them in other situations, such as extinction and reward magnitude shifts, support for the sequential view will be provided.

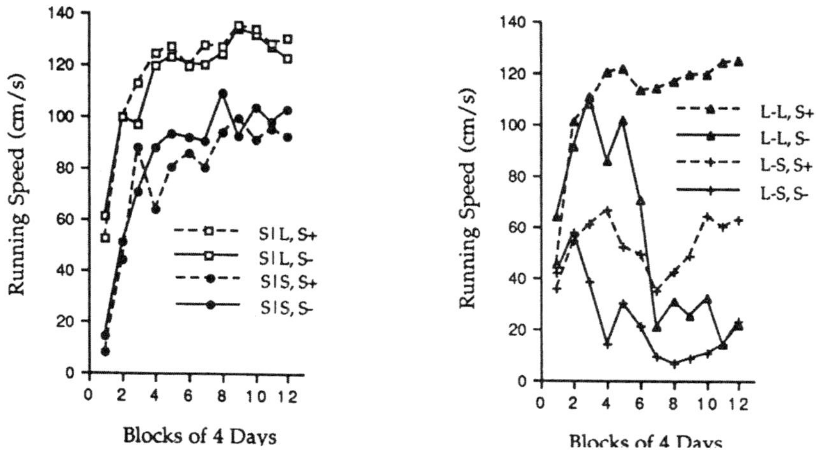

Figure 5. Running speed in blocks of four days on S+ and S- trials in Groups S-L and S-S (left panel) and Group L-L and L-S (right panel).

Experiment 2

The four groups were trained either 0-2 and 2-0 (Group S-S), 0-2 and 10-0 (Group L-S), 0-10 and 2-0 (Group S-L), or 0-10 and 10-0 (Group L-L). In each of the groups the memory of 0 pellets signaled reward on the next trial and thus the memory of 0 pellets was the S+ cue in all groups. The memory of 2 pellets signaled nonreward in Groups S-S and S-L (2-0 pair) while the memory of 10 pellets signaled nonreward in Groups L-S and L-L. Thus, the memory of 2 pellets was the S- cue in Groups S-S and S-L while the memory of 10 pellets was the S- cue in Groups L-S and L-L.

Figure 5 shows the running speed of the four groups on S+ and on S- trials in blocks of four days. Groups having the easier discrimination learning problem, i.e., that between S^0+ and S^{10}- (L-S and L-L) are shown in the right panel of Figure 5. Groups having the more difficult discrimination problem, i.e., that between S^0+ and S^2- (Groups S-S and S-L) are shown in the left panel of Figure 5. The results were quite clear: Groups L-L and L-S came to run faster on S+ than on S- trials while Group S-L and S-S did not. Also obvious in Figure 5 is that large reward on S+ trials (Groups L-L and S-L) produced faster running than small reward on S+ trials (Groups L-S and S-S). Finally of interest is the fact that Group L-L acquired discriminative responding more slowly than Group L-S. The major reason for this was that Group L-L required more training than Group L-S to run slowly on S- trials.

The groups employed in Experiments 1A and 1B had an equally

difficult discrimination learning problem. Experiments 1A and 1B showed that under those conditions it is more difficult for rats to run slowly to nonreward following larger than smaller reward. Thus, all else equal in Experiment 2, it should have been more difficult to run slowly on S- trials for rats trained 10-0 than for rats trained 2-0. Yet, the opposite was found. This should not be surprising. In Experiment 2, the discriminative and response enhancing functions of reward produced memories were in opposition. In this particular case the discriminative function played the greater role. Another way of putting it is to say the groups trained 0-2, 2-0 and 0-10, 2-0 had such a difficult discrimination learning problem, the response enhancing factor, the tendency to run slowly following S^2, was unable to manifest itself.

The data shown in Figure 5 indicate clearly that it is much easier for rats to discriminate between S^{10} and S^0 than between S^2 and S^0. The failure of Groups S-S and S-L to show any discriminative responding at all by the end of training is very interesting. That finding could mean that rats are unable to discriminate between S^2 and S^0. Why then would rats show extinction after being trained with small rewards of 1 or 2 .045 g pellets? The answer to that question would require conjecture of a sort not appropriate here. However, whatever may be the case, rats appear to have a most difficult time discriminating between small reward and nonreward. And they discriminate readily between large reward and nonreward.

Experiment 3

There were two groups in Experiment 3. Group Monotonic received the pairs 7-5 and 5-0 and a single 1 pellet trial. Group Nonmonotonic received the pairs 1-5, 5-0 and a single trial of 7 pellets. Thus the two groups received the same reward magnitudes on Trial 1 and on Trial 2. The rule learning view expects, if it expects anything, that 0 pellets should be better anticipated by Group Monotonic than by Group Nonmonotonic. Group Monotonic has a single rule to learn, reward magnitudes decrease monotonically (7-5 and 5-0), whereas Group Nonmonotonic must learn two rules, reward magnitude sometimes increases (1-5 pair) and sometimes decreases (5-0). The prediction of sequential view is opposite to that of rule learning. It should be easier to discriminate between the more dissimilar memories S^1+ and S^5- (Group Nonmonotonic) than between the more similar memories S^7+ and S^5- (Group Monotonic), a proposition consistent with Weber's law.

Figure 6 presents the running speeds of the two groups on S+ trials and on S- trials in blocks of five days. On S+ trials, the two groups ran at about the same speed. But on S- trials, Group Nonmonotonic ran much more slowly than Group Monotonic. These findings support the sequential

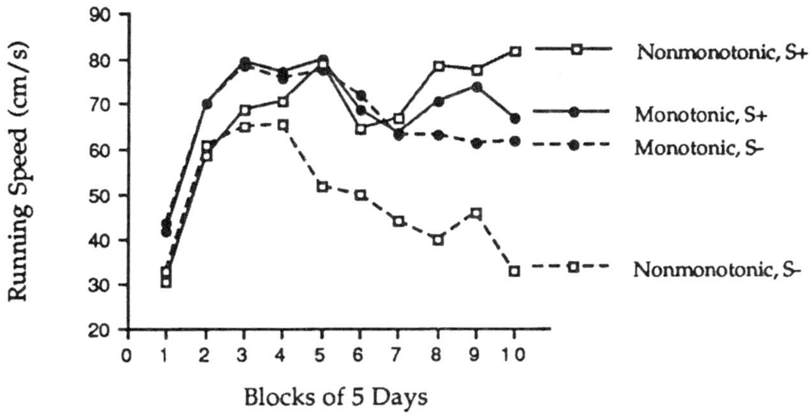

Figure 6. Running speed in blocks of 5 days on S+ and S- trials in Group Monotonic and Group Nonmonotonic on Trial 2 of the 5-0 pair.

view. They fail to support rule-learning, a matter considered more fully later.

Reward Produced Memories and Pavlovian Conditioning

A widely held view is that in instrumental tasks, Pavlovian conditioning occurs giving rise to the anticipation of reward (e.g., Amsel, 1958, 1992; Rescorla & Solomon, 1967; Spence, 1956; Trapold & Overmier, 1972). Apparatus and other cues are usually identified as the conditioned stimuli (CSs) in this Pavlovian arrangement. However, it has been shown elsewhere that in many, perhaps most learning situations, goal events are more validly predicted by reward produced memories than by other cues, e.g., apparatus or other exteroceptive cue (see Capaldi, 1994; Capaldi et al., 1995). For example, in Experiment 1A here, 0 and 10 pellets on Trial 2 of the pairs were better predicted by S^2 and S^{10}, respectively, (Groups 2-0 and 10-0) than by apparatus cues. Accordingly, I have suggested that it is not uncommon in instrumental tasks, indeed it may be common, for reward produced memories to function as CSs (Capaldi, 1994; Capaldi et al., 1995). This being so, reward produced memories should have the same properties as other CSs. Two characteristics of conventional CSs are (a) they enter into association with the unconditioned stimulus (US, see e.g., Mackintosh, 1983) and (b) more intense CSs elicit stronger conditioned responding than less intense CSs (see e.g., Kamin, 1965). I suggest that reward produced memories have both of these characteristics. That is, reward produced memories enter into association with goal events (discriminative stimulus or directive function) and instrumental behavior is more strongly elicited by memories of larger than by memories of smaller reward magnitudes (response enhancing function). The response enhancing capacity of exteroceptive CSs has been explained in two ways. According to one view as CSs become more intense they elicit or energize behavior more strongly by virtue of their intensity (e.g., Hull, 1949). According to another view, more intense stimuli are more discriminable than less intense stimuli from other cues such as background cues (e.g., Logan, 1954; Perkins, 1953). According to this view, it is the discriminative properties of the cues which regulates their response enhancing property. It may be that both of the mechanisms described above are involved (see Logan, 1954).

CONCLUSIONS

The experimental data reported here indicate that reward produced memories elicit and direct responding (discriminative stimulus function) and that memories associated with larger rewards enhance responding more strongly than memories associated with smaller rewards or nonreward (response enhancing function). Rats in a runway were presented with trials in pairs. Reward magnitude differed on Trial 1 of the pair and Trial 2 terminated either in reward or nonreward. The reward magnitude associated with Trial 1 of the pair could not be predicted because the pairs were presented irregularly. However, the reward produced memory stored on Trial 1 and retrieved on Trial 2 was a reliable indicator of whether Trial 2 would be rewarded or nonrewarded. In the absence of employing the reward produced memories as S+ and S- cues, the discrimination problem on Trial 2 of the pairs could not be learned. The problems were learned.

Experiment 1A provided evidence for the response enhancing properties of reward produced memories by showing that rats discriminating between the memories of ten (S^{10}) and two (S^{2}) pellets ran faster to nonreward when S^{10} was the S- cue than when S^{2} was the S- cue. Experiments 2 and 3 showed that rats acquired discriminative responding more easily the greater the difference in the reward produced memories which served as the S+ and S- cues. For example, in Experiment 3 rats discriminated more easily between S^{1}+ and S^{5}- than between S^{7}+ and S^{5}-.

In addition to supporting the writer's sequential view (e.g., Capaldi, 1994), the data reported here present difficulties for four alternative views described earlier. Experiment 1 creates difficulties for two of the major theories which have attempted to explain how learning and performance is regulated by reward and reward magnitude. Experiments 2 and 3 creates difficulties for two major theories of serial learning. First I will show why the present findings create problems for these alternative theories. Then I will outline some of the implications the present results have for the writer's view.

Theories of Reward

As is well known, larger reward magnitudes produce more vigorous responding (see e.g., Wagner, 1961). This is the result of response contingent reinforcement according to the alternative views to be examined. For example, according to Hull (1943, see also Rescorla & Wagner, 1972), larger contingent reward results in stronger associations. And according to a variety of theorists, larger contingent rewards produce distinctive expectancies which in one way or another promote more vigorous responding than expectancies associated with smaller contingent rewards (e.g., Amsel, 1992; Rescorla & Solomon, 1967; Spence, 1956; Trapold & Overmier, 1972).

Experiment 1A demonstrated (see also Experiment 1B; Bloom & Capaldi, 1961; Capaldi et al., 1975) that the larger the remembered reward the more vigorous the responding. In other words, responding is affected not only by the reward event contingent upon responding but also the prior or remembered reward event. Experiment 1B provided data indicating that responding is a resultant of both the expected reward event and the remembered reward event. Responding was most vigorous when 20 pellets was both remembered and expected. It was least vigorous when 0 pellets was both remembered and expected. And responding was intermediate when 20 pellets was expected and 0 pellets was remembered or vice-versa. According to the findings presented here, any theory purporting to explain the effects of reward and reward magnitude on learning and performance cannot ignore reward produced memories and how such memories interact with expectancies.

Theories of Serial Learning

According to the position-item view, distinctive cues associated with the position of an item in the series may become associated with items. Any number of such cues could be identified. As one example, item 2 of a series occurs later in the series than item 1 and thus, e.g., temporal cues can be employed to predict item 2 and other items. This sort of position-item analysis cannot explain the findings of any of the three experiments reported here. This is because the reward outcomes on Trial 2 of the pairs were not preceded by any conceivable position cues which would support discriminative responding. The pairs, as indicated, were presented irregularly and so Trial 2 of the pair, considering other than memory cues, receive 50% irregular reinforcement. In each of the three experiments reported here, rats were able to correctly anticipate Trial 2 reward outcomes, running faster when Trial 2 was rewarded than when it was nonrewarded. Only a memory analysis can explain those findings.

The rule encoding view cannot explain why in Experiment 3 Group Nonmonotonic exhibited better discriminative responding than Group Monotonic. This is because, as previously indicated, the reward events received by Group Monotonic occurred according to a simpler rule than those received by Group Nonmonotonic. Simpler rules should be more easily learned than more complicated ones according to the rule encoding view (see e.g., Hull & Dorsky, 1977). In Group Monotonic, Trial 2 reward magnitude was always smaller than Trial 1 magnitude, whereas in Group Nonmonotonic Trial 2 reward magnitude was sometimes larger and sometimes smaller than Trial 2 reward magnitude. According to Hulse (1980), the results obtained in Experiment 3 lie outside the scope of the rule learning view because of the short series employed (only two items). Hulse (1980) mentioned several other variables which, in his view, lay outside the scope of the rule learning view. In any event, subsequent animal learning data suggest that Hulse's view may be incorrect in two respects (see e.g., Capaldi, Nawrocki, Miller, & Verry, 1986). On the one hand, findings inconsistent with rule learning were obtained employing variables which supposedly encourage rule learning. On the other hand, findings compatible with rule learning were obtained employing variables which supposedly discourage rule learning. Experiment 3 is a more effective refutation of the rule learning view than some previous findings. In those earlier experiments, it is true, findings were obtained which are incompatible with rule learning but compatible with the sequential view. However, those findings, unlike those obtained in Experiment 3, could be explained in other ways, e.g., by position-item associations (see e.g., Capaldi & Molina, 1979).

The findings obtained in Experiment 1A pose additional difficulties for the rule learning view. In Experiment 1A, it was found that all else equal, rats respond less vigorously to 0 pellets following smaller than following larger reward. Consider that one of the major findings reported in support of the rule learning view is that rats anticipate 0 pellets more quickly (slow running) when given the series of decreasing food pellets 14-7-3-1-0 than when given the series 14-1-3-7-0. Rules aside, a major reason for that result according to the findings reported in Experiment 1A, is that rats will run more slowly to 0 pellets following 1 pellet than following 7 pellets. Similar data, which suffer from the same defect, have been employed to support the position-item view (Roitblat et al., 1983).

The failure to recognize that rats might respond differently following 7 pellets vs. 1 pellet illustrates a general problem associated with the rule learning and position-item views. Both views treat series composed of different reward magnitude as similar in most, if not all, respects to series composed of other items such as colors, forms, etc. (see e.g., D'Amato, 1991; Hulse & Dorsky, 1977). Essentially both views have all but ignored the fact that reward events have a directive function and a response enhancing function. It may be possible to present animals with different forms or colors which are equally response enhancing. However, this is not possible employing different reward magnitudes. Future investigations purporting to support rule learning or position-item learning cannot ignore the response enhancing function of reward magnitudes.

The Sequential View

According to the view outlined here, distinctive reward events produce distinctive memories which can acquire discriminative control over responding and enhance it. As support for these positions the sequential model relied on indirect evidence for many years. For example, the sequential view assumed that rats were capable of determining how many nonrewarded (or rewarded) trials occurred in a row and that the distinctive memories associated with different strings of rewarded or nonrewarded trials could become discriminative stimuli for reward or nonreward. The major evidence for these assumptions was indirect. The evidence was essentially, that if memories associated with long strings of nonrewarded trials occurred on rewarded trials in acquisition, resistance to extinction would be increased. This is because animals so trained would come to expect reward on late extinction trials and so would respond vigorously (see e.g., Capaldi, 1964, 1966, 1994). Animals trained with short strings of nonrewarded trials in acquisition would not expect reward late in extinction and so would run slowly.

Direct evidence for the ability of rats to remember how many rewarded or nonrewarded trials occurred in a runway was not obtained until many years later. This evidence consisted of rats learning that, e.g., while memories associated with one or two rewards would be rewarded (S+ cue), the memory associated with three rewards would be nonrewarded (S- cue). These numerical discriminations as I characterized them, were not as difficult for rats as perhaps might be supposed (see e.g., Capaldi, 1993; Capaldi & Miller, 1988; Capaldi, Miller, & Alptekin, 1988).

In the investigation reported here rather than employing different numbers of rewarded or nonrewarded events as the S+ and S- cues, I employed different magnitudes of reward in that capacity. This approach had two consequences of importance within the sequential model. It allowed the direct test of reward magnitude assumptions which previously were based on indirect evidence. And it allowed a convincing demonstration of the proposition that reward produced memories enhance responding, memories associated with greater reward magnitudes producing greater response enhancement.

An example of such indirect evidence, it was indicted earlier, is the finding that extinction is more rapid the larger the magnitude of consistent reward (e.g., Wagner, 1961). I interpreted this finding to mean that memories associated with nonreward are more easily discriminated from the memory of large reward than from the memory of small reward (see Capaldi, 1967). Experiment 2 here provided direct evidence for that assumption by demonstrating that the discrimination between the memory of small reward (S- cue) and nonreward (S+ cue) is much more difficult than discriminating between memory of large reward (S- cue) and nonreward (S+).

Of course, reward outcome serial learning is but one example of the various phenomena that the sequential approach interprets in terms of the discriminative properties of reward produced memories. Capaldi (1994) has recently described a number of other findings which can be understood in terms of the discriminative properties of reward produced memories. These include not only reward schedule effects, such as have been described so far, but conventional forms of discrimination learning as well, i.e., cases in which the explicit S+ and S- cues are exteroceptive stimuli.

It has been clear for some time that rats respond more vigorously as the magnitude of remembered reward increases (see e.g., Bloom & Capaldi, 1961; Capaldi et al., 1975). Direct evidence for this proposition was presented in this chapter (Experiments 1A and 1B). An interesting implication of the findings obtained in Experiment 1B is that memories and expectancies of reward interact to determine performance. They may both prompt the same vigor of responding or one may prompt more vigorous responding than the other. Another interesting implication of Experiment 1B is that the response enhancing properties of reward produced memories, unlike their discriminative properties, may not be modified by training. That is, in Experiment 1B the memories of nonreward and reward enhanced responding late in training (40 days) as they had early in training. This conclusion may not be contradicted by the finding obtained in Experiment 1A that rats eventually ran as slowly to nonreward when remembering large reward as when remembering small reward. This finding may represent a floor effect. Another interesting implication of Experiment 1B is that responding on a current trial may be enhanced not only by the reward outcome on the immediately preceding trial but the reward outcomes associated with even earlier trials.

There are many interesting ideas one may entertain about the response enhancing capacity of reward produced memories, but these are highly speculative at this time. For example, is the response enhancing function of reward produced memories a learning factor or a performance factor? What is the effect of retention interval of the response enhancing function of reward produced memories? These are only a few of the questions which may be raised in connection with the ability of reward produced memories to energize responding.

REFERENCES

Amsel, A. (1958). The role of frustrative nonreward in noncontinuous reward situations. *Psychological Bulletin*, **55**, 102-119.

Amsel, A. (1992). *Frustration theory: An analysis of dispositional learning and memory*. New York: Cambridge University Press.

Bloom, J. M., & Capaldi, E. J. (1961). The behavior of rats in relation to complex patterns of partial reinforcement. *Journal of Comparative and Physiological Psychology*, **54**, 261-265.

Capaldi, E. J. (1964). Effect of N-length, number of different N-lengths, and number of reinforcements on resistance to extinction. *Journal of Experimental Psychology*, **68**, 230-239.

Capaldi, E. J. (1966). Partial reinforcement: A hypothesis of sequential effects. *Psychological Review*, **73**, 459-477.

Capaldi, E. J., (1967). A sequential hypothesis of instrumental learning. In K. W. Spence & J. T. Spence (Eds.), *Psychology of learning and motivation* (Vol. 1, pp. 67-156). New York: Academic Press.

Capaldi, E. J. (1992). Levels of organized behavior in rats. In W. K. Honig & G. Fetterman (Eds.), *Cognitive aspects of stimulus control* (pp. 385-404). Hillsdale, NJ: Erlbaum.

Capaldi, E. J. (1993). Animal number abilities: Implications for a hierarchical approach to instrumental learning. In S. T. Boysen & E. J. Capaldi (Eds.), *The development of numerical competence: Animal and human models* (pp. 191-209). Hillsdale, NJ: Lawrence Erlbaum Associates.

Capaldi, E. J. (1994). The sequential view: From rapidly fading stimulus traces to the organization of memory and the abstract concept of number. *Psychonomic Bulletin & Review*, **1**, 156-181.

Capaldi, E. J., Berg, R., & Morris, M. D. (1975). Stimulus control of responding in the early trials of differential conditioning. *Learning & Motivation*, *6*, 217-229.

Capaldi, E. J., Birmingham, K. M., & Alptekin, S. (1995). Memories of reward events and expectancies of reward events may work in tandem. *Animal Learning & Behavior*, **23**, 40-48.

Capaldi, E. J., & Miller, D. J. (1988). Counting in rats: Its functional significance and the independent cognitive processes that constitute it. *Journal of Experimental Psychology: Animal Behavior Processes*, **14**, 3-17.

Capaldi, E. J., Miller, D. J., & Alptekin, S. (1988). Numerical aspects of nonreinforcement: The same phase nonreinforcement procedure. *Animal*

Learning & Behavior, **16**, 411-416.

Capaldi, E. J., & Molina, P. (1979). Element discriminability as a determinant of serial pattern learning. *Animal Learning & Behavior*, **7**, 318-322.

Capaldi, E. J., Nawrocki, T. M., Miller, D. J., & Verry, D. R. (1986). Time between events as a retrieval cue: Recall and the temporal similarity between the storage and retrieval intervals. *Journal of Experimental Psychology: Animal Behavioral Processes*, **12**, 258-269.

Capaldi, E. J., Nawrocki, T. M., & Verry, D. R. (1983). The nature of anticipation: An inter- and intraevent process. *Animal Learning & Behavior*, **11**, 193-198.

D'Amato, M. R. (1991). Comparative cognition: Processing of serial order and serial pattern. In L. Dachowski, & C. F. Flaherty (Eds.), *Current topics in animal learning: Brain, emotion, and cognition* (pp. 165-185). Hillsdale, NJ: Erlbaum.

Domjan, M. (1993). *Domjan and Burkhard's The Principles of Learning and Behavior*. Pacific Grove, CA: Brooks/Cole Publishing.

Fountain, S. B. (1990). Rule abstraction, item memory, and chunking in rat serial pattern tracking. *Journal of Experimental Psychology: Animal Behavior Processes*, **16**, 96-105.

Hull, C. L. (1931). Goal attraction and directing ideas conceived as habit phenomena. *Psychological Review*, **38**, 487-506.

Hull, C. L. (1943). *Principles of behavior*. New York: Appleton-Century-Crofts.

Hull, C. L. (1949). Stimulus intensity dynamism (*V*) and stimulus generalization. *Psychological Review*, **56**, 67-76.

Hulse, S. H. (1980). The case of the missing rule: Memory for reward vs. formal structure in serial-pattern learning by rats. *Animal Learning & Behavior*, **8**, 689-690.

Hulse, S. H., & Dorsky, N. P. (1977). Structural complexity as a determinant of serial pattern learning. *Learning & Motivation*, **8**, 488-506.

Kamin, L. J. (1965). Temporal intensity characteristics of the conditioned stimulus. In W. F. Prokasy (Ed.), *Classical conditioning: A symposium* (pp. 118-147). New York: Appleton-Century-Crofts.

Logan, F. A. (1954). A note on stimulus intensity dynamism (V). *Psychological Review*, **61**, 77-80.

Mackintosh, N. J. (1983). *Conditioning and associative learning*. Oxford: Clarendon Press.

Perkins, C. C., Jr. (1953). The relation between conditioned stimulus intensity and response strength. *Journal of Experimental Psychology*, **46**, 225-231.

Rescorla, R. A., & Solomon, R. L. (1967). Two-process learning theory: Relationships between Pavlovian conditioning and instrumental learning. *Psychological Review*, **74**, 151-182.

Rescorla, R. A., & Wagner, A. R. (1972). A theory of Pavlovian conditioning: Variations in the effectiveness of reinforcement and nonreinforcement. In A. H. Black & W. F. Prokasy (Eds.), *Classical conditioning: Current research and theory* (pp. 64-99). New York: Appleton-Century-Crofts.

Roitblat, H. L., Polage, B., & Scopatz, R. A. (1983). The representation of items in serial position. *Animal Learning & Behavior*, **11**, 489-498.

Spence, K. W. (1956). *Behavioral theory and conditioning.* New Haven, CT: Yale University Press.

Terrace, H. A. (1991). Chunking during serial learning by a pigeon: I. Basic evidence. *Journal of Experimental Psychology: Animal Behavior Processes*, **17**, 81-93.

Trapold, M. A., & Overmier, J. B. (1972). The second learning process in instrumental learning. In A. H. Black & W. F. Prokasy (Eds.), *Classical conditioning II: Current theory and research* (pp. 427-452). New York: Appleton-Century-Crofts.

Wagner, A. R. (1961). Effects of amount and percentage of reinforcement and number of acquisition trials on conditioning and extinction. *Journal of Experimental Psychology*, **62**, 234-242.

6

THE ORGANIZATION OF SEQUENTIAL BEHAVIOR

Stephen B. Fountain, Douglas G. Wallace, and James D. Rowan
Kent State University

For decades, beginning with Lashley's (1951) celebrated paper or even earlier with the work of Ebbinghaus and others (Hunter, 1920; Skinner, 1934), sequential learning has repeatedly aroused debate over the fundamental nature of learning, memory, and representation in humans and animals. In recent years, debate has centered on whether nonhuman animals can use nonassociative symbolic processes such as rule-induction to learn about the structure of patterned sequences ("serial patterns") of events. The work by Hulse and colleagues on rats' serial-pattern learning (SPL) of food reward magnitude in runways, in particular, supported a Rule-Learning (RL) theory (e.g., Hulse & Campbell, 1975; Hulse & Dorsky, 1977; Hulse, 1978; Hulse & Dorsky, 1979; Fountain & Hulse, 1981; Fountain, Evensen, & Hulse, 1983). The RL theory proposed that rats learned some representation of the abstract rules that described organized sequences (Fountain, 1986; Roitblat, Pologe, & Scopatz, 1983; Wathen & Roberts, 1994). The implication was that rats did not have to rely on chaining or remote associations alone to master sequences. Later work in our laboratory involved a somewhat different SPL paradigm employing patterns of stimuli drawn from another stimulus dimension (viz., a spatial dimension rather than food reward magnitude). This work also supported the RL view of serial-pattern learning in both rats and mice (e.g., Fountain, 1990; Fountain

& Rowan, 1995b; Fountain & Rowan, 1995a).

Very early on, however, Capaldi and colleagues contested the RL view, proposing as a competing view that SPL in rats could be accounted for by associative mechanisms alone (e.g., Capaldi & Molina, 1979; Capaldi, Verry, & Davidson, 1980a; Capaldi, Verry, & Davidson, 1980b; Capaldi, Nawrocki, & Verry, 1982; Capaldi, Nawrocki et al., 1985; Capaldi, 1986). Capaldi likened sequential learning to other discrimination learning problems. However, as Capaldi himself admitted at the time (Capaldi, Verry, & Davidson, 1980a), his "sequential memory" model lacked sufficient parametric power to make strong predictions regarding the outcome of sequential learning tasks involving long patterns like those used by Hulse and colleagues. Adding to the complexity of the issue, the RL view has frequently failed to predict outcomes for shorter patterns, and even Hulse (Hulse, 1980) conceded that Capaldi's sequential memory theory was a better account in that domain because, according to Hulse, short patterns approximate the paradigm of paired-associates learning rather than that of rule learning.

Roitblat, Pologe, and Scopatz (1983) proposed that serial position played an important role in rat serial-pattern learning. Serial position models assume that sequence elements become associated with their position in the sequence, not with other sequence items (Burns & Gordon, 1988; Burns, Hulbert, & Cribb, 1990; Chen et al., 1997; Roitblat et al., 1983). Although serial position models may have much in common with Capaldi's sequential memory view, the critical difference is that sensitivity to serial position, the central construct here, seems to imply the additional cognitive ability of counting or timing.

Although the foregoing models were originally proposed by some as mutually exclusive accounts of sequential learning and memory, recent behavioral and psychobiological work has produced evidence to support all of these positions, sometimes suggesting that rats may employ more than one strategy or type of information to encode and reproduce a single complex behavioral sequence (Fountain, 1990; Fountain & Rowan, 1995a; Fountain & Rowan, 1998). In this chapter, we consider how best to achieve a rapprochement of these sometimes competing views of SPL. We will describe recent work in our laboratory involving behavioral, computational, and psychobiological approaches to understanding SPL. Our conclusion, for now, is that no general process theory is currently adequate to describe all SPL phenomena, and that in all likelihood SPL involves multiple cognitive and brain processes.

THE SUBSYMBOLIC ORGANIZATION OF SEQUENTIAL BEHAVIOR: ITEM ASSOCIATION

Capaldi and colleagues have championed the idea that rats learn to anticipate sequences of events by learning a series of stimulus-response or stimulus-stimulus associations relating successive events of the series. Sequential learning using this method is essentially discrimination learning (cf. Capaldi & Molina, 1979). Although Restle and his colleagues rejected similar ideas as being too simplistic to explain human SPL (Restle & Brown, 1970a; Restle & Burnside, 1972), a possible objection to their analysis is that the range of associative hypotheses they considered as potential mechanisms of sequential learning and memory were overly narrow. In work on rat sequential learning, for example, sequences have been composed of quantities of food presented sequentially in runway tasks. The difficulty of anticipating the successive quantities and the vigor of rats' response in anticipation of a given quantity can be influenced by a number of associative factors under appropriate conditions. These associative factors include generalization of "reward signal capacity" between similar quantities of the sequence (cf. Capaldi & Molina, 1979) and "remote anticipations" of quantities to be received beyond the immediate trial (cf. Capaldi, Nawrocki, & Verry, 1983; Capaldi & Miller, 1988), to name but two.

Although our work has tended to look for evidence of RL processes, we have repeatedly encountered evidence that discrimination learning also plays a significant role in SPL even under conditions thought to be maximally conducive for observing RL processes. First, we discuss how a subsymbolic associative computational model has been successful in describing some classical RL phenomena and some new phenomena in terms of discrimination learning processes. Later, we discuss evidence that discrimination learning also plays a significant role in how extra-sequence cues, particularly temporal "phrasing cues," affect serial-pattern learning.

Reward Magnitude SPL as Discrimination Learning: An Associative Computational Model

Several studies have been taken as strong evidence that rats are sensitive to the formal structure of the serial patterns they learn. For example, Hulse and Dorsky (1977, 1979) created patterns from a stimulus "alphabet" (Jones, 1974) of various quantities of food. The patterns they created varied in formal structure. A strongly monotonic sequence of 14-7-3-1-0 food pellets was formally simple in structure, requiring a single "less than" rule to describe the relationships of all successive pairs of quantities. A weakly monotonic 14-5-5-1-0 pattern and a nonmonotonic 14-1-3-7-0 pattern were progressively more complex, requiring combinations of "less than," "greater than," and "equal" rules to adequately describe the patterns. Rats learned to track the formally simple monotonic pattern most rapidly and to the most proficient level of tracking performance of the three patterns, followed by the weakly monotonic and the nonmonotonic patterns. Hulse and Dorsky ruled out a variety of pattern-learning strategies the rats might have adopted to learn these patterns in favor of the hypothesis that rats are sensitive to pattern structure (see Hulse and Dorsky, 1977).

Two different studies were used as evidence that the rules rats used in serial-pattern learning were, in fact, abstract. In one study (Hulse & Dorsky, 1979), rats demonstrated the ability to generalize a rule from one set of patterns to a different pattern. Rats were pre-trained with a set of sequences that were either consistently monotonically decreasing or randomly increasing and decreasing. Later, rats were trained on one of two sequences composed of new quantities, either monotonically decreasing (16-9-3-1-0) or nonmonotonic (16-1-3-9-0). The differential pattern tracking observed for the four groups during the training phase indicated that pre-training produced positive and negative transfer as predicted by RL theory. Specifically, monotonic pre-training produced better tracking in the training phase than random pre-training when both received a monotonic pattern in the training phase. Monotonic pre-training produced poorer tracking in the training phase than random pre-training when both received a nonmonotonic pattern in the training phase. In a second study, Fountain and Hulse (1981) evaluated rats' ability to extrapolate patterns of varied structural complexity. Rats were better able to predict a new 0-pellet element added to a structurally simpler 14-7-3-1 pattern than 0 pellets added to a structurally complex 14-3-7-1 pattern. The authors reasoned that the monotonic pattern produced better anticipation of "0" than the nonmonotonic pattern because the monotonically decreasing 14-7-3-1 sequence was described by a single "less than" rule that would lead naturally to the added "0" after "1".

Capaldi and colleagues (Capaldi & Molina, 1979) challenged the RL theory by proposing that stimulus associations and "generalized reward signal capacity" (GRSC) mediate pattern tracking when rats learn sequences of food reward quantities. Simply put, this view supposed that each food quantity in a sequence serves as a cue for the next. Acquisition depends in part on discriminability and salience of pattern elements, but tracking performance is also determined by generalization between food quantity signals. For example, the GRSC theory would describe performance on a 14-3-7-1-0 sequence of food quantities as follows. Rats would anticipate "0" based on memory of "1" that came to signal it, but anticipation of "0" would be modulated by the fact that "1", by generalization from the similar item "3", would also lead to some anticipation of "7." Other interactions would also be expected, where the amount of generalization would depend, as it does in traditional discrimination tasks, on the relative similarity of cues. This kind of generalization between signals would be expected based on traditional notions of discrimination learning as well as by Capaldi's sequential memory model.

Evidence supporting Capaldi's GRSC theory has been generated from various runway studies. First, Capaldi and Molina (1979) demonstrated that a structurally complex sequence constructed from highly discriminable elements (1-29-0) produced better pattern tracking relative to a structurally simple sequence constructed of less discriminable elements (20-10-0). Second, in response to Hulse and Dorsky's (1977) rule learning theory, Haggbloom (1985) demonstrated that pattern tracking was disrupted upon transfer only when associative information was removed, but was unaffected by manipulations that violated rule or serial position information. Finally, Haggbloom and Brooks (1985) showed that discriminability of pattern elements, not pattern structure, was the best predictor of pattern extrapolation.

The foundation for the RL versus GRSC debate was that GRSC theory as described by Capaldi and colleagues (Capaldi & Molina, 1979; Capaldi, Verry, & Davidson, 1980a) could not make strong predictions concerning patterns that contained more than a few stimuli. For example, Capaldi, Verry, and Davidson (1980) stated that the generalized reward-signal capacity received by "1" in sequences such as 14-7-3-1-0 or 14-5-5-1-0 is "completely indeterminate." In other words, even though "1" signaled nonreward ("0") by way of direct association, GRSC theory predicted that "1" should also acquire the ability to signal reward through generalization from other similar items in series. It was thought that the amount of generalization was crucial in that it determines how much "1" will signal reward versus nonreward. However, the amount of reward-signal capacity "1" received through generalization was indeterminable presumably because, as suggested by Roitblat (1982), Capaldi et al. could not explicitly describe the requisite generalization functions. For that reason, Capaldi et al. (1980) were unable to make strong predictions regarding learning for 5-item series that would have allowed them to claim that GRSC could describe the available behavioral data. Instead, they were forced to adopt the weaker position that the GRSC interpretation "cannot be excluded" (Capaldi, Verry, & Davidson, 1980).

We have recently completed work in our lab with a computational model of SPL we have called the Sequential Pairwise Associative Memory (SPAM) model (Wallace & Fountain, in press). SPAM includes all of the essentials of the TODAM model described by Murdock (1982; Murdock, 1983) and Metcalfe (Eich, 1982; Metcalfe, 1990) and subsequently used by

Metcalfe (1990) as the model known as CHARM in her simulations of eyewitness misinformation effects and "blending" of memories. The model is described in great detail by Murdock (1982; 1983; 1985) and Metcalfe (1982; 1990). SPAM is also one possible instantiation of GRSC because encoding is based on pairwise associations of items and because generalization occurs between similar cues.

The SPAM program was written in C (Turbo C, Borland International, Scotts Valley, CA) for PC-compatible hardware. As in the model described by Metcalfe (1982; 1990) and Murdock (1982; 1983; 1985), items to be stored in memory in SPAM were represented as 63-element random vectors, where each element of the vector is termed a "feature" of the item. The features of item vectors were random normal deviates so that features were statistically independent. Deviates were restricted to the range -2.0 to 2.0, and item vectors were normalized so that the dot product of each vector with itself is 1. The dot product of any two unrelated item vectors should approach 0. An item file of 100 unrelated item vectors of this sort was generated.

Because the simulations required using items that varied systematically in similarity to each other, item *dimensions* (sets of items with step-wise increments/decrements in similarity) were also set up from which the items to be used as "reward magnitudes" (food quantities) in the simulations were drawn. The item dimensions were constructed in the same manner as the one used by Metcalfe (1990) in her simulation of color-shift phenomena in memory (so-called "memory blends"). Each dimension was composed of 11 items. To produce such a dimension, we first created a set of independent (unrelated) items, such as Item Vectors ***20-30***, using our standard item generation routine. Item Vectors ***21-29*** were then replaced by Item Vector ***20***. Then, the first 6 features of Items ***21-29*** were set equal to the same features of Item ***30***. Next, the adjacent 6 features of Items ***22-29*** were set equal to those of Item ***30***. This process continued until the ninth set of 6 features was replaced for Item ***29*** only. [Note that Metcalfe (1990) achieved the same result by randomly choosing the 6 features—without replacement—to be modified on each step of this routine. Because features are independent, the result is the same using either method.] An item in such a dimension differs from its neighbors by approximately 10% per step away in the dimension, and item vectors at the extremes of the dimension (i.e., Items ***20*** and ***30***, in this case) are unrelated.

One assumption of the model is that perceived food quantity is a log-linear dimension. This assumption, based on the view that Weber's Law should apply to the perception of food quantity as it does to the perception of quantity in other stimulus dimensions, is common to theories of generalization and traditional notions of discrimination learning (Spence, 1937; Spence, 1936; Blough, 1975). This assumption was also implicitly accepted by Roitblat (1982), who attempted to apply Blough's (1975) model of generalization to the serial-pattern learning phenomena reported by Hulse and Dorsky (1979). To create a log-linear food quantity dimension, food quantities were represented by vectors in an approximate log-linear mapping that expected larger just noticeable differences as food quantity increased. In the following simulations, stimulus dimension Vector ***0*** represented 0 food pellets, Vector ***1*** represented 1 food pellet, Vector ***3*** represented 3 pellets, Vector ***5*** represented 7 pellets, Vector ***6*** represented 10 pellets, Vector ***7*** represented 14 pellets, Vector ***8*** represented 20 pellets, and Vector ***9*** represented 29 pellets.

In SPAM, two item vectors such as "14 pellets" (represented by the vector $\underset{\sim}{A}$) and "7 pellets" (represented by the vector $\underset{\sim}{B}$) may be associated by the process of *convolution*, which is computed according to the equation:

$$(A * B)_m = \sum_{(ij)\in S(m)} a_i b_j$$

where ***A*** and ***B*** are the item vectors ($a_{-31}, a_{-30}, \ldots, a_{30}, a_{31}$) and ($b_{-31}, b_{-30}, \ldots, b_{30}, b_{31}$), respectively, that are being associated, and $S(m) = \{(i, j)| - (n-1)/2 \leq i, j \leq (n-1)/2$, and $i + j = m$. If $n = 3$, that is, if item vectors are 3 elements (features) in length, it is straightforward to compute that the convolution of ***A*** and ***B***, called ***A*B***, will result in the 5-element vector, $\boldsymbol{A*B} = (a_{-1}b_{-1}, a_0b_{-1} + a_{-1}b_0, a_1b_{-1} + a_0b_0 + a_{-1}b_1, a_1b_0 + a_0b_1, a_1b_1\}$, following Metcalfe (1990).

"Encoding" is accomplished by adding the resulting trace vector and item vectors that were components of the association to the composite trace vector, ***T***, which is the composite memory trace:

$$\boldsymbol{T_n} = \alpha \boldsymbol{T}_{n-1} + \gamma_1 \boldsymbol{A} + \gamma_2 \boldsymbol{B} + \omega(\boldsymbol{A*B})$$

where $\boldsymbol{T}_{n-1}$ and $\boldsymbol{T}_n$ represent the trace before and after encoding information for a new event, respectively, ***A*** and ***B*** are vectors representing the individual items contributing to the association, and ***A*B*** is the convolution of the vectors representing 14 pellets and 7 pellets. Following Murdock (1983), the Greek letter α represents a retention parameter that determines how much of prior learning will be retained. [Note that Murdock called this parameter a "forgetting" parameter.] The symbols γ_1 and γ_2 are weights for the item vectors, and ω is an encoding parameter that represents the strength of the convolution vector (the pairwise association) that is to be added to the composite trace vector. Encoding new information is accomplished simply by adding vectors to the trace vector, ***T***. That is, the model assumes only one composite trace, trace vector ***T***, that contains all of the individual pairwise associations (convolutions) of items that compose a sequence of events.

"Retrieval" is accomplished by the process of *correlation*, which is represented by #, where a cue item is correlated with the memory trace. Correlation is defined as

$$R_m = \sum_{(ij)\in S(m)} a_i b_j$$

where $S(m) = \{(i, j)| - (n - 1)/2 \leq i, j \leq (n - 1)/2$, and $i - j = m$, following Metcalfe (1990). So, for example, ***A*** (the vector representing 14 pellets) might be correlated with the trace, ***T***. This retrieval would be depicted as ***A#T***, and it would recover a vector representation of the item previously associated with ***A***, in this case, the vector ***B*** (which represents 7 pellets). The recovered item vector will be similar to the originally encoded item vector, but will be somewhat degraded by noise inherent in the convolution and correlation processes. When the composite trace contains multiple convolutions with the same cue item (e.g., both ***A*B*** and ***A*C***, where ***C*** represents "3 pellets"), the result of correlation of ***A*** with the composite trace (i.e., ***A#T***) will be a vector bearing information about each item originally associated with the cue, ***A***. In this case, the recovered vector (***A#T***) would contain noisy representations of both vector ***B*** (7 pellets) and vector ***C*** (3 pellets) in composite form. To determine how a subject with such a composite memory trace would perform on a test, the recovered vector is compared to each possible item in the dimension (that is, to each possible food quantity in the dimension), and the most similar item is taken to be the item "anticipated" by the subject. For purposes of comparing item vectors, the similarity of two vectors is defined as the dot product of the vectors (cf. Metcalfe, 1990). In this example, similarity comparisons would be made by calculating the dot product of the recovered vector with each vector representing a food quantity in the dimension.

Figure 1 provides a schematic of the general steps involved in an iteration of the computational model. For example, a simulation of performance for a 20-10-0 sequence would include three associations, hereafter called "convolutions". The first convolution would be ***Start*8***, where "***Start***" is an independent item vector representing distinctive cues signaling the first trial of the sequence and Item Vector ***8*** represents 20 pellets of food. Later convolutions would be ***8*6*** and ***6*0***. The convolution process is labeled "(1)" in Figure 1. The item that was convolved with ***0***, namely, ***6***, was later used to cue the composite memory trace, labeled "(2)" in Figure 1. This resulted in the retrieved item, labeled "(3)", that was compared to the items in the lexicon. The lexicon was "dimensional," that is, the lexicon contained all the items of the stimulus dimension. The comparison process generated dot product values that reflected the degree of similarity between the retrieved item and each individual item in the lexicon, as indicated by "(4)" in Figure 1. The lexical item with the highest dot product value was chosen as the item best predicted by the cue for that iteration, shown as "(5)" in Figure 1. In order to obtain a good representation of the central tendency of the program's performance, 1000 iterations were performed for each sequence simulation.

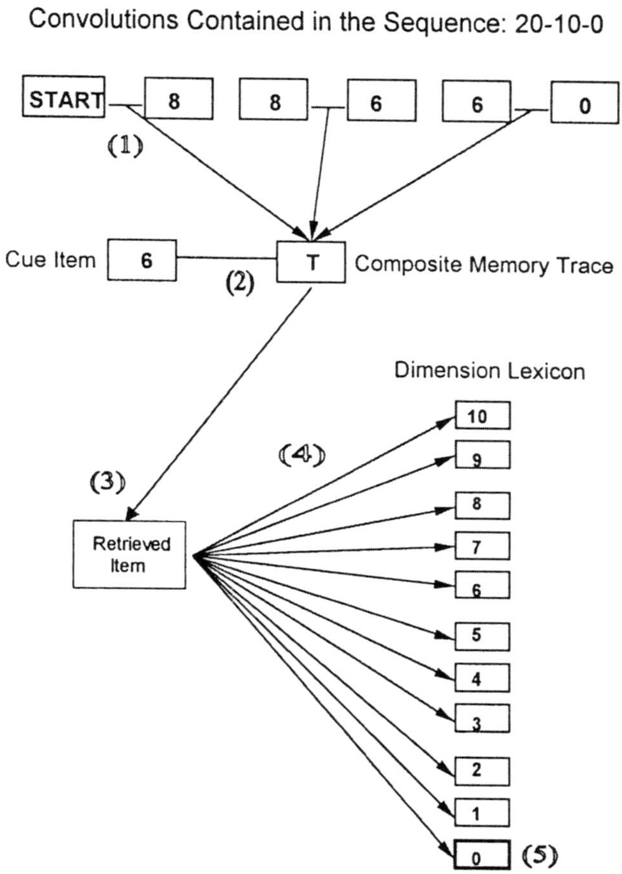

Figure 1. Flow diagram for one iteration of the SPAM model for the sequence 20-10-0 (coded as vector ***8*** → vector ***6*** → vector ***0***) with "6" as a cue for the anticipated item. The schematic depicts the following steps: (1) association by convolution, (2) storage in a composite memory trace, (3) retrieval by correlation of a cue item with the composite memory trace, (4) determining dot products (similarity) of the retrieved item vector and each individual lexicon item vector, and (5) recognition by selecting the lexicon item with the highest similarity to the retrieved item.

We used SPAM to test whether simple associative processes could account for some of the fundamental behavioral evidence for the GRSC and RL theories. In two studies (Wallace & Fountain, in press), SPAM predicted that a short pattern with highly discriminable elements would be learned better than one with less discriminable elements even though the former was monotonic and the latter was nonmonotonic (cf. Capaldi & Molina, 1979). This prediction from the model might be expected given the similarities of SPAM and GRSC theory, but SPAM also predicted that a long, strongly monotonic pattern of food quantities should be tracked better than a long nonmonotonic pattern, as observed by Hulse and Dorsky (1977). In two other experiments (Wallace & Fountain, in press), SPAM also successfully simulated behavioral "rule generalization," "extrapolation," and associative transfer data (cf. Fountain & Hulse, 1981; Haggbloom, 1985; Haggbloom & Brooks, 1985; Hulse & Dorsky, 1979). Thus, to our surprise, SPAM provided an existence proof for a purely associative GRSC mechanism that can account for several critical behavioral phenomena previously taken as strong evidence for rule learning in rats. SPAM shows that whereas both GRSC and RL theory can account for a subset of the serial-pattern learning literature, as claimed by Capaldi for a number of years, GRSC theory can also account for a number of phenomena outside the scope of RL theory.

Although rats may indeed use rule-induction to learn serial patterns, SPAM simulations indicate that this is not a necessary conclusion from the classical behavioral studies on reward magnitude SPL. SPAM's success also argues that computational models can play a role in animal cognition research. In this case, SPAM demonstrated that complex behavior can sometimes result from the aggregate effects of simple behavioral processes acting in complex settings, and that computational modeling may be required to demonstrate that simpler processes are viable alternatives to more complex processes as explanations. Computational modeling provides one means of maintaining our vigilance regarding the ever-present dangers of the argument from design (Blumberg & Wasserman, 1995) in animal cognition theorizing (cf. Church, 1997, 2001).

In order to determine if novel predictions of the SPAM model would parallel behavioral data observed in rat serial pattern learning, seven 3-element sequences were selected for study that: 1) systematically varied sequence discriminability, 2) systematically varied reward magnitude, and 3) systematically manipulated the order of food quantities. The chosen series directly pitted earlier hypotheses against each other in a way not attempted before in a single study. First, the sequences 1-1-0, 1-4-0, and 1-16-0 were included because they contain successively more discriminability between the Trial 1 and 2 sequence elements. Capaldi's idea that greater discriminability should facilitate learning would predict successively better anticipation of the Trial 3 0-pellet quantity for these sequences. However, in the foregoing sequences, discriminability is confounded with increases in Trial 2 food quantity. To evaluate the effects of food quantity under conditions where discriminability of Trial 1 and 2 quantities is held constant, the sequences 1-1-0 (aforementioned), 2-2-0, and 4-4-0 were included. The sequences 4-1-0 and 16-1-0 were included to examine, by comparison to sequences 1-4-0 and 1-16-0, respectively, how the sequential order of elements affects sequence tracking when element discriminability is conserved. Comparisons of anticipation for 0 pellets between the 4-1-0 and 1-4-0 sequences or between the 16-1-0 and 1-16-0 sequences can also be used to assess the importance of response enhancement in sequential learning.

The foregoing seven 3-element sequences were selected for study in part because SPAM made some predictions not anticipated by any of the earlier hypotheses alone. As one particularly striking example, the SPAM model predicted *a priori* that rats in the 16-1-0, 1-16-0, and 1-1-0 pattern groups should all run slowly for the terminal 0-pellet element, that is, they should have approximately equally good anticipation of "0". This result is not expected based on the element discriminability hypothesis because of differential levels of discriminability between Elements 1 and 2 (high discriminability for 16 vs. 1 pellets in 16-1-0 and 1-16-0, and low discriminability for 1 vs. 1 pellets in 1-1-0). Discriminability alone would predict that "1" as Element 2 in 1-1-0 should be a poor cue for anticipating "0" whereas "1" in 16-1-0 and "16" in 1-16-0 should be good cues for anticipating "0". Likewise, the result predicted by SPAM is not expected based on response enhancement because of expected differential enhancement from Element 2 in the three sequences; "1" as Element 2 in 16-1-0 and 1-1-0 should produce slower speeds on the following "0" when compared to the effect of "16" on "0" in 1-16-0. Thus, this experiment directly pits these earlier hypotheses against the SPAM model. Both computer simulations and behavioral testing were conducted with these sequences to evaluate the correspondence of the behavioral data with earlier theoretical ideas and to evaluate the correspondence between the behavioral data and the predictions of the SPAM model.

As shown in Figure 2, predictions from the SPAM model correlated well with the observed behavioral data; the overall correlation was $r = .94$ for data collapsed across all blocks of training. What are the processes instantiated in SPAM that produced accurate prediction of the behavioral data? We propose that pairwise associations and generalization are the necessary mechanisms, with stimulus coding scheme and retention rate as parameters that influence, the predictions of the model quantitatively rather than qualitatively. In this case, sequential behavior in tracking 3-element sequences is largely accounted for by pairwise associations between successive elements and generalization between event memories. For example, the sequence 1-16-0 is represented as the two associations 1-16 and 16-0. Anticipation of a given element is then determined by the amount of generalization that occurs between the Elements 1 and 2. Considering that 1 and 16 food pellets are very discriminable, little generalization occurs between the different signaling food quantities. Therefore, anticipation of the terminal 0-pellet element is relatively easy, and the same argument holds

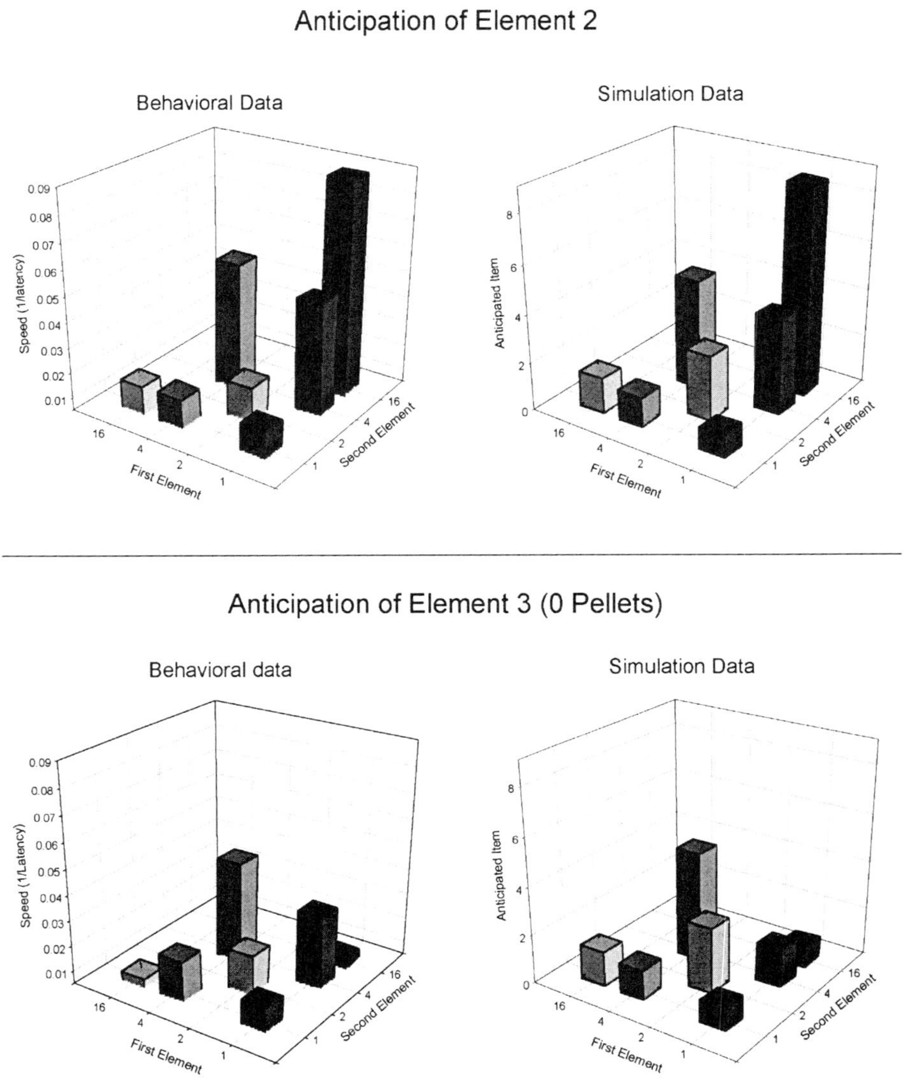

Figure 2. A comparison of behavioral data and SPAM model predictions. The top panel presents behavioral data (running speeds as 1/latency) and SPAM model predictions (anticipated item scores) for anticipation of Element 2 of 3-element sequences, which varied according to the sequence. The bottom panel presents behavioral data and SPAM model predictions for Element 3, which was 0 pellets in all 3-element sequences. Behavioral data and SPAM model predictions are presented in 3-dimensional graphs as a function of the two food quantities serving as Elements 1 and 2.

for the 16-1-0 sequence. However, for the 4-4-0 sequence, "4" signals both "4" for Element 2 and "0" for Element 3. Consequently, maximal generalization of Elements 1 and 2 (because they were the same quantity) as cues for Elements 2 and 3, respectively, leads to anticipation of a larger quantity on Element 3 compared to other conditions, and thus poorer anticipation reflected in faster running speeds as observed in both the behavioral and simulation data. As might be expected, because of complete generalization between Elements 1 and 2 as cues in the 4-4-0 pattern, running speeds on Elements 2 and 3 were observed to be very similar, as predicted by the SPAM simulations. According to SPAM, the progressively worse anticipation of the terminal element found in the sequences, 1-1-0, 2-2-0, and 4-4-0, is the result of maximal generalization of signals for "0" and progressively larger food quantities; across sequences, Element 1 cues progressively larger food quantities, whereas Element 2 always cues the terminal 0-pellet element. Since the Elements 1 and 2 are identical in each sequence, maximum generalization is observed, and the net result is progressively poorer anticipation of the terminal element. In fact, this case reflects the results of the classical "branching" problem, where the same stimulus signals different events at different times and leads to performance based on anticipating both outcomes simultaneously. This idea should be compared to similar phenomena described by the term "memory blends" in the human memory literature by Metcalfe (1990) and to Capaldi's idea of generalized reward signal capacity in rat serial learning (Capaldi, Verry, & Davidson, 1980). Thus, SPAM accounts for differential element anticipation in different sequences by appealing to pairwise association of sequence events and generalization between cues as the principle factors mediating pattern tracking.

Two interesting points can be made here. First, the sequences 1-1-0, 2-2-0, and 4-4-0 share equivalent amounts of discriminability between Elements 1 and 2, so Capaldi and Molina's (Capaldi & Molina, 1979) idea of discriminability alone would tend to incorrectly predict no differences between sequences in terms of anticipation of 0-pellets on Element 3. However, the sequences 1-1-0, 2-2-0, and 4-4-0 presumably have successively increasing amounts of response enhancing properties of larger rewards on the anticipation of 0 pellets, so differences in anticipation of the final element could be attributed to increasing levels of response enhancing properties of the stimuli preceding 0 pellets. Do the data support the notion of response enhancement as a separate behavioral process? The SPAM model has no separate response enhancement process and thus, to the extent that response enhancement is a separate process, SPAM should fail to account for the data. The fact that SPAM can predict effects attributed to response enhancement indicates that response enhancement is likely an epiphenomenon resulting from pairwise association and generalization, that it is not a separate process or independent factor.

A second point is that a similar argument applies to the idea that multiple hedonic memory is required for serial pattern learning. Since SPAM does not include "remote associations" or similar multiple-association processes, the model would be expected to fail to account for the behavioral data to the extent that these are necessary for serial pattern learning. Given the high correlation between the SPAM simulation data and the observed behavior, the results are consistent with the idea that rats did not use multiple hedonic memory in this task. This is not to say that rats cannot use multiple hedonic memory when it is necessary to do so, only that this construct is not necessary to account for the present data.

SPAM captures the relationship between reward magnitude and generalization in the context of pairwise association that has eluded other theories. As reward magnitude increases in sequences with low discriminability held constant, anticipation of the terminal, 0-pellet, element is progressively more difficult due to generalization from increasingly larger rewards. Reducing generalization by increasing discriminability between sequence elements can overcome the disruptive effects of large reward magnitude on anticipation of nonreward. Finally, intermediate levels of reward magnitude and generalization produce intermediate levels of anticipation. The correspondence between the predictions from SPAM simulations and the behavioral data identify the processes that mediate serial behavior in 3-element sequences, namely, pairwise association and generalization. These ideas have been foreshadowed by earlier investigators, particularly by Capaldi, Verry, and Davidson's (1980) idea of generalized reward signal capacity. However, the SPAM model is the first to formalize the ideas of pairwise association and generalization in such a way as to make possible novel predictions regarding pattern tracking in serial patterns and to account for effects such as response enhancement without appeal to independent processes.

THE SYMBOLIC ORGANIZATION OF SEQUENTIAL BEHAVIOR: EVIDENCE FOR RULE-LEARNING

Although the foregoing experiments and discussion might seem to argue for a strong refutation of RL theory in rat serial-pattern learning, several additional phenomena deserve mention because they are consistent with RL theory and we have not yet been able to simulate them with SPAM. These include some aspects of data from studies of pattern phrasing and chunking (Fountain, Henne, & Hulse, 1984), pattern learning for hierarchical versus "linear" serial patterns (Fountain & Rowan, 1995a), and interleaved pattern learning in rats (Fountain, Rowan, & Benson, 1999).

Evidence that Phrasing Facilitates Chunking and Rule-Learning

Pattern phrasing and chunking have received significant attention in the literature on animal sequential learning and memory (Fountain, Henne, & Hulse, 1984; Fountain, 1990; Capaldi, Verry et al., 1984; Terrace, 1987; Terrace, 1991; Terrace & Chen, 1991a, 1991b). Depending on cue salience and placement within the pattern, phrasing can facilitate or retard pattern learning. Cues can be lights, tones, temporal intervals, runway brightness (color), or spatial cues.

Fountain, Henne, and Hulse (1984) argued that phrasing cues speed learning by facilitating chunking processes. That is, phrasing cues facilitate learning by marking significant features of pattern structure. According to this view, phrasing cues facilitate the process of detecting and abstracting structure from a sequence. Work showing that phrasing cues need not involve temporal intervals to facilitate pattern learning in rats (Fountain, Henne, & Hulse, 1984) is also consistent with this hypothesis. Finally, another set of studies showed that "misplaced" phrasing cues, that is, so-called "bad" phrasing (Restle, 1972; Fountain, Henne, & Hulse, 1984), impaired learning relative to no phrasing in both humans and rats (Restle, 1972; Fountain, Henne, & Hulse, 1984). These results, too, fit well with this hypothesis. In the latter studies, it was suggested, bad phrasing cues (temporal or otherwise) obscure pattern organization by distracting attention from significant features of pattern structure.

Capaldi, Verry, Nawrocki, and Miller (1984) argued that phrasing effects are the result of associative processes; phrasing cues overshadow associations between successive pattern elements by establishing associations between the phrasing cue and the next sequence item. In studies of extinction and spontaneous recovery, Capaldi, Verry, Nawrocki, and Miller (1984) found evidence that when temporal pauses were used as phrasing cues, they indeed overshadowed inter-item associations. We have also found evidence to support this view (Stempowski, Carman, & Fountain, 1999).

One phenomenon not easily accounted for by this latter overshadowing view of phrasing is the fact that when congruent "good" phrasing cues are removed in a transfer test, savings are sometimes observed in tracking performance. For example, when temporal or spatial phrasing cues were positioned at chunk boundaries, they both facilitated learning. However, when cues were removed, rats trained with temporal phrasing cues showed deficits in performance whereas those trained with spatial phrasing cues showed savings (Fountain, Henne, & Hulse, 1984). The interpretation of this result by RL theory is that spatial phrasing facilitated chunking, rule-induction, and, thus, encoding, not simply the establishment of associations involving phrasing cues and sequence items. The overshadowing view proposed by Capaldi and associates would seem to predict a deficit following cue removal, not savings. This phenomenon, then, suggests that under some conditions it is possible to show that rats do indeed chunk patterns and induce rules. It would be surprising if SPAM could predict savings after cue-removal transfer of this sort, though SPAM's surprising ability to simulate pattern "generalization" and "extrapolation" phenomena in rats in the experiments we report earlier in this chapter suggests that we should not prejudge the outcome.

Rats are Sensitive to Pattern Structure

One prediction from the rule-learning view is that a highly organized, hierarchically structured sequence should be easier to learn than a sequence having little or no higher-order structure. We designed a study to explore whether pattern structure would determine the ease or difficulty of learning long and elaborate patterns. Our method for rats is a functional analogue of human pattern learning tasks that require subjects to learn to press buttons in an array in the proper sequential order (Restle, 1970; Restle & Brown, 1970b; Restle & Brown, 1970c; Restle, 1973). Rats learned to press levers in an array in the proper sequential order. They were tested in an octagonal operant chamber (see Figure 3) equipped with a retractable lever mounted on each wall (Fountain & Rowan, 1995b; Fountain & Rowan, 1995a). All levers were presented at the beginning of each trial and the rat could press any of the 8 levers. If the correct lever was pressed, the rat received hypothalamic brain-stimulation reward. If an incorrect lever was pressed, all levers except the correct lever were removed from the box and the rat was not reinforced until the correct response was emitted. This method a) is easily learned by the rat without pretraining procedures, b) provides the rat with a continuous (circular) stimulus array, and c) allows us to record response accuracy measures on a trial-by-trial basis while the rat performs the task at its own pace.

We tested whether pattern structure would determine the ease or difficulty of pattern learning by developing patterns with hierarchical structure, then reordering chunks of the pattern to produce "linear" structure, that is, a sequence of unrelated chunks. The Hierarchical (H) and Linear (L) patterns were:

H Pattern 123 234 345 456 567 876 765 654 543 432...
L Pattern 123 234 <u>543</u> 456 567 876 765 654 <u>345</u> 432...

For both groups, the digits indicate the clockwise position of the correct response on successive trials and spaces indicate brief pauses.

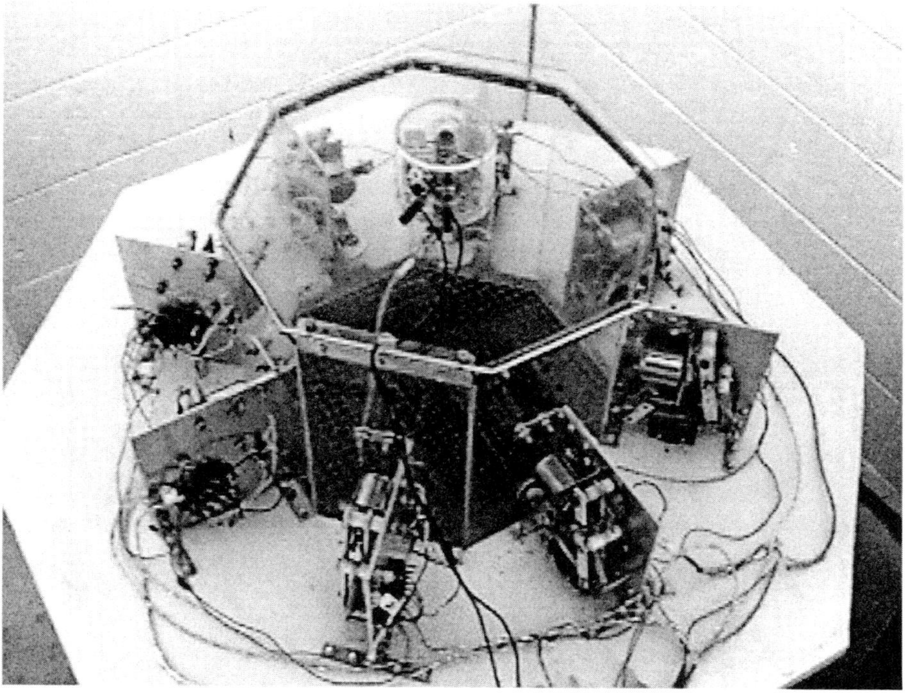

Figure 3. An octagonal operant chamber equipped with a retractable lever on each wall. Rats were reinforced with pulses of brain-stimulation reward for correct responses.

The completely nested H pattern is described by a simple hierarchical rule structure: elements within 3-element chunks are related by first-order rules, chunks within the first and second halves of the pattern, respectively, are related to each other by second-order rules, and the first half of the pattern is related to the second half of the pattern by a third-order rule. A formal description of this pattern is

$$(M(T_{+1}{}^{4}(T_{+1}{}^{2}(1))))$$

where "1" refers to the starting lever, T_{+n} represents a "transpose" rule (i.e., to move n units in the indicated direction, where + indicates clockwise), M represents a "mirror image" rule, and superscripts reflect the number of repeated applications of the rule that are required. Because of the nested structural organization, the second-order T_{+1} rule applies a "+1" rule to each item in the first chunk to generate the second chunk, and so on. The third-order M rule produces a "mirror image" (a more complex form of a "reverse" rule) of the first half of the pattern to generate the entire second half of the pattern.

The incompletely nested L pattern was generated by exchanging the two underlined 3-element chunks in the H pattern. In so doing, however, it should be noted that all pairwise associations were maintained; rats were always required to press a lever immediately to the left or right of the last correct response in both patterns, and the number of transitions from a given lever to any other was conserved across patterns. In this structure, elements within any chunk are related by a rule, but chunks are not related to each other in any systematic way. A formal description of this pattern is

$$(T_{+1}(T_{+1}^{2}(1))) - (T_{-1}^{2}(5)) - (T_{+1}(T_{+1}^{2}(4))) - (T_{-1}^{2}(T_{-1}^{2}(8))) - (T_{+1}^{2}(3)) - (T_{-1}^{2}(4))$$

where T_{+n} and T_{-n} represent rules indicating to “transpose” clockwise and counter-clockwise, respectively. Dashed lines indicate connections that must be learned by non-hierarchical rules or non-rule-based associations between rule-governed chunks or chunk subsets. Note that this complicated structure resulted from changing the serial positions of only 4 of 30 pattern elements compared to the H pattern.

The results showed that, for rats, pattern complexity predicted pattern learning difficulty (Fountain & Rowan, 1995a). The formally simpler H pattern was easier to learn than the formally more complex L pattern. In addition, rats in H were sensitive to the hierarchical structure of their pattern. For rats, as in humans, in the H pattern groups, the difficulty of learning to respond appropriately on any trial was a function of the hierarchical level of the rule required to predict the item. Figure 4 shows rats' group mean element-by-element error rates for Week 1 of the experiment for the H group (top panel) and the L group (bottom panel). As shown in Figure 4, during Week 1, rats produced significantly more errors on the first trial of Chunks 1 and 6 than on all other trials. These trials corresponded to the highest-order rule transitions in the pattern structure (i.e., third-order rule transitions). Fewer errors were observed on the first trial of other chunks; these trials corresponded to second-order rule transitions. The fewest errors occurred within chunks, where trials corresponded to first-order rule transitions. Thus, in the completely hierarchical pattern, the difficulty of learning to respond appropriately on any trial was a function of the hierarchical level of the rule required to predict the item.

Rats in L did not show the 3-level hierarchical pattern of errors observed for H rats. L rats found trials within chunks easier than the first trial of each chunk, but their response to the first trial of chunks was disorganized. That is, L rats, unlike H rats, showed no differential response for the first trial of Chunks 1 and 6 (corresponding to third-order rule transitions in the completely nested pattern) versus the first trials of other chunks (corresponding to second-order rule transitions in the completely

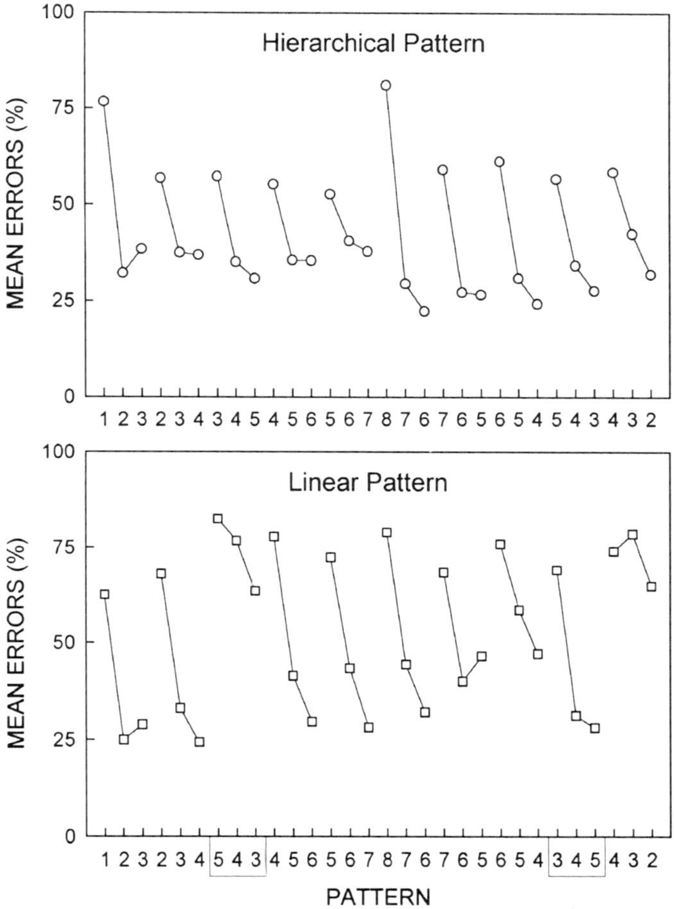

Figure 4. Rats' mean percentage of pattern tracking errors for hierarchical (top panel) and linear (bottom panel) patterns as a function of the 30 items of the patterns. Mean percentage of errors are shown for the first week of training. Pattern complexity predicted pattern learning difficulty (Fountain & Rowan, 1995a).

nested pattern). However, error rates for the second and third elements of each chunk (with the exception of the second element of Chunks 3 and 10) were significantly lower than for the first element of each chunk. These results support the view that L rats were sensitive to the actual pattern structure; they recognized that elements within a given chunk were related by a single rule, but that chunks were somewhat haphazardly arranged.

In the hierarchical versus linear structure experiment just described, rats demonstrated sensitivity to multi-level hierarchically-organized pattern structure. Rats found learning completely nested hierarchical patterns easier than learning less organized patterns even when pairwise associations and pattern length were conserved across patterns. In another study from the same series (Fountain & Rowan, 1995a), a 3-level hierarchy was easier to learn than a 4-level hierarchy when pattern length was conserved across patterns. As a rule, then, pattern complexity was a better predictor of acquisition difficulty in these studies than was pattern length. These acquisition results alone are strong evidence that pattern organization, that is, pattern complexity, was the primary determinant of pattern difficulty, as argued by a rule-learning view of sequential learning.

Detecting Pattern Structure in Interleaved Patterns

Humans have the ability to chunk together information from nonadjacent serial positions in sequential patterns. For example, human subjects can extrapolate the pattern, A-M-B-N-C-O-D-P-E-_, to find the missing element, "Q," by sorting pattern elements into two component interleaved subpatterns: A-B-C-D-E and M-N-O-P-_. Fountain and Annau (1984) earlier reported evidence that rats spontaneously detected relations among nonadjacent items in series under conditions where the rats were not provided unique cues associated with different pattern components. We conducted two experiments to investigate further rats' ability to reorganize pattern elements from nonadjacent serial positions into chunks not presented by the experimenter (Fountain, Rowan, & Benson, 1999). Rats learned either a structured or unstructured sequence interleaved with elements of a repeating sequence or an alternation sequence. The question was whether rats would learn the interleaved subpatterns at different rates as a function of subpattern complexity.

The first experiment sought to determine whether rats would show signs of being sensitive to the organization of nonadjacent items from interleaved subpatterns when one subpattern was a composed of simple, repeating element and the second subpattern was either highly structured or not. For rats in the Structured (ST) subpattern condition, a 123 234 345 456 567 subpattern was interleaved with a Repeating (R) subpattern, 888 888 888 888 888, resulting in the ST-R pattern that rats were required to learn:

182838 283848 384858 485868 586878.

For rats in the Unstructured (UNST) subpattern condition, a 153 236 345 426 547 subpattern was interleaved with the same R subpattern to create the UNST-R pattern in the same manner. For both patterns, integers represent the clockwise position of correct levers in the octagonal chamber on successive trials and spaces represent pauses that served as phrasing cues.

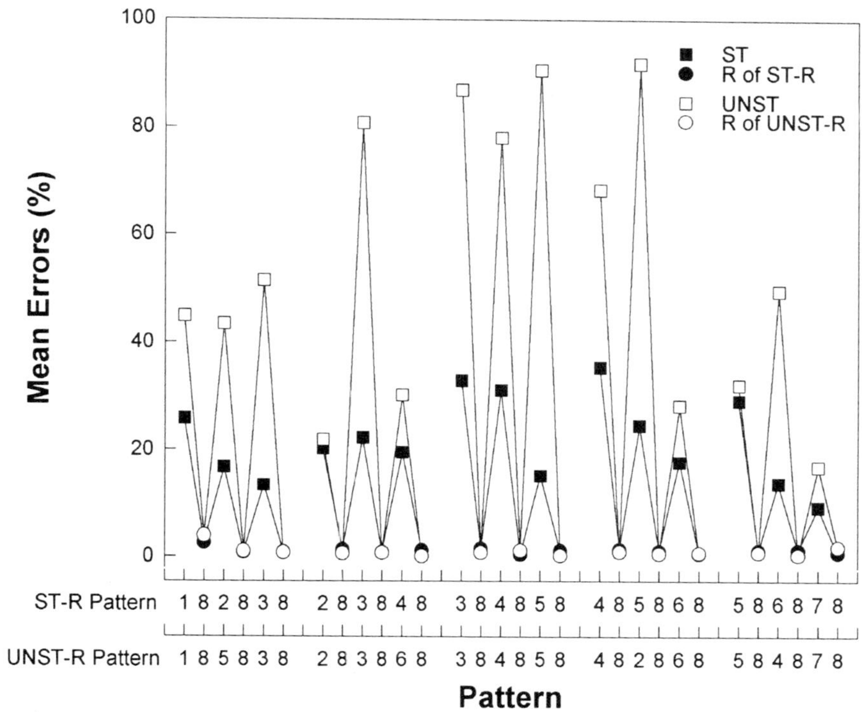

Figure 5. Group mean element-by-element errors for the ST-R and UNST-R interleaved patterns averaged across Week 3 of training (Fountain, Rowan, & Benson, 1999).

Acquisition of the interleaved structured pattern (i.e., ST-R) was significantly faster than for the interleaved unstructured pattern (i.e., UNST-R). The unstructured pattern was generated by exchanging only two pairs of elements in the structured pattern, as described above. In so doing, however, all pair-wise associations in the interleaved patterns were maintained because all of the relocated items were preceded by "8" trials. Nevertheless, Figure 5 shows that the effects of disrupting pattern structure were apparent throughout the pattern. This was so even in the third (middle) chunk that was not altered in producing the unstructured pattern; rats found this chunk, 384858, harder to learn in the context of the UNST-R pattern than in the ST-R pattern.

In the second experiment, rats learned two interleaved sequences where both were created from sets composed of more than one element. As before, longer patterns were composed of two interleaved subpatterns; either a structured or unstructured subpattern was interleaved with a subpattern of two alternating elements. For one group of rats, the structured (ST) subpattern, 1 2 3 4 5 6, was interleaved with the alternating (A) subpattern, 7

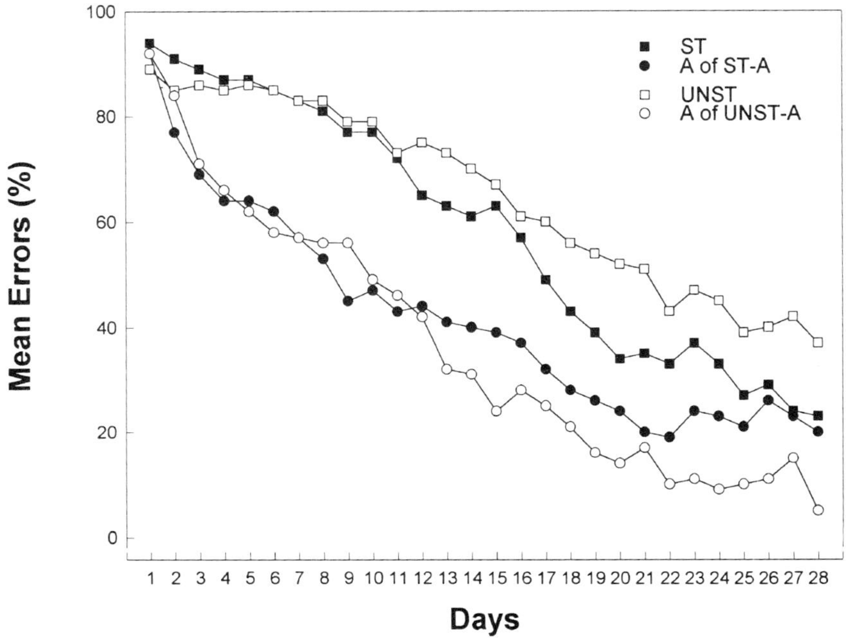

Figure 6. Acquisition of component structured (ST) or unstructured (UNST) and alternating (A) subpatterns for groups learning the ST-A and UNST-A interleaved patterns over 28 days of training (Fountain, Rowan, & Benson, 1999). Daily mean errors were averaged across elements of the subpatterns. (Error bars: ±S.E.M.)

8 7 8 7 8 to create the ST-A pattern. For another group of rats, the unstructured (UNST) subpattern, 1 5 3 4 2 6, was likewise interleaved with the same alternating subpattern to produce the UNST-A pattern. Note that the unstructured subpattern was generated by exchanging two items of the structured subpattern.

As shown in Figure 6, rats learned the subpatterns of their interleaved patterns at different rates both within and between pattern groups. As predicted based on subpattern structure, in the case of the UNST-A pattern, the A subpattern was acquired faster than the UNST subpattern. The A subpattern would be expected to be acquired faster because it is formally simple whereas the UNST subpattern has little structure (Hulse, 1978; Hulse & Dorsky, 1977; Jones, 1974). Based on similar reasoning, it was expected that the ST subpattern should be easier to learn than the UNST subpattern, and this result was obtained. In the case of the ST-A pattern, since both subpatterns were structured, it might be difficult to predict in advance based on subpattern structure alone whether rats should find either the ST or A subpattern easier to learn than the other. However, if structural complexity is equated (i.e., if the same number of rules are needed to describe subpattern structure), rats might show the same predisposition that humans do (Kotovsky & Simon, 1973) to detect repeating items before other structural features of patterns. In fact, evidence for the latter assertion was obtained in this experiment. Rats in the ST-A pattern group showed better acquisition for A with its repeating "7" and "8" elements than ST subpatterns of their interleaved pattern despite the fact that both ST and A subpatterns have simple structure that can be described by a single rule (viz., a "+1" rule for the 123456 ST subpattern versus an "alternate" rule for the 787878 A subpattern). The results of differential acquisition of ST and UNST subpatterns support the notion that accurate performance on these interleaved subpatterns was dependent on a mnemonic representation characterized by relations for nonadjacent events (Roitblat et al., 1991; Roitblat et al., 1987). The results indicate that rats are sensitive to the organization of nonadjacent elements in serial patterns and that they can detect and sort structural relationships in interleaved patterns. Pattern and subpattern structure appear to drive how animals sort, chunk together, and represent nonadjacent pattern elements that are related by common rules.

THE TEMPORAL ORGANIZATION OF SEQUENTIAL BEHAVIOR

Although we have not focused on the temporal organization of rat serial pattern learning, recent evidence from our lab indicates that timing is also an important factor in rat sequential behavior. By this we mean that rats are sensitive to the relative timing or serial position of items in sequence, even when other cues should be perfectly adequate for anticipating sequential events. In one recent study, rats learned a 24-element pattern that contained one "violation" element, that is, an element that cannot be predicted from pattern structure. The pattern was:

123 234 345 456 567 678 781 81$\underline{8}$ ...

where integers represent the clockwise position of correct levers in the octagonal chamber on successive trials and spaces represent pauses that served as phrasing cues. After rats were trained to a high criterion of accurate performance, they were transferred to one of three patterns that varied in length from 18 elements to 30 elements. These transfer patterns also contained chunks with as few as 1 element and as many as 5 elements between phrasing cues. However, in each case, the violation element was the last element of the pattern and remained in the same spatial location in the chamber (i.e., on Lever 8). Response requirements after phrasing cues and within chunks were conserved; phrasing cues continued to signal left

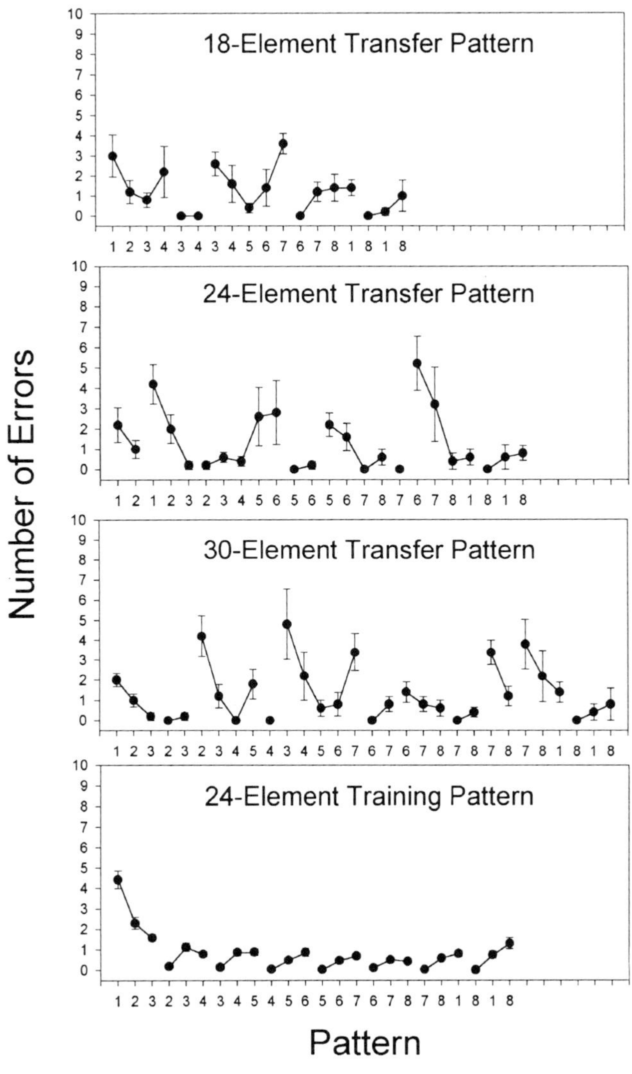

Figure 7. Rats' mean number of pattern tracking errors for the successive elements of the 18-, 24-, and 30-element transfer patterns and the 24-element training pattern experienced in transfer.

turns in transfer, and within-chunk responses were always right turns, as in training. The transfer patterns were:

18-element transfer pattern
1234 34 34567 6781 81<u>8</u>
24-element transfer pattern
12 123 23456 56 5678 7 6781 81<u>8</u>
30-element transfer pattern
123 23 2345 4 34567 67 678 78 78 781 81<u>8</u>

Rats received one of each transfer pattern interspersed with normal training for 10 days.

Figure 7 shows rats performance during the transfer phase on each of the three transfer patterns and for the 24-element training patterns among which the transfer patterns were scattered. The results show clearly that rats were concurrently using multiple cues to anticipate pattern elements (Muller, Wallace, Smith, & Fountain, 2000). First, one should note that rats' performance on the violation element (the last element of each pattern) was no poorer in the transfer patterns than in the training pattern despite large changes in the pattern leading up to the violation and in serial position. This pattern of result is consistent with the idea that rats were using spatial cues to anticipate the violation element. Second, it should be noted that performance on the first element of chunks—elements that should have been cued by phrasing cues—varied from excellent to quite poor. A close perusal of the data indicates that phrasing cues were effective as a cue for the next response whenever the preceding chunk was composed of 3 elements or more. When phrasing cues followed chunks that were 1 or 2 elements long, performance on the trial after the phrasing cue was poor relative to performance after phrasing cues in the training pattern. Thus, in the transfer phase rats were using phrasing cues to anticipate chunk boundary elements, but only when chunks were as long or longer than they were in training. Although it is clear that the phrasing cue should have been sufficient to cue what to do at chunk boundaries ("always turn left"), rats learned about chunk length even when other cues should have been perfectly adequate for anticipating forthcoming events.

THE NEURAL ORGANIZATION OF SEQUENTIAL BEHAVIOR

Several recent studies have begun to explore the neurobiological basis of serial learning and memory in rats. They suggest how serial learning processes might be integrated into more general neurobiologically based models of learning and memory. In one study, Olton, Shapiro, and Hulse (1984) tested rats' sequential memory. Four quantities of food 14, 7, 1, and 0 pellets of food were placed in the goal boxes at the ends of the four arms of a plus maze. Rats were allowed to choose freely among the arms and over a period of days learned to choose the large quantities of food first and the smallest quantities of food last. Thus rats had encoded a stimulus alphabet of four elements and had also learned an orderly response to the four elements as they were distributed in four spatial locations represented by the four goal boxes of the plus maze. Once rats had learned this task, they were given lesions of the fimbria-fornix (FFx), the major extrinsic pathway of the hippocampus. Subsequent testing showed that the rats could remember to search out the quantities in the order in which they had previously learned them. In other words, rats would go first to 14, then to 7, then to 1, then to the arm containing 0 pellets of food. However, if before given a free choice in the maze the rats were required to sample one or more of the quantities out of order in a forced choice procedure, they subsequently failed to remember having sampled the quantities when they were tested in the free choice test. For example, if a rat were allowed to retrieve the 1-pellet quantity before being given the free choice test, when the rat was allowed to make a free choice among the arms, the rat went first to 14, then to 7, then to 1 just as if it had never sampled the 1-pellet quantity in the preexposure. These kinds of mistakes indicated that rats had no memory for previously sampling food quantities from the maze before the free choice. However, subsequent tests showed that rats could remember elements sampled in preexposure as long as the quantities were received in the order in which they had originally learned them. For example, if a rat was first preexposed to the arms containing 14 and 7 pellets of food, when given a free choice, the rat would not run down the arms previously containing 14 and 7, but would go immediately to 1 and then 0. Thus FFx-lesioned rats could remember elements presented in order, but could not remember elements presented out of the order originally learned. These results are consistent with Olton et al.'s (1984) interpretation that the impairment produced by FFx lesions was an impairment of working memory, but not reference memory. They are also consistent with Eichenbaum et al.'s (1992) idea that hippocampus mediates *representational flexibility*, the ability to use declarative memories flexibly in new configurations, situations, or tasks. This idea predicts that FFx-lesioned rats should be inflexible in their use of sequential information learned before surgery, and therefore they should be impaired in their ability to respond to probe situations where patterns differed from the training pattern. Yet another interpretation of Olton et al.'s (1984) results is consistent with the view that hippocampus mediates item associative processes in SPL, but not rule-induction and memory for pattern structure (Fountain, Schenk, & Annau, 1985). According to this latter

RL view, FFx lesions spare information about pattern structure that mediates responding according to the rule learned in training prior to surgery.

A second experiment shows that hippocampal lesions produce results predicted by the RL view of serial-pattern learning if one views rule-induction as a process potentially dissociable from item association formation. Fountain, Schenk, and Annau (1985) (1985) trained rats with long monotonic and nonmonotonic patterns created from quantities of brain-stimulation reward. The monotonic pattern was 18-10-6-3-1-0 and the nonmonotonic pattern was 18-1-3-6-10-0. Prior to training, one group of rats was exposed to trimethyltin (TMT), a neurotoxic organometal which produces damage in the limbic system, primarily in the hippocampus. TMT-exposed rats learned the formally simple monotonic 18-10-6-3-1-0 pattern as fast as control rats, but learned the nonmonotonic 18-1-3-6-10-0 pattern slower than controls.

According to Olton et al. (1984), hippocampal damage should have impaired working memory, but for both groups of rats reference memory should have been intact. However, the results indicate differential impairment for two different kinds of patterns despite the fact that reference memory should have been intact and available for learning both kinds of patterns. Similarly, according to Eichenbaum et al. (1992), flexible declarative memory should have been impaired, but inflexible nondeclarative memory should have been spared. This view suggests that since learning for both groups involved learning to respond to a consistent repeating pattern, learning for both should have been spared following TMT damage to the hippocampus. The results fit best with the notion that rule-induction processes were spared following TMT damage, whereas item associative processes were impaired by TMT damage. These, along with other data, suggested that item association formation is a hippocampal-dependent process, whereas rule induction is not. A dissociation of this sort may be difficult to model with SPAM, though Metcalfe (1993) has succeeded in simulating characteristics of Korsakoff's amnesia using a closely related model, CHARM.

In a third series of experiments (Fountain & Rowan, 2000), we sought additional evidence for this distinction between item associative and rule induction processes. In the first study of the series, rats were trained on two patterns, one which was structurally "perfect" and a second virtually identical to the first, but containing a single element that violated the otherwise simple structure. The Perfect (P) and Violation (V) patterns were:

P Pattern: 123 234 345 456 567 678 781 812

V Pattern: 123 234 345 456 567 678 781 818

As before, the digits indicate the reinforced lever for successive trials. The last "8" item of the V pattern (underlined) was the violation element. Rats from one group for each pattern condition were injected with MK-801 daily before training. MK-801 is a systemically administered NMDA receptor antagonist that blocks neuronal plasticity, known as long-term potentiation,

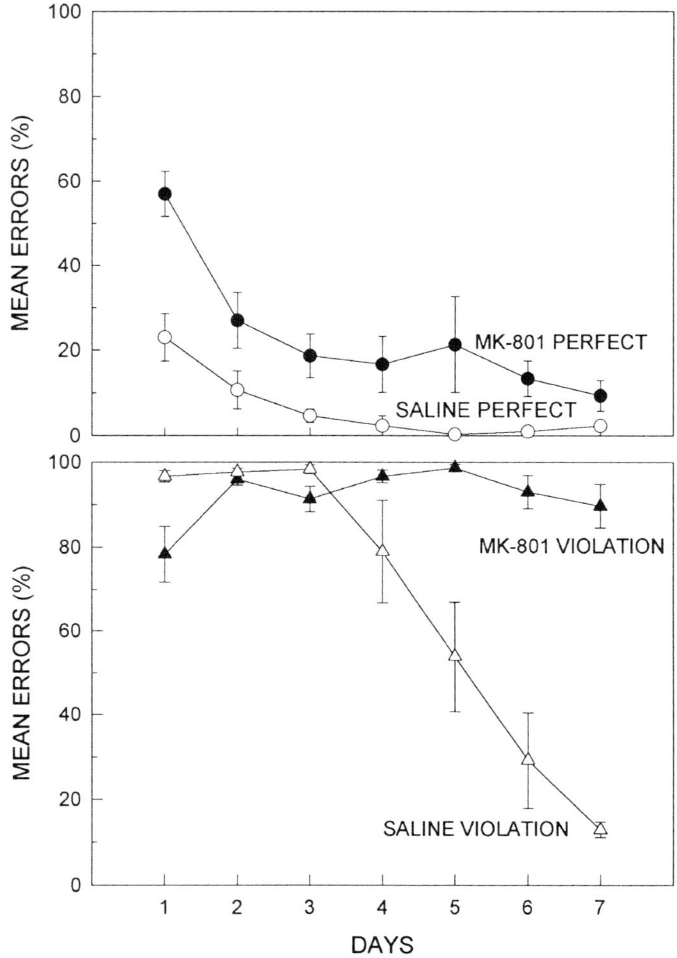

Figure 8. Acquisition of the last element of the perfect pattern (top panel) and violation pattern (bottom panel) over the 7 days of training for the Saline and MK-801 groups. The last element was structurally consistent in the perfect pattern and it was the violation element in the violation pattern. Daily mean errors are shown for the last element of the pattern only (Fountain & Rowan, 2000).

in the hippocampus. It is thought that MK-801 should impair any hippocampal-dependent learning. As shown in Figure 8, MK-801 had little effect on learning to respond to rule-based items within chunks. However, it did impair responding at points where rules were violated, namely, on the first trial of each new chunk and, most dramatically, for the violation element. Although rats showed no signs of learning to respond to the violation element, throughout the experiment they produced rule-based errors on the violation trial by responding “2” instead of “8” at the end of the sequence (Fountain & Rowan, 2000). The results are strong evidence that hippocampal damage impaired learning the item associations necessary to track violations of pattern structure while sparing the rule induction processes necessary to induce pattern structure and extrapolate the sequence on the violation trial.

In a later study in the same series (Fountain & Rowan, 2000), we examined the role of hippocampus when new serial pattern information is added to old. Rats were first trained to a high criterion on a pattern consisting of the first 7 chunks of the P pattern above: 123 234 345 456 567 678 781. After rats learned the pattern, they were transferred to one of two new patterns that contained all elements of the first pattern and an additional chunk of three additional elements. The three added elements were either structurally consistent with the first pattern (viz., 812), making it structurally “perfect” (P), or they contained a violation (V) of the pattern structure learned in training (viz., 818). On the day of transfer, half the rats were injected with MK-801 to determine the effects of hippocampal dysfunction on the rats’ ability to integrate structurally consistent or inconsistent new information with an already learned pattern.

As shown in Figure 9, when a structurally consistent chunk was added in the P transfer, the effects of MK-801 were very similar to the effects of the drug on acquisition (Fountain & Rowan, 2000). That is, the drug produced a selective decrease in the animals' accuracy on the first elements of each chunk of the original pattern, but produced virtually no change in accuracy on the remaining two elements of the 3-element chunks. The most interesting result occurred when a structurally inconsistent chunk was added in the V transfer. As shown in Figure 9, although saline controls showed difficulty in learning the new chunk, there was little effect on the rest of the pattern. However, MK-801 dramatically disrupted performance for elements both in the new chunk to be learned and throughout the rest of the pattern (Fountain & Rowan, 2000). When this effect is compared to the effects of MK-801 in the P transfer, the effect can only be accounted for by the addition of the terminal violation element. One interpretation of these results is that adding new information to a pattern representation is possible under MK-801, but only if the information is consistent with pattern structure that has already been encoded. In fact, this initial evidence indicates that for rats with hippocampal impairment, new information that is structurally inconsistent can disrupt previously well-learned response patterns. This suggests that, in intact animals, nonhippocampal systems mediate rule induction whereas hippocampus may play a role in the successful integration of new rule-inconsistent SPL information with already encoded information about pattern structure. These ideas are

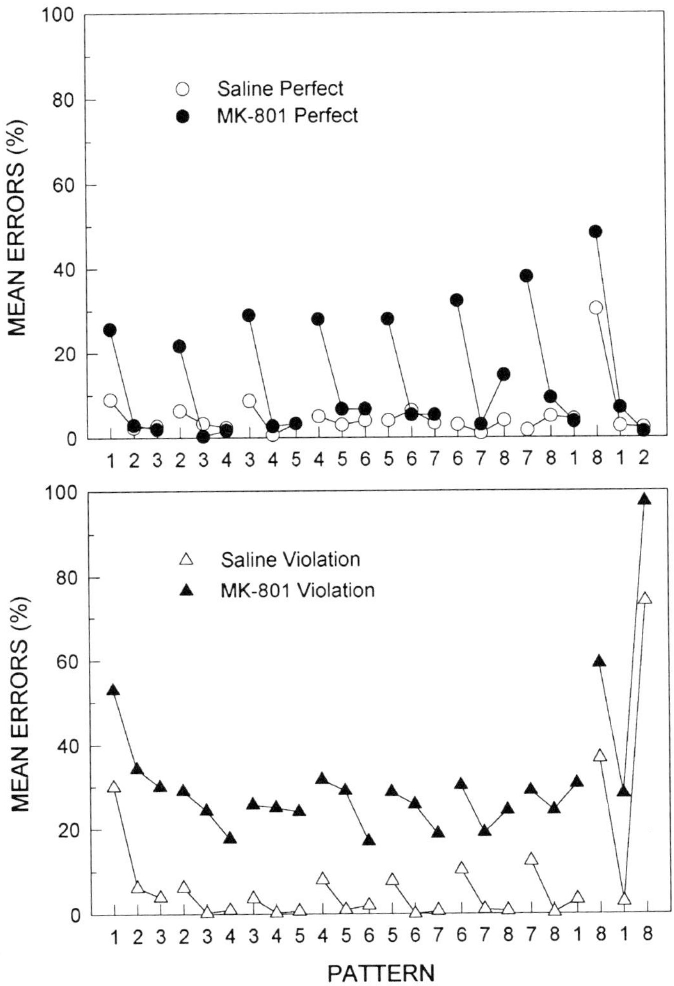

Figure 9. Rats' mean percentage of pattern tracking errors for the perfect (top panel) and violation (bottom panel) patterns as a function of the 24 items of the patterns on the day of transfer when the eighth 3-element chunk was added to the previously learned 7-chunk pattern. On the day of transfer, rats were injected with either saline or MK-801, an NMDA receptor antagonist (Fountain & Rowan, 2000).

reminiscent of the distinction between flexible and inflexible memory processes proposed by Eichenbaum et al. (1992), but our MK-801 results suggest that what constitutes "representational flexibility" is far from resolved. Under MK-801, rats were able to add a rule-consistent chunk to their already learned pattern with relatively little difficulty, but not a rule-inconsistent chunk.

CONCLUSIONS: AN INTEGRATIVE VIEW OF SEQUENTIAL BEHAVIOR

This chapter describes rats as multi-modality information processors; rats appear able to use multiple cues from different modalities and cognitive processes to navigate sequential patterns. We have described evidence that rats used spatial cues for anticipating violation elements in patterns, temporal phrasing cues to anticipate chunk boundaries, serial position cues (as compound cues with phrasing cues), and proprioceptive cues or rules to perform within-chunk responses. It should also be noted that sensitivity to serial position implies counting, timing, or some related process for discriminating chunk length. [One might also wonder how the rat concurrently tracks temporal pauses between trials to identify phrasing cues while concurrently tracking chunk length.] In addition, we have reported evidence that at least two brain systems are recruited in serial-pattern learning, and recent evidence in our lab suggests that more than two dissociable brain processes can be demonstrated. All-in-all, our current view of the rat performing sequential tasks is a complicated one, and it is no longer proper to think of the rat's performance or behavioral systems as "simple." The implication is that rats have sophisticated cognitive processes for online monitoring of internal and external information, for cue selection, and for efficiently representing sequential episodes and problems. The challenge will be to determine how multiple competing cognitive and brain processes work together to produce organized and adaptive sequential behavior.

ACKNOWLEDGMENTS

This work was supported by the National Institute of Mental Health Grant MH48402. Douglas Wallace is now at Department of Psychology, University of Lethbridge. James Rowan is now at Department of Psychology, Chowan College. Correspondence should be addressed to Stephen B. Fountain, Department of Psychology, Kent State University, Kent, Ohio 44242-0001. E-mail may be addressed to sfountai@kent.edu.

REFERENCES

Blumberg, M.S., & Wasserman, E.A. (1995). Animal mind and the argument from design. *American Psychologist*, **50**, 133-144.

Burns, R. A., & Gordon, W. U. (1988). Some further observations on serial enumeration and categorical flexibility. *Animal Learning & Behavior*, **16**, 425-428.

Burns, R. A., Hulbert, L. G., & Cribb, D. (1990). A test for order relevance in a three-element serial learning task. *Journal of General Psychology*, **117**, 91-98.

Capaldi, E.J. (1986). Serial learning and trimethyltin: An unfortunate case of ad hoc conclusions. *Physiological Psychology*, **14**, 71-72.

Capaldi, E.J., & Miller, D.J. (1988). The rat's simultaneous anticipation of remote events and current events can be sustained by event memories alone. *Animal Learning & Behavior*, **16**, 1-7.

Capaldi, E.J., & Molina, P. (1979). Element discriminability as a determinant of serial-pattern learning. *Animal Learning & Behavior*, **7**, 318-322.

Capaldi, E.J., Nawrocki, T.M., Miller, D.J., & Verry, D.R. (1985). An examination into some variables said to affect serial learning. *Animal Learning & Behavior*, **13**, 129-136.

Capaldi, E.J., Nawrocki, T.M., & Verry, D.R. (1982). Difficult serial anticipation learning in rats: Rule-encoding vs. memory. *Animal Learning & Behavior*, **10**, 167-170.

Capaldi, E.J., Nawrocki, T.M., & Verry, D.R. (1983). The nature of anticipation: An inter- and intraevent process. *Animal Learning & Behavior*, **11**, 193-198.

Capaldi, E.J., Verry, D.R., & Davidson, T. (1980a). Memory, serial anticipation pattern learning, and transfer in rats. *Animal Learning & Behavior*, **8**, 575-585.

Capaldi, E.J., Verry, D.R., & Davidson, T. (1980b). Why rule encoding by animals in serial learning remains to be established. *Animal Learning & Behavior*, **8**, 691-692.

Capaldi, E.J., Verry, D.R., Nawrocki, T.M., & Miller, D.J. (1984). Serial learning, interitem associations, phrasing cues, interference, overshadowing, chunking, memory, and extinction. *Animal Learning & Behavior*, **12**, 7-20.

Chen, S., Swartz, K.B., & Terrace, H.S. (1997). Knowledge of the ordinal position of list items in rhesus monkeys. *Psychological Science*, **8**, 80-86.

Church, R.M. (1997). Quantitative models of animal learning and cognition. *Journal of Experimental Psychology: Animal Behavior Processes*, **23**, 379-389.

Church, R.M. (2001). A Turing test for computational and associative theories of learning. *Current Directions in Psychological Science*, **10**, 132-136.

Eich, J.M. (1982). A composite holographic associative recall model. *Psychological Review*, **89**, 627-661.

Eichenbaum, H., Otto, T., & Cohen, N.J. (1992). The hippocampus—What does it do? *Behavioral and Neural Biology*, **57**, 2-36.

Fountain, S.B. (1986). Serial-pattern learning: A unitary process? *Physiological Psychology*, **14**, 67-70.

Fountain, S.B. (1990). Rule abstraction, item memory, and chunking in rat serial-pattern tracking. *Journal of Experimental Psychology: Animal Behavior Processes,* **16**, 96-105.

Fountain, S.B., & Annau, Z. (1984). Chunking, sorting, and rule-learning from serial patterns of brain-stimulation reward by rats. *Animal Learning & Behavior,* **12**, 265-274.

Fountain, S.B., Evensen, J.C., & Hulse, S.H. (1983). Formal structure and pattern length in serial pattern learning by rats. *Animal Learning & Behavior,* **11**, 186-192.

Fountain, S.B., Henne, D.R., & Hulse, S.H. (1984). Phrasing cues and hierarchical organization in serial pattern learning by rats. *Journal of Experimental Psychology: Animal Behavior Processes,* **10**, 30-45.

Fountain, S.B., & Hulse, S.H. (1981). Extrapolation of serial stimulus patterns by rats. *Animal Learning & Behavior,* **9**, 381-384.

Fountain, S.B., & Rowan, J.D. (1995a). Coding of hierarchical versus linear pattern structure in rats and humans. *Journal of Experimental Psychology: Animal Behavior Processes,* **21**, 187-202.

Fountain, S.B., & Rowan, J.D. (1995b). Sensitivity to violations of "run" and "trill" structures in rat serial-pattern learning. *Journal of Experimental Psychology: Animal Behavior Processes,* **21**, 78-81.

Fountain, S.B., & Rowan, J.D. (2000). Differential impairments of rat serial-pattern learning and retention induced by MK-801, an NMDA receptor antagonist. *Psychobiology*, **28**, 32-44.

Fountain, S.B., Schenk, D.E., & Annau, Z. (1985). Serial-pattern-learning processes dissociated by trimethyltin exposure in rats. *Physiological Psychology,* **13**, 53-62.

Haggbloom, S.J. (1985). Serial learning and transfer in rats: effects of changes in stimulus-stimulus associations, pattern structure, and serial position information. *Animal Learning & Behavior,* **13**, 370-374.

Haggbloom, S.J., & Brooks, D.M. (1985). Serial anticipation and pattern extrapolation in rats as a function of element discriminability. *Animal Learning & Behavior,* **13**, 303-308.

Hulse, S.H. (1978). Cognitive structure and serial pattern learning by animals. In S. H. Hulse, H. Fowler, & W. K. Honig (Eds.), *Cognitive processes in animal behavior* (pp. 311-340). Hillsdale,NJ: Erlbaum.

Hulse, S.H. (1980). The case of the missing rule: Memory for reward vs. formal structure in serial-pattern learning by rats. *Animal Learning & Behavior,* **8**, 689-690.

Hulse, S.H., & Campbell, C.E. (1975). "Thinking ahead" in rat discrimination learning. *Animal Learning & Behavior,* **3**, 305-311.

Hulse, S.H., & Dorsky, N.P. (1977). Structural complexity as a determinant of serial pattern learning. *Learning and Motivation*, **8**, 488-506.

Hulse, S.H., & Dorsky, N.P. (1979). Serial pattern learning by rats: Transfer of a formally defined stimulus relationship and the significance of nonreinforcement. *Animal Learning & Behavior,* **7**, 211-220.

Hunter, W.S. (1920). The temporal maze and kinaesthetic sensory processes in the white rat. *Psychobiology,* **2**, 1-17.

Jones, M.R. (1974). Cognitive representations of serial patterns. In B. Kantowitz (Ed.), *Human information processing: Tutorials in performance and cognition*. Hillsdale,NJ: Erlbaum.

Lashley, K.S. (1951). The problem of serial order in behavior. In L. A. Jeffress (Ed.), *Cerebral mechanisms in behavior* (pp. 112-146). New York: Wiley.

Metcalfe, J. (1990). Composite holographic associative recall model (CHARM) and blended memories in eyewitness testimony. *Journal of Experimental Psychology: General,* ***119***, 145-160.

Metcalfe, J. (1993). Novelty monitoring, metacognition, and control in a composite holographic associative recall model: Implications for Korsakoff amnesia. *Psychological Review,* **100**(1), 3-22.

Muller, M.D., Wallace, D.G., Smith, D.P., & Fountain, S.B. (2000). Predicting pattern violations using serial position in rat sequential learning. *Society for Neuroscience Abstracts*, **26**, 2244.

Murdock, B.B. (1982). A theory for the storage and retrieval of item and associative information. *Psychological Review,* **89**, 609-626.

Murdock, B.B. (1983). A distributed memory model for serial-order information. *Psychological Review,* **90**, 316-336.

Murdock, B.B. (1985). Convolution and matrix systems: A reply to Pike. *Psychological Review,* **92**, 130-132.

Olton, D.S., Shapiro, M.L., & Hulse, S.H. (1984). Working memory and serial patterns. In H. L. Roitblat, T. G. Bever, & H. S. Terrace (Eds.), *Animal cognition* (pp. 171-182). Hillsdale, NJ: Lawrence Erlbaum Associates.

Restle, F. (1970). Theory of serial pattern learning: Structural trees. *Psychological Review,* **77**, 481-495.

Restle, F. (1972). Serial patterns: The role of phrasing. *Journal of Experimental Psychology,* **92**, 385-390.

Restle, F. (1973). Serial pattern learning: Higher order transitions. *Journal of Experimental Psychology,* **99**, 61-69.

Restle, F., & Brown, E.R. (1970a). Organization of serial pattern learning. In G. H. Bower (Ed.), *Psychology of learning and motivation.* New York: Academic Press.

Restle, F., & Brown, E.R. (1970b). Serial pattern learning. *Journal of Experimental Psychology,* **83**, 120-125.

Restle, F., & Brown, E.R. (1970c). Serial pattern learning: Pretraining of runs and trills. *Psychonomic Science,* **19**, 321-322.

Restle, F., & Burnside, B.L. (1972). Tracking of serial patterns. *Journal of Experimental Psychology,* **95**, 299-307.

Roitblat, H.L. (1982). The meaning of representation in animal memory. *Behavioral and Brain Sciences,* **5**, 353-406.

Roitblat, H.L., Bever, T.G., Helweg, D.A., & Harley, H.E. (1991). On-line choice and the representation of serially structured stimuli. *Journal of Experimental Psychology: Animal Behavior Processes,* **17**, 55-67.

Roitblat, H.L., Pologe, B., & Scopatz, R.A. (1983). The representation of items in serial position. *Animal Learning & Behavior,* **11**, 489-498.

Roitblat, H.L., Scopatz, R., & Bever, T. (1987). The hierarchical representation of three-item sequences. *Animal Learning & Behavior,* **15**, 179-192.

Skinner, B.F. (1934). The extinction of chained reflexes. *Proceedings of the National Academy of Sciences,* **20**, 234-237.

Stempowski, N.K., Carman, H.M., & Fountain, S.B. (1999). Temporal phrasing and overshadowing in rat serial-pattern learning. *Learning and Motivation*, **30**, 74-100.

Terrace, H.S. (1987). Chunking by a pigeon in a serial learning task. *Nature,* **325**, 149-151.

Terrace, H.S. (1991). Chunking during serial learning by a pigeon: I. Basic evidence. *Journal of Experimental Psychology: Animal Behavior Processes,* **17**, 81-93.

Terrace, H.S., & Chen, S. (1991a). Chunking during serial learning by a pigeon: II. Integrity of a chunk on a new list. *Journal of Experimental Psychology: Animal Behavior Processes,* **17**, 94-106.

Terrace, H.S., & Chen, S. (1991b). Chunking during serial learning by a pigeon: III. What are the necessary conditions for establishing a chunk? *Journal of Experimental Psychology: Animal Behavior Processes,* **17**, 107-118.

Wallace, D.G., & Fountain, S.B. (In press). What is learned in sequential learning? An associative model of reward magnitude serial-pattern learning. *Journal of Experimental Psychology: Animal Behavior Processes*.

Wathen, C.N., & Roberts, W.A. (1994). Multiple-pattern learning by rats on an eight-arm radial maze. *Animal Learning & Behavior,* **22**, 155-164.

Part 3: Counting and Timing

7

MECHANISMS OF "COUNTING" IN ANIMALS

William A. Roberts
University of Western Ontario

For about the last 100 years, scientists interested in animal behavior and cognition have addressed the question of numerical competence in animals (Rilling, 1993; W. A. Roberts, 1998). As we shall see, there are a number of research procedures by which one can address this question. One thing that all of these approaches have in common is that number is responded to as a *concept*. A concept is a response or tag that is given to a group of stimuli that are perceptually distinguishable from one another but have some common property. Considerable interest in animal concept learning has arisen in recent years, with the discovery that pigeons and nonhuman primates readily learn to discriminate visual concepts, such as pictures of trees, people, flowers, cats, cars, and chairs (Herrnstein, Loveland, & Cable, 1976; W. A. Roberts & Mazmanian, 1988; Wasserman, 1995). Perhaps less well recognized has been the fact that all well controlled numerical studies with animals also involve concept learning. For example, if we wish an animal to learn to make a particular response that identifies a stimulus as containing five objects, we would not wish to present the same picture of five objects each time the response is to be made for reward. If we did present the same stimulus each time, a critic would rightly complain that we have not taught the animal to respond to quantity but only to a particular visual pattern. In order to insure that our subject is learning to respond to number, we would present stimuli that contain five items which vary in size, shape, color and the pattern by which they are displayed. If the subject can consistently make the same response to these stimuli, we can safely make the argument that the subject is responding to the concept of five. Although the stimuli may vary along many dimensions, they all have the property of fiveness in common. In order to find out if an animal can learn to respond to number as a concept, experiments in numerical discrimination must be performed.

NUMERICAL DISCRIMINATION

A basic question about the numerical competence of an animal is whether it can discriminate between quantities of items. Numerical discrimination studies may require discrimination based on either *relative numerousness* or *absolute number* (Davis & Perusse, 1988). In a relative numerousness discrimination, an animal must choose the stimulus that has the most or fewest items, even though the stimuli may contain different numbers of items from one trial to the next. In an absolute number discrimination, a stimulus containing a precise number of items is always correct.

Relative Numerousness Discrimination

In a nicely controlled study of relative numerousness, Thomas, Fowlkes, and Vickery (1980) trained two squirrel monkeys to discriminate between cards that contained different numbers of black circles. The monkey's task was to choose the card with the smaller number of circles, for which it was rewarded with fruit. Up to ten circles could be presented on a card, but the sizes of the circles and their pattern of presentation varied from card to card, with 100 different patterns used for each number of circles. Monkeys could not make choices based on the card with the smaller blackened area or the larger background area, because the card with the smaller number of circles sometimes had a larger surface area covered with black than the card with the larger number of circles. Both subjects learned to choose the smaller number with combinations of circles that differed by two or more circles, such as 2:6, 2:7, 3:7, 4:7, and 4:6. Eventually, both monkeys chose correctly above chance accuracy on an 8:9 discrimination, and one monkey was able to master a 9:10 problem. The data strongly suggest that these monkeys used relative number as the basis for their choice.

Although the items to be discriminated were presented *simultaneously* as an array in the Thomas, et al. experiment, animals' ability to discriminate between stimuli presented *successively* also may be tested. W. A. Roberts and Mitchell (1994; Experiment 1) trained pigeons to

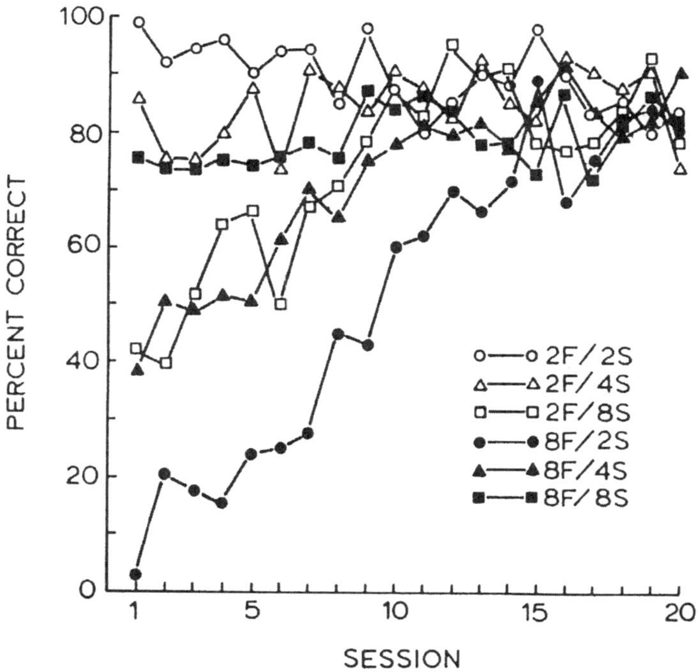

Figure 1. Acquisition of a numerical discrimination by pigeons between 2 and 8 light flashes. (From W. A. Roberts & Mitchell, 1994).

discriminate between two successive light flashes and eight successive light flashes. Pigeons were tested in an operant chamber, and each flash consisted of a 200-ms presentation of an overhead red light followed by 800 ms of darkness. Therefore, flashes occurred at the rate of one per second, with 2 flashes in 2 s (2F/2S) and 8 flashes in 8 s (8F/8S). Two keys were lit after each sequence; a peck on the left key produced grain reward after the 2F/2S sequence, and a peck on the right key produced reward after the 8F/8S sequence. Six pigeons learned to discriminate these sequences at 80% accuracy within an average of 15 sixty-trial sessions.

It should be clear that we cannot conclude that pigeons learned to discriminate more from less light flashes based on this experiment. Since two and eight light flashes occupied different durations, time is an obvious confounding variable. Pigeons may have been judging longer versus shorter lengths of flash sequences. In order to rule sequence duration out as a source of discrimination, the pigeons were transferred to sequences of 2 and 8 200-ms flashes that lasted for equal durations (W. A. Roberts & Mitchell, 1994; Experiment 2). These sequences were 2F/2S, 8F/2S, 2F/4S, 8F/4s, 2F/8s and 8F/8s. Performance over 20 sessions with these six sequences is shown in Figure 1. The fact that pigeons did not show immediate transfer to novel sequences suggests that the original discrimination between 2F/2S and

8F/8s was not based solely on number. In particular, the 8F/2s sequence begins at near zero accuracy, suggesting that duration of this sequence may have formed the basis for initial response to it. Over 20 sessions, however, pigeons learned to respond at 80% accuracy or higher to all six sequences, indicating response to number of flashes.

In experiments examining numerousness discrimination between sequences of events, it is very difficult to control all possible sources of confound. With regard to the data presented in Figure 1, it could be objected that pigeons might have summed the durations of light flashes and used that difference in time to make the discrimination. Although this possibility was not controlled by Roberts and Mitchell, Fernandes and Church (1982) performed a similar experiment with rats and varied both the sequence length and the duration of sound bursts. Rats learned to press one lever after 2 bursts and another lever after 4 bursts, even when the 2-burst sequence had a longer sound-on period than the 4-burst sequence.

Absolute-Number Discrimination

In an absolute-number discrimination, an animal must learn to make a specific response when only one number of items is presented. The Roberts and Mitchell (1994) experiment just described may have involved absolute-number discrimination, because pigeons were reinforced for making distinct responses to 2 and 8 light flashes. However, it is also possible that pigeons learned to respond to just fewer versus more flashes. An absolute-number discrimination is most clearly shown when an animal learns to respond in one way to a particular number and to respond differently to numbers higher and lower than that value. For example, Hicks (1956) trained rhesus monkeys to discriminate between cards that contained 1 to 5 geometric forms. The forms took different shapes, such as triangles, squares, and circles, and varied in size, color and pattern of distribution. Monkeys always were rewarded for choosing the card containing 3 items. After considerable training, monkeys learned to choose whichever card contained 3; thus, the card containing 3 items was chosen when cards containing 2 and 3 items were shown and when cards containing 3 and 4 items were shown.

Davis and Albert (1986) demonstrated absolute-number discrimination in rats when stimuli were successive white noise bursts. Series of 2, 3, and 4 bursts of white noise were played, but only bar pressing after 3 bursts was reinforced. To control for the use of time cues, the length of each noise burst was varied between 0.2 and 0.5 s, and the interval between bursts was varied between 0.5 and 0.9 s. All three rats tested learned to press the bar more after 3 noise bursts than after 2 or 4 noise bursts. Rats apparently learned to discriminate precisely the number 3.

CAN ANIMALS COUNT?

Human counting systems involve the application of a set of number tags to a collection of items to be counted (Gelman & Gallistel, 1978). These tags or number symbols have the properties of *cardinality* and *ordinality*. Cardinality refers to the unique symbols assigned to counted items, with only one symbol applied to each item. The items counted may be assigned symbols in any order (the order-irrelevance principle), as long as the symbols are assigned in a fixed order (the stable-order principle). The final symbol assigned to a set of items is the *cardinal number* and defines that set of items.

Whether animals can use a number tagging system with these properties is a controversial topic. Both Capaldi (Capaldi, 1993; Capaldi & Miller, 1988) and Gallistel (1990, 1993) have argued that animals count routinely and automatically. The symbols used to count may be hidden within an animal's nervous system; these hypothesized symbols have been referred to as *numerons*. Numerons form the basis for internal counting in both people and animals, but humans have, in addition, associated verbal labels with numerons (Gallistel & Gelman, 1992).

By contrast, Davis and Perusse (1988) argue that the best evidence indicates only *protocounting* in animals. Protocounting suggests that animals show numerical competence in a number of types of numerical processing tasks but fall short of the process of counting found in people. The fact that animals can make relative numerousness and absolute number discriminations does not necessarily mean that they can count. Mechanisms other than the assignment of symbols to items could allow an animal to distinguish among quantities of items. Human subjects can *subitize* or identify the number of items shown in a pattern presented for only a fraction of a second (Kaufman, Lord, Reese, & Volkmann, 1949). The speed with which subitizing occurs appears to preclude the sequential assignment of cardinal symbols to items, although there is some debate on this issue (Mandler & Shebo, 1982; Miller, 1993). Davis and Perusse suggest that animals may use subitizing to make numerical discriminations when different quantities of items are shown simultaneously and a related process of *rhythm discrimination* to distinguish among different numbers of items presented sequentially.

Another issue regarding numerical competence in animals is transfer of training. Humans use counting systems to count virtually anything (the abstract principle), including objects, events, and hypothetical ideas. However, evidence for transfer of numerical discriminations learned in one situation to another situation is fairly rare in animal research. Lack of evidence for transfer does not rule out the possibility that animals are capable of counting. As Broadbent, Church, Meck, and Rakitin (1993) have

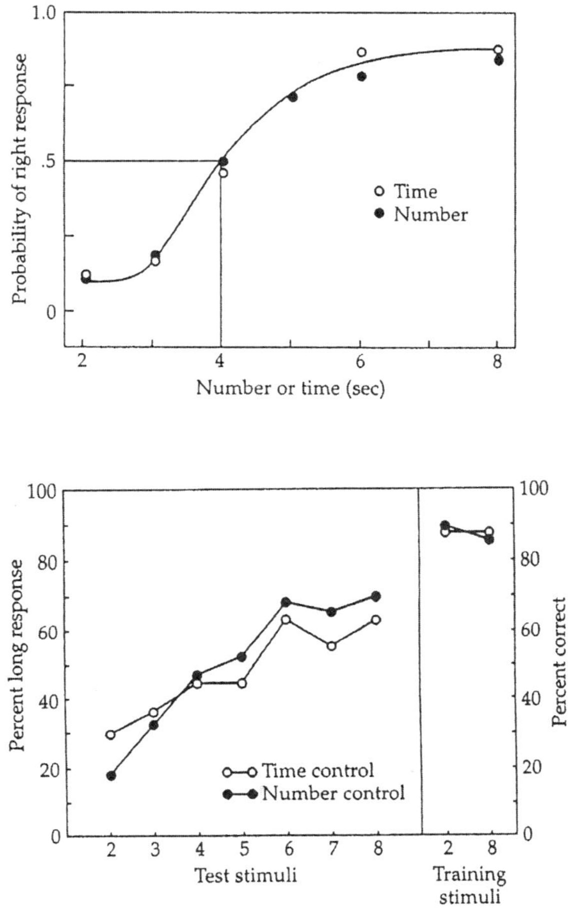

Figure 2. Psychophysical curves from rats (upper panel) and pigeons (lower panel) showing equivalent control by time and number after animals were trained to discriminate between two sound bursts or light flashes in 2 s and eight sound bursts or light flashes in 8 s. (From Meck & Church, 1983, and W. A. Roberts & Mitchell, 1994).

pointed out, an animal may be able to count only within a limited domain. There is also the possibility that, with sufficient training, an animal might learn to use a counting system in several different situations.

A MODEL FOR COUNTING AND TIMING

The scalar-timing model is a well-known model of timing in animals. In a dual-mode version of this model, Meck and Church (1983) showed how the same mechanism that accounted for precise interval timing could also account for counting. The model was based on an experiment in which rats

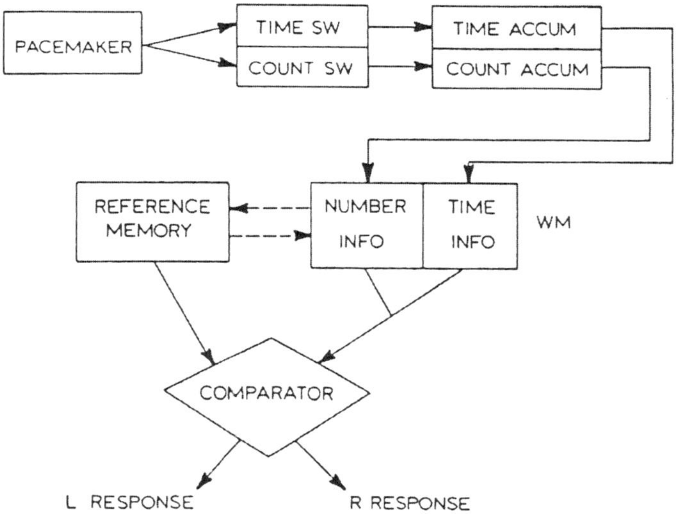

Figure 3. The dual-mode model accounts for timing and counting by separate time and count switches that operate in the run and event modes, respectively. (From Roberts & Mitchell, 1994).

were reinforced for disciminating between sequences of sound bursts by pressing left and right levers. For example, a press on the left lever would be reinforced after a sequence of 2 sound bursts that lasted 2 s, and a press on the right lever would be reinforced after a sequence of 8 sound bursts that lasted 8 s. When rats responded accurately to these sequences, tests were carried out in which one dimension was held constant at an intermediate value and the other dimension was varied. Thus, time was held constant at 4 s, and number was varied from 2 to 8 bursts; on other tests, number was held constant at four bursts, and time was varied from 2-8 s. In a similar experiment later performed with pigeons by W. A. Roberts and Mitchell (1994; Experiment 3), the same procedures were followed, but pigeons discriminated between 2 light flashes in 2 s and 8 light flashes in 8 s.

The psychophysical curves produced by these experiments are shown in Figure 2. The important finding is that rats in the upper panel and pigeons in the lower panel show essentially the same curves when time and number were varied. The findings suggest that both species show equal control by time and number. The dual-mode model that provides an account of these findings is shown in Figure 3. At the top level of the model, a clock process is shown in which a pacemaker transmits pulses at a constant rate to time and number accumulators. The important aspect of the model that allows it to both time and count are the time and count switches placed between the pacemaker and the accumulators. When a trial begins with the presentation of a stimulus event (sound burst or light flash), these switches close and allow pulses to flow into the accumulators. However, the switches operate in different modes. The time switch operates in a *run mode*, which means that it stays closed throughout the presentation of successive stimuli and only opens at the end of the sequence. By constrast, the count switch operates in an *event mode*, which means that it closes briefly for a short constant interval each time an event occurs. At the end of the stimulus presentation, the total pulses in the time accumulator will represent the total time to present the sequence of events, and the total pulses in the count accumulator will represent the number of events in the sequence. At the second level of the model, these accumulated pulses are sent to time and number working memories. If the correct response is made (right or left bar press or key peck), that number of pulses is stored in a reference memory along with the appropriate response. The third level of the model contains a comparator that compares number and time criteria retreived from reference memory with pulses accumulated in the working memories on new trials. If the value in working memory closely matches the time or number value retrieved from reference memory for a left response, the left lever or key will be chosen; if the working memory value more closely matches values that dictate a right response, the right lever or key will be chosen. The accumulated pulses then correspond to the numerons suggested by Gallistel and Gelman (1992; 2000). Animals can thus count or track numbers by using a neural process (accumulated pulses) that does not require the assignment of learned symbols to numbers of events.

THE PRODUCTION OF NUMBER

The Peak Procedure in Timing

In a method referred to as the *peak procedure* (S. Roberts, 1981), animals are initially trained to respond to a fixed-interval (FI) schedule of reinforcement signaled on discrete trials by the onset of a stimulus. A tone or houselight may initiate a trial, signaling that the first response made after an FI will be reinforced. On occasional test trials inserted among the FI trials, the signal is presented for a period far longer than the FI, and no reinforcer is delivered for responding; these test trials are often referred to as *empty trials*. The data accumulated from a number of empty trials are then plotted as the rate of pecking in successive time bins. The peak-time curve approximates a Gaussian distribution and shows a peak near the FI (See Figures 5, 6, and 7). The subject then has produced through its behavior an estimate of the time at which it expected to be reinforced.

The scalar-timing model has been particularly successful in accounting for peak-time data. The model assumes that the pulse total

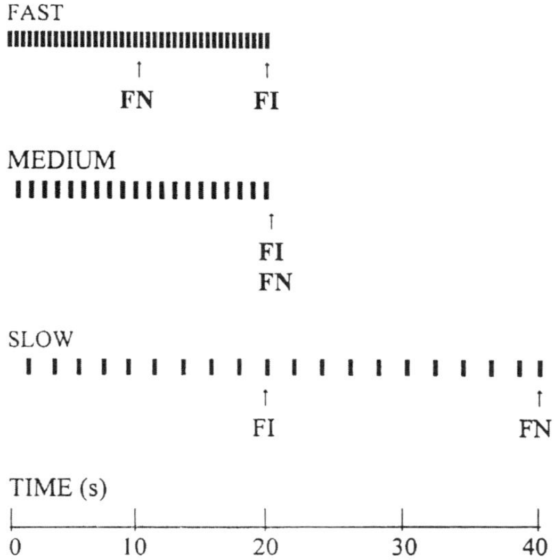

Figure 4. The spaced dark bars stand for light flashes presented at fast (2/s), medium (1/s), or slow (0.5/s) rates. Birds reinforced on a fixed interval 20 s schedule are reinforced at the FI arrows, and birds reinforced on a fixed number 20 flashes schedule are reinforced at the FN arrows. (From W. A. Roberts & Boisvert, 1998).

progressively accumulating in the time working memory is regularly compared with a criterion drawn from reference memory that would represent the FI. When the accumulating pulses become sufficiently close to the criterion number, an animal begins to lever press or key peck and continues to do so on empty trials until the accumulating pulses exceed the criterion. The average peak of responding appears right around the FI.

The Peak Procedure Applied to Counting and Timing

W. A. Roberts and Boisvert (1998) investigated the possibility that pigeons could produce estimates of number as well as time. According to the dual-mode model shown in Figure 3, the same process of pulse accumulation is used to track both time and number. The model then predicts that the peak procedure should reveal evidence of accurate counting as well as timing. Two groups of pigeons were trained on sessions within which discrete trials began with the illumination of a white center key in a Skinner box. The key periodically flashed red for 200 ms. The rate at which the key flashed varied randomly between three rates. The key flashed at a fast rate of two times per s on a third of the trials, at a medium rate of one time per s on another third of the trials, and at a slow rate of once every 2 s on the remaining third of the trials. Pigeons in one group were always reinforced according to an FI 20 s schedule; the first key peck made after 20

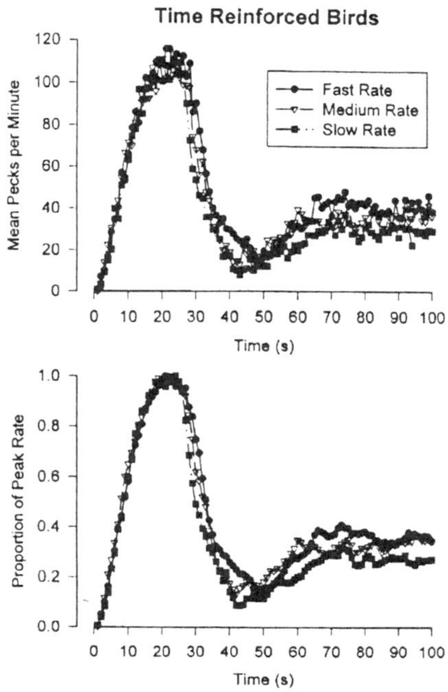

Figure 5. Peak-time curves obtained from FI trained pigeons on empty trials at three flash rates. All three curves peak near the training FI of 20 s. (From W. A. Roberts & Boisvert, 1998).

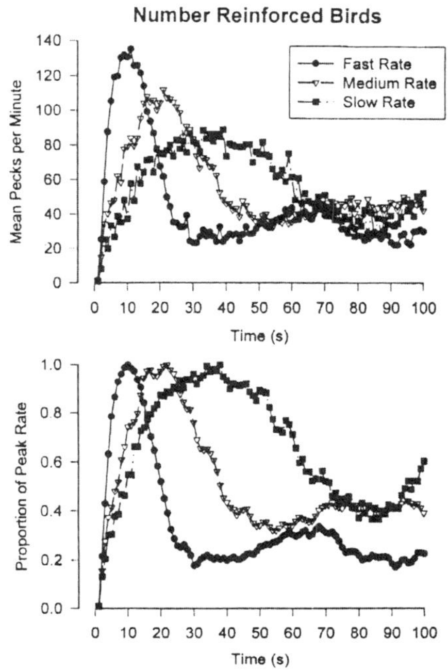

Figure 6. Peak-time curves obtained from FN trained pigeons on empty trials at three flash rates. The curves peak near the times when the FN of 20 flashes was reached. (From W. A. Roberts & Boisvert, 1998).

s had gone by since the key light came on was reinforced regardless of the flash rate. Pigeons in the other group were reinforced on a fixed number (FN) 20 flashes schedule; the first peck made after 20 flashes had occurred was reinforced. This meant that the first response after 10 s since the key light came on would be reinforced at the fast flash rate; at the medium flash rate, the first response after 20 s was reinforced, and, at the slow flash rate, the first response after 40 s was reinforced (See Figure 4).

Empty test trials were occasionally given to each group, in which the key flashed at any one of the three flash rates for 100 s, and no reinforcer was delivered. The peak time curves for the FI-reinforced group are shown in Figure 5, and those for the FN-reinforced group are shown in Figure 6. In each figure, the upper panel shows mean pecks per min, and the lower panel shows curves that have been smoothed and normalized by plotting proportions of each curve's peak rate of pecking. The FI-reinforced group's curves for different flash rates superpose on one another, and all produce peak times near 20 s. The FN-reinforced group's curves, on the other hand, are clearly separated, with the peaks at the fast, medium, and slow flash rates approximating 10, 20, and 40 s. These data then suggest that the group reinforced according to time kept track of time, and the group reinforced according to number of flashes kept track of number. These finding then are clearly in agreement with expectations generated by the dual-mode model.

Could a pigeon learn to either time or count on cue? This question was addressed in experiments performed by W. A. Roberts, Coughlin, and Roberts (2000). A single group of pigeons was trained on trials that began with the onset of a white key half the time and with the onset of blue key half the time. On both white-key and blue-key trials, the key periodically flashed red for 200 ms at the fast, medium, or slow rate. If the key color was blue, reinforcement was always delivered on an FI 20 s schedule. If the key color was white, however, reinforcement was always delivered on an FN 20 flashes schedule. After considerable training on these mixed schedules sessions, empty trials were introduced on which the key was either white or blue, and the each of the three flash rates was tested. Figure 7 shows the normalized curves generated by this procedure. The peak time curves found on trials cued by the blue FI cue appear in the top panel; these curves superpose on one another and yield mean peak times very close to 20 s. The peak time curves yielded by white key FN cued empty trials are shown in the bottom panel. The counting cued curves peak near 10 s at the fast flash rate and near 20 s at the medium rate; at the slow rate, the curve rises and appears to plateau, finally reaching its highest point at 39-40 s. These findings are similar to the W. A. Roberts and Boisvert (1998) findings but show that pigeons can flexibly switch between timing and counting processes when cued to do so.

A Control Experiment

A critic of these experiments might object that it could be possible for a pigeon to appear to be counting when in fact it was timing. The argument is that pigeons might detect flash rate without counting flashes and then use flash rate as a conditional cue for different length FIs. Thus a pigeon might learn that a fast flash rate cues a 10-s FI, a medium flash rate cues a 20-s FI, and a slow flash rate cues a 40-s FI. Of course, in the within-subjects design, pigeons would have to use both key color and flash rate as conditional cues indicating the FI.

In a control experiment, the pigeons that yielded the evidence for cue-controlled counting and timing shown in Figure 7 were tested with empty trials on which pigeons were cued to time or count and the rate of flashing was irregular. On all of these test trials, the flash rate was made aperiodic by randomly intermixing interflash intervals of 0.5, 1, and 1.5 s. However, the relative proportions of these interflash intervals were varied

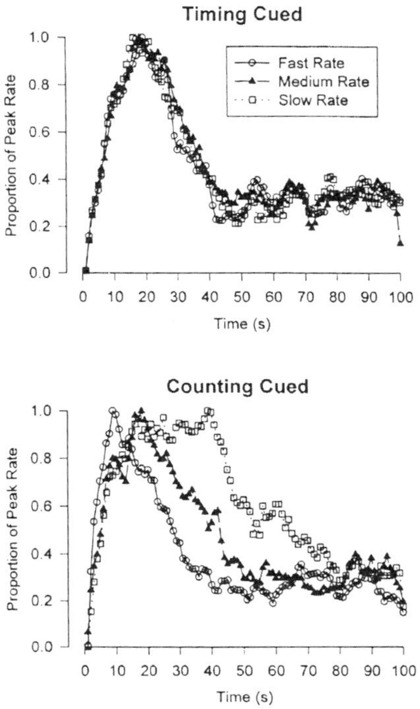

Figure 7. Peak-time curves obtained from pigeons cued to either time or count on different trials. The curves from timing cued trials tend to all peak near the 20-s FI, but the curves from counting cued trials tend to peak near the time when 20 flashes has occurred. (From W. A. Roberts, Coughlin, & Roberts, 2000).

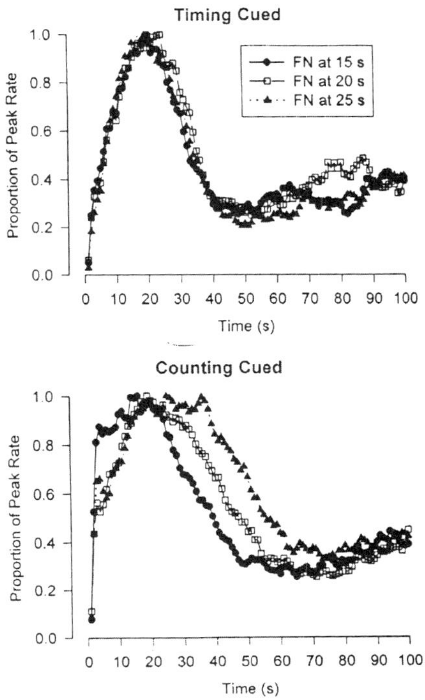

Figure 8. Peak-time curves obtained from a control experiment in which flashes on empty test trials occurred at irregular intervals and the 20th flash occurred after either 15, 20, or 25 s. On timing cued trials, all curves peak near 20 s, but on counting cued trials, the curves tend to peak at the times when 20 flashes are reached.

over three conditions so that the 20th flash was reached after either 15, 20, or 25 s. If birds used flash rate on FN-cued trials to time the point of reinforcement, they should show general confusion when aperiodic flash rates occurred. On the other hand, if pigeons were counting to the 20th flash, they should show peak time curves that peak near the time the 20th flash occurs, regardless of the irregular flash rate.

The peak-time curves from this control experiment are shown in Figure 8. The timing-cued trials shown in the upper panel are almost identical regardless of the time at which 20 flashes was reached, and they all peak at about 20 s. The counting-cued curves shown in the lower panel are clearly sorted out by the time the FN was reached, with the mean peak times being 16.35, 23.21, and 27.67 s at FN = 15, 20, and 25 s, respectively. These means differed significantly and clearly suggest that pigeons were counting number of flashes and not using flash rate as a conditional cue to track time.

IS THE SAME MECHANISM RESPONSIBLE FOR TIMING AND COUNTING?

Simultaneous Timing and Counting?

The dual-mode model clearly suggests that the same process, pulse accumulation, is responsible for timing and counting. The model also implies that both of these processes should occur simultaneously. That is, in the experiments just described, pigeons were recording pulses both in the run mode in the time accumulator and in the event mode in the number accumulator. W. A. Roberts et al. (2000) asked whether pigeons would keep track of time when instructed to count and would keep track of number when instructed to time. The same pigeons used in the preceding two experiments continued to be trained on FI 20 s and FN 20 flashes schedules. Occasional empty trials now occurred on which either the FI blue key cue or the FN white key cue appeared for the first 10 s of the trial and the key flashed red at the medium rate. The key then changed to the other color for the remaining 90 s of the trial (from blue to white or from white to blue), and the flash rate was fast, medium, or slow. By examining where pigeons peaked at different flash rates, it was possible to infer whether they had been counting while instructed to time and whether they had been timing while instructed to count. On trials when pigeons were switched from the counting cue to the timing cue, the curves at all three flash rates superpose and peak near 20 s from the beginning of the trial. These data suggest that pigeons did keep track of time while cued to count. When switched from the timing cue to the counting cue, however, the peak times obtained suggested that pigeons did not begin to count until the counting cue was presented. In other words, pigeons did not keep track of number when instructed to keep track of time. It was suggested that this asymmetry may point to a basic difference between timing and counting mechanisms. Timing may be an automatic process that can occur simultaneously with other processes, such as counting. Counting, on the other hand, may be a more effortful process and may be more closely related to behavior. The possibility that animals might count behaviorally has been suggested before, but has received little empirical support (Boysen, Berntson, Shreyer, & Hannan, 1995; Davis & Perusse, 1988; W. A. Roberts, 1998).

This finding represents an initial failure for prediction from the dual-mode model and suggests the possibility that different mechanisms control timing and counting. Another recent report from Breukelaar and Dalrymple-Alford (1998) suggested a nonequivalence of timing and counting. When rats were trained to discriminate between short and long sequences of sound bursts, it was found that time showed better control of discriminative behavior than number. They suggested that rats showed a time bias or a preference for using the time dimension over the number dimension.

Some Evidence for Different Mechanisms of Timing and Counting

The discovery that pigeons time while cued to count but do not count while instructed to time led us to look for evidence that birds might be coding number differently from time. We were particularly interested in the possibility that a behavioral code might underlie counting but not timing. In the previous studies discussed, rate of pecking had been measured in 1-s time bins. In a set of new experiments (W. A. Roberts, Roberts, & Kit, 2001), peck rates were measured in 0.5-s time bins in order to more carefully see changes in the peck rate within successive flash cycles. Six new pigeons were trained on mixed FI 20 s and FN 20 flashes schedules, with the FI schedule signaled by a blue key and the FN schedule signaled by a white key. On both schedules, the key flashed red for 200 ms at the fast, medium, and slow rates on different trials.

When rate of pecking was plotted against 0.5-s time bins for the period up to the FI or FN, very clear differences in response to the flash appeared. These can be seen most clearly on trials in which the flash occurred at the slow rate. Since the flash appeared every 2 s, there were four 0.5-s time bins in each cycle. The flash occurred in the first bin, and the remaining three bins contained only the blue timing cue or the white counting cue. The data for FI-cued trials at the slow flash rate are shown in Figure 9. The curve shows the typical rise in pecking over the 20-s FI. However, a clear pattern of falling and rising pecking is superimposed on the overall rise in the curve. Pecking during the 0.5 s when a flash occurred is shown as the filled circles; it can be seen that the pecking rate dropped at each flash and rose again over the three following 0.5-s bins when the flash was absent. Data from the FN 20 flashes schedule are shown in Figure 10. The pattern seen here is directly opposite that seen during the FI schedule. Each time a flash occurred, the pigeons' rate of pecking increased on both the 0.5-s bin that contained the flash and the 0.5-s bin that followed the flash. Rate of pecking then declined on the two bins that completed the cycle.

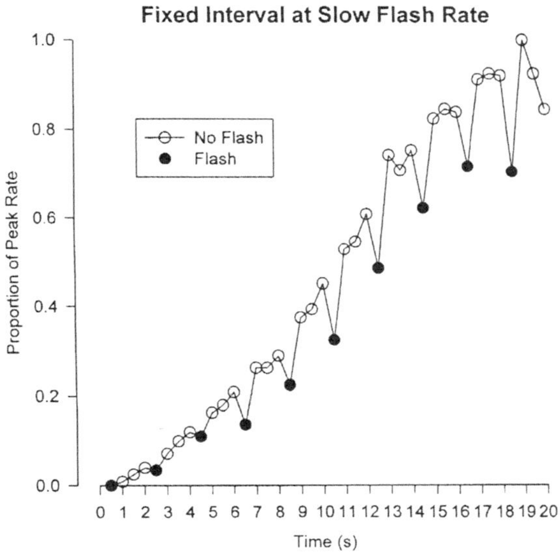

Figure 9. Pigeons' proportional rate of pecking plotted over 0.5-s time bins for trials on which an FI 20 s schedule of reinforcement was signalled. Rate of pecking drops in each time bin containing a flash. (From W. A. Roberts, Roberts, & Kit, 2001).

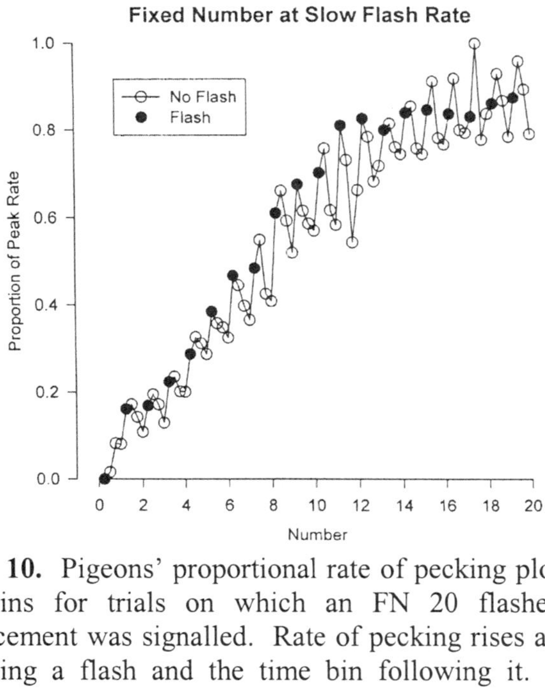

Figure 10. Pigeons' proportional rate of pecking plotted over 0.5-s time bins for trials on which an FN 20 flashes schedule of reinforcement was signalled. Rate of pecking rises at each time bin containing a flash and the time bin following it. (From W. A. Roberts, Roberts, & Kit, 2001).

The fact that pigeons responded very differently to the flash on FI and FN schedules suggests that the flash signaled quite different things to pigeons under the two schedules. In particular, the acceleration of pecking to the flash on FN-cued trials might represent an excitatory response to the flash that could act as a code for counting. The decline in pecking at the flash seen on FI-cued trials is interesting but may not be functional in timing, since previous experiments show that pigeons' peak time curves tend to peak near 20 s, regardless of the flash rate and hence regardless of how many times a momentary decline in pecking may have occurred.

If response to the flash on FN-cued trials acts as a counting code, it should be possible to interfere with the counting process by interfering with the pattern of pecking shown on FN-cued trials. An experiment was carried out in which opportunity to maintain a regular pecking pattern was prevented by periodically turning off the key light and leaving the pigeons in the dark. The six pigeons that participated in the preceding experiment continued to be trained on cued FI and FN schedules. Occasional empty test trials were presented on which the key was either blue or white and the key flashed red at the slow rate of once every 2 s. On odd-numbered sessions, performance was measured on control empty trials in which pecking proceeded with no interruptions. On even-numbered sessions, empty trials contained blackouts. On every fourth flash cycle, the key light containing the the blue FI cue or the white FN cue was turned off. Thus during these blackout periods, the flash occurred during the first 200 ms of the cycle and the key was then dark during the remaining 1.8 s of the cycle. Since pigeons could not peck during this blackout, the regular pattern of pecking was seriously disrupted. If the maintenance of a regular pattern of pecking is necessary for counting, the peak time curve on FN-cued blackout trials should show either a marked shift of the peak to a time when more than 20 flashes occurs or a complete absence of counting. If timing is a more automatic process that is little dependent on overt behavior, we would expect to find evidence of accurate timing even on blackout trials.

The upper panel in Figure 11 shows pigeons' rate of pecking on FI-cued empty trials when the key was periodically blacked out and when no blackouts occurred. The no-key-blackouts control curve shows the typical rise in response rate to a peak near 20 s. The effect of blackouts can be seen by the periodic drop in pecking to near zero in the key-blackouts curve. The important observation here is that, despite these periodic interruptions of the pecking pattern, pigeons still tended to show peak responding near the FI. The lower panel in Figure 11 presents mean peck rate over the periods when the key was illuminated. The rise in pecking around 20 s is seen both with and without blackouts. Analysis of variance revealed significant variation across time periods in both conditions, and the calculated mean peaks of

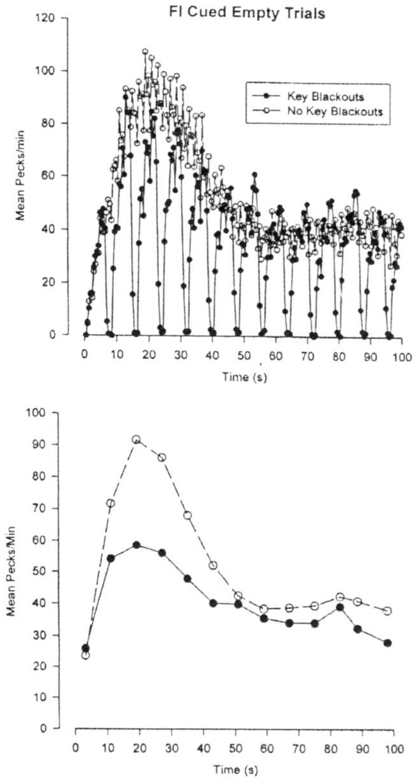

Figure 11. Response rates on FI-cued empty trials containing key blackouts or no key blackouts. The upper curve plots rate at each 0.5-s time bin, including blackouts, and the lower curve plots mean rate over successive time blocks when the key was lit. (From W. A. Roberts, Roberts, & Kit, 2001).

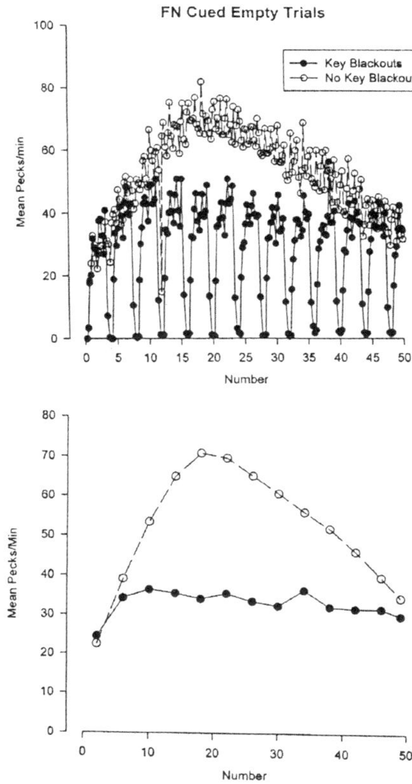

Figure 12. Response rates on FN-cued empty trials containing key blackouts or no key blackouts. The upper curve plots rate at each 0.5-s time bin, including blackouts, and the lower curve plots mean rate over successive time blocks when the key was lit. (From W. A. Roberts, Roberts, & Kit, 2001).

these curves were 19.25 s on blackout trials and 23.25 s on control trials. These means did not differ significantly from one another or from 20 s.

Data from FN-cued empty trials are shown in the upper panel of Figure 12, plotted as a function of number of flashes. The control curve with no key blackouts tends to peak around 20 flashes. The key-blackout curve shows the periodic interruptions of pecking caused by the blackouts but further shows no peak during the periods when birds pecked at the key. The lower panel of Figure 12 shows performance during time periods when the key was illuminated. Notice the clear peak in the control curve and the complete absence of a peak in the blackouts curve. When subjected to analysis of variance, the control curve yielded significant variation across time periods. Further, the mean peak number found for the control curve was 22.29 flashes and did not differ significantly from 20 flashes. Analysis of the blackouts curve found no significant variation across time blocks, showing that this curve is basically flat.

The findings of this experiment agree with the hypothesis that pecking pattern is important for counting but not for timing. Blacking out the response key at regular intervals during empty trials completely eliminated any evidence of counting on FN-cued trials but left the timing process intact on FI-cued trials. These finding also help us to better understand the finding that pigeons could time while instructed to count but could not count while instructed to time. If pigeons were cued to track time, they would have shown a decrease in peck rate during each flash, as revealed in Figure 9. Since this behavior was just the opposite of the increase in peck rate needed to code each flash (as shown in Figure 10), no flashes should have been counted when pigeons were responding to the timing cue. When pigeons were cued to count, their rate of pecking increased each time a flash occurred. Although this behavior is opposite the deceleration of peck rate seen on FI-cued trials, timing is not dependent upon behavior and should thus take place regardless of how pigeons respond to the flash.

A MODEL FOR COUNTING LIGHT FLASHES

In Figure 13, a behavioral model is presented to explain how pigeons count light flashes and to explain the different responses to light flashes when FI and FN cues are presented. The model suggests that responses to the flash represent a form of classically conditioned or autoshaped behavior. Furthermore, it is suggested that the FN and FI key cues act as occasion setters for differential responding to flashes. The argument is that when the FN cue is presented, birds learn that the number of flashes is perfectly correlated with the delivery of reinforcement; the first peck made after 20 flashes always delivers a reinforcer. When the FI cue is presented, however, pigeons learn that the number of flashes is noncorrelated with the delivery of the reinforcer; the reinforcer can come after 10, 20, or 40 flashes. As a consequence, pigeons response to flashes in the presence of the FN occasion setter is excitatory and leads to an acceleration of pecking at the key. Pigeons response to flashes in the presence of the FI occasion setter is inhibitory and leads to just the opposite behavior of a decline in peck rate. However, this decline in peck rate is not functional in the timing process, since timing is not controlled by overt

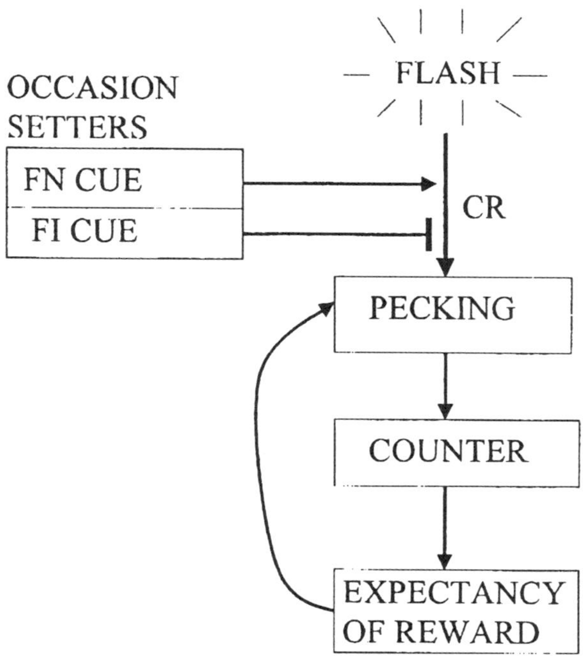

Figure 13. A model showing how the FN and FI cues may act as excitatory and inhibitory occasion setters, respectively, for pecking at light flashes. Accelerated pecking at the flash on FN-cued trials is assumed to increment a counter that in turn raises the expectancy of reward and increases the overall rate of pecking. (From W. A. Roberts, Roberts, & Kit, 2001).

behavior. An internal, automatic process such as the accumulation of pulses suggested by the scalar-timing theory may best account for timing.

The model clearly views the tendency of pigeons to peck faster at each flash as a causal factor in counting. It suggests that each increase in peck rate adds to a counter. As the number on the counter approaches a criterion value (20 in these experiments), the expectancy of reward increases, and increased expectancy then increases the overall rate of pecking. Thus, the periodic flashes drive up the rate of pecking at the flash through the excitatory influence of the flash in the presence of the FN occasion setter and drive up the overall rate of pecking by increasing the value on the counter and thus the expectancy of reward.

It should be pointed out that the use of FI and FN cues as occasion setters in this context is unusual. Occasion setters are usually ambient stimuli (sound or light) that signal that a specific conditioned stimulus (CS) will be followed by an unconditioned stimulus (US). In these experiments, there is no specific CS. Rather, the occasion setter indicates whether the number of light flashes will or will not be a reliable indicator of the US (food delivery). Each flash itself is not a unique CS but only a member of the number dimension that predicts the occurrence of the US. Nevertheless, the data indicate that pigeons learn these relationships and come to respond to each flash as a good or bad predictor of the US.

The experiments and model presented suggest that at least within the context of a flashing light signal, pigeons use different mechanisms to track time and number. An internal, automatic process appears to be used to keep time, while counting appears to be dependent upon the animal's own behavior. Although the dual-mode model accounts for a number of the counting phenomena reported here, it appears that it cannot explain the behavioral nature of counting found in these experiments. Animals appear to be able to accurately respond to number in a number of different procedures, as described earlier in this chapter. That behavioral counting would be responsible for number discrimination in all of these cases seems doubtful. It may turn out that there are a number of ways in which animals can count both simultaneously and successively presented events. The experiments reported here have revealed one way in which they do it, a way that is particularly related to the animal's own behavior.

ACKNOWLEDGMENTS

Preparation of this chapter was supported by a Research Grant from the Natural Sciences and Engineering Research Council of Canada to William A. Roberts.

REFERENCES

Boysen, S.T., Berntson, G.G., Shreyer, T.A., & Hannan, M.B. (1995). Indicating acts during counting by a chimpanzee (*Pan troglodytes*). *Journal of Comparative Psychology,* **109**, 47-51.

Breukelaar, J. W. C., & Dalrymple-Alford, J. C. (1998). Timing ability and numerical competence in rats. *Journal of Experimental Psychology: Animal Behavior Processes,* **24**, 84-97.

Broadbent, H.A., Church, R.M., Meck, W.H., & Rakitin, B.C. (1993). Quantitative relationships between timing and counting. In S.T. Boysen & E.J. Capaldi (Eds.), *The development of numerical competence: Animal and human models* (pp. 171-187). Hillsdale, NJ: Erlbaum.

Capaldi, E.J. (1993). Animal number abilities: Implications for a hierarchical approach to instrumental learning. In S.T. Boysen & E.J. Capaldi (Eds.), *The development of numerical competence: Animal and human models* (pp. 191-209). Hillsdale, NJ: Erlbaum.

Capaldi, E.J., & Miller, D.J. (1988). Counting in rats: Its functional significance and the independent cognitive processes that constitute it. *Journal of Experimental Psychology: Animal Behavior Processes,* **14**, 3-17.

Davis, H., & Albert, M. (1986). Numerical discrimination by rats using sequential auditory stimuli. *Animal Learning & Behavior,* **14**, 57-59.

Davis, H., & Memmott, J. (1982). Counting behavior in animals: A critical evaluation. *Psychological Bulletin,* **92**. 547-571.

Davis, H., & Perusse, R. (1988). Numerical competence in animals: Definitional issues, current evidence, and a new research agenda. *Brain and Behavioral Sciences,* **11**, 561-579.

Fernandes, D.M., & Church, R.M. (1982). Discrimination of the number of sequential events by rats. *Animal Learning & Behavior,* **10**, 171-176.

Gallistel, C.R. (1990). *The organization of learning.* Cambridge, MA: The MIT Press.

Gallistel, C.R. (1993). A conceptual framework for the study of numerical estimation and arithmetic reasoning in animals. In S.T. Boysen & E.J. Capaldi (Eds.), *The development of numerical competence: Animal and human models* (pp. 211-223). Hillsdale, NJ: Erlbaum.

Gallistel, C.R., & Gelman, R. (1992). Preverbal and verbal counting and computation. *Cognition,* **44**, 43-74.

Gallistel, C. R., & Gelman, R. (2000). Non-verbal numerical cognition: from reals to integers. *Trends in Cognitive Sciences,* **4**, 59-65.

Gelman, R., & Gallistel, C.R. (1978). *The child's understanding of number.* Cambridge, MA: Harvard University Press.

Herrnstein, R.J., Loveland, D.H., & Cable, C. (1976). Natural concepts in pigeons. *Journal of Experimental Psychology: Animal Behavior Processes,* **2**, 285-302.

Hicks, L.H. (1956). An analysis of number-concept formation in the rhesus monkey. *Journal of Comparative and Physiological Psychology,* **49**, 212-218.

Kaufman, E.C., Lord, M.W., Reese, T.W., & Volkmann, J. (1949). The discrimination of visual number. *American Journal of Psychology,* **62**, 498-525.

Mandler, G., & Shebo, B.J. (1982). Subitizing: An analysis of its component processes. *Journal of Experimental Psychology: General,* **111**, 1-22.

Meck, W.H., & Church, R.M. (1983). A mode control model of counting and timing processes. *Journal of Experimental Psychology: Animal Behavior Processes,* **9**, 320-334.

Miller, D.J. (1993). Do animals subitize? In S.T. Boysen & E.J. Capaldi (Eds.), *The development of numerical competence: Animal and human models* (pp. 149-169). Hillsdale, NJ: Erlbaum.

Rilling, M. (1993). Invisible counting animals: A history of contributions from comparative psychology, ethology, and learning theory. In S. T. Boysen & E. J. Capaldi (Eds.), *The development of numerical competence: Animal and human models* (pp. 3-37). Hillsdale, NJ: Erlbaum.

Roberts, S. (1981). Isolation of the internal clock. *Journal of Experimental Psychology: Animal Behavior Processes,* **7**, 242-268.

Roberts, W.A. (1998). *Principles of animal cognition.* Boston: McGraw-Hill.

Roberts, W. A., & Boisvert, M. J. (1998). Using the peak procedure to measure timing and counting processes in pigeons. *Journal of Experimental Psychology: Animal Behavior Processes,* **24**, 416-430.

Roberts, W. A., Coughlin, R., & Roberts, S. (2000). Pigeons flexibly time or count on cue. *Psychological Science,* **11**, 218-222.

Roberts, W.A., & Mazmanian, D.S. (1988). Concept learning at different levels of abstraction by pigeons, monkeys, and people. *Journal of Experimental Psychology: Animal Behavior Processes,* **14**, 247-260.

Roberts, W.A., & Mitchell, S. (1994). Can a pigeon simultaneously process temporal and numerical information? *Journal of Experimental Psychology: Animal Behavior Processes,* **20**, 66-78.

Roberts, W. A., Roberts, S., & Kit, K. A. (2001). Different mechanisms control timing and counting of light flashes by pigeons. *Manuscript submitted for publication.*

Thomas, R.K., Fowlkes, D., & Vickery, J.D. (1980). Conceptual numerousness judgments by squirrel monkeys. *American Journal of Psychology,* **93**, 247-257.

Wasserman, E.A. (1995). The conceptual abilities of pigeons. *American Scientist,* **83**, 246-255.

8

DISTORTIONS IN THE CONTENT OF TEMPORAL MEMORY

Neurobiological Correlates

Warren H. Meck
Duke University

Distortions in timing and time perception are presented by a number of different neurological disorders (e.g., Bünning, 1973; Corkin, 1984; Fraisse, 1984; Gloning & Hoff, 1969; Goody, 1969; Malapani, Pillon, Lo Russo, DuBois, Munari, & Agid, 1992; Malapani, Rakitin, Meck, Deweer, Dubois, & Gibbon, 1998; Meck & Benson, 2001; Nichelli, Venneri, Molinari, Tavani, & Grafman, 1993; Penney, Meck, Roberts, Gibbon, & Erlenmeyer-Kimling, 2001; Richards, 1973; Williams, 1969) as well as normal aging (e.g., Lustig & Meck, 1998, 2001; Meck, Church, & Wenk, 1986), pharmacological manipulations (e.g., Levin, Conners, Sparrow, Hinton, Erhardt, Meck, Rose, & March, 1996; Levin, Conners, Silva, Hinton, Meck, March, & Rose, 1998; Meck, 1983, 1986; Meck & Williams, 1997a; Paule, Meck, McMillan, Bateson, Popke, Chelonis, & Hinton, 1999) and stimulus modality effects (e.g., Meck, 1991; Penney, Gibbon, & Meck, 2000) in both animals and humans. Normal subject populations often show large and consistent individual differences in the remembered times of events that are not a direct consequence of the time of feedback or signal duration (Church & Meck, 1988; Lejeune, Ferrara, Soffie, Bronchart, & Wearden, 1997). The primary focus of this chapter is on factors that define or produce such systematic changes (discrepancies) in the content of temporal memory. The findings obtained from this research will be used to augment a quantitative

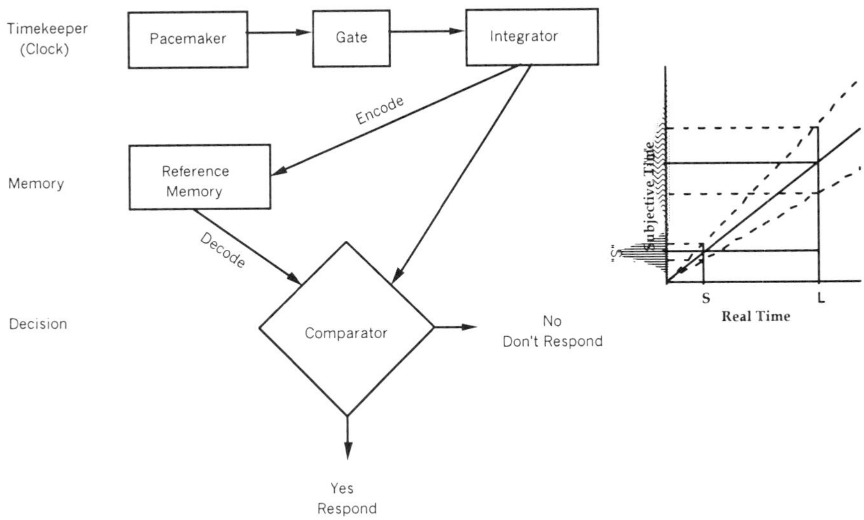

Figure 1. Information-processing (IP) model of interval timing developed to represent scalar timing theory. The three levels of processing include clock, memory, and decision stages.

theory of scalar timing in memory (Church, Meck, & Gibbon, 1993; Gibbon, 1977; Gibbon & Church, 1990, 1992; Gibbon, Church, & Meck, 1984) by providing the appropriate physiological underpinnings.

Scalar timing theory is represented by a mathematical model that describes the formal properties of the cognitive processes that operate when an individual solves a variety of duration discriminations that rely on the storage and retrieval of clock readings maintained in temporal memory (Gibbon & Church, 1984). Consequently, an information-processing (IP) model of this theory can provide us with both the conceptual framework and the experimental procedures to develop quantitative models of duration discrimination. Such models are necessary to investigate the neural mechanisms involved in timing and time perception so that real-time physiological models can eventually be developed (see Gibbon, Malapani, Dale, & Gallistel, 1997). A possible flow-diagram of a cognitive model showing the three stages of information processing (clock, memory, and decision stages) is illustrated in Figure 1.

In this IP model of interval timing, a *pacemaker* emits pulses that are gated by a switch into an *accumulator* (cf., Matell & Meck, 2000). This accumulator integrates these pulses in a linear fashion over time. The value in the accumulator is then compared to a sample taken from the appropriate distribution of the expected times of occurrence of relevant events (e.g., reinforcement) stored in *reference memory*. If the clock reading and the memory sample are close enough according to a specific decision rule applied by a *comparator*, a response is made. If a response is reinforced, the value in the accumulator is transferred to reference memory. In this model, reference memory is used to store information about the expected time of an event, gained from experience with many trials. *Working memory* is used to store information about the durations of sequential events comprising a single trial. Neurobiological support for these types of timing models comes from brain imaging (e.g., Hinton, Meck, & MacFall, 1996; Meck, Hinton, & Matell, 1998; Rao, Mayer, & Harrington, 2001); lesion (e.g., Meck & Benson, 2001), pharmacological (e.g., Meck, 1996; Malapani et al., 1998), and electrophysiological recording and simulation studies (e.g., Matell & Meck, 2000).

Scalar timing theory combines this IP analysis with specific assumptions about distribution forms and sources of variability (Gibbon, Church, & Meck, 1984) to account for most of the variability in performance of duration discriminations. For example, both scalar timing theory and the IP model have been extended to account for the behavioral patterns that are produced by changes in the speed of the internal clock used to measure signal durations and by changes in the speed of the memory processes used to store these measured durations (Meck, 1983; 1996). There have also been applications to situations where attentional bias toward different stimulus modalities occurs (Hinton & Meck, 1997a; Meck, 1984, 1991; Penney, Holder, & Meck, 1996).

Variations of the standard discrete-trials fixed-interval (FI) schedule of reinforcement have allowed selective tests of the processes described by scalar timing theory. One of these modifications is the peak-interval (PI) procedure (Church, 1984; Church, Miller, Meck, & Gibbon, 1991; Roberts, 1981). In this task, signals (e.g., white noise) are separated by a 130-s intertrial interval (ITI). On a random half of the trials food is made available after the signal has been present for a fixed duration (e.g., 50-s). The rat's first lever press after the fixed duration causes the termination of the signal and the delivery of a 45 mg food pellet. On the remaining trials (peak trials) no food is given and the signal continues for 130 s. On peak trials, the rat's probability of responding (averaged over many trials) increases as a function of time since signal onset until the time that food is sometimes made available. The time of the maximal response probability for this function is referred to as the *peak time* and the response rate at the peak time is referred to as the *peak rate*. Once the maximal level of responding is obtained, the rat's probability of responding declines in a fairly symmetrical manner when response probability is plotted on a linear time scale. Peak functions averaged over many trials often resemble Gaussian distributions that are centered around the time of reinforcement. The peak time measure serves as an indicator of the rat's remembered time of reinforcement, a value that is stored in reference memory. For illustrative purposes, response functions obtained from a single rat trained on the "tri-peak" procedure (e.g., Matell & Meck, 1999; Matell, Meck, & King, 1996) with FI values of 10-s, 30-s, and 90-s are presented in Figure 2 and individual peak functions for three different rats trained on the standard PI procedure with a 20-s criterion are shown in Figure 3. The individual data depicted in Figure 3 were selected from a sample of approximately 100 male control rats (4-12 months of age) in order to illustrate the potential range of peak times exhibited by stable response functions. As one might imagine, the peak functions centered around 15.5 s and 26 s are unusual for "normal" adult rats trained with a 20-s criterion. The peak function centered around 20 s is fairly standard.

Current evidence suggests that the frontal and temporal lobes are both involved in the reference memory of the duration of an event, but in complementary ways. A double dissociation in brain function resulted from lesions of the FC and the nucleus basalis magnocellularis (NBM) as compared with those of the HIP, fimbria-fornix (FFx), and the medial septal area (MSA). Lesions in the frontal system (NBM, FC) produced a rightward shift (overestimation) of the expected time of reinforcement, while lesions in the hippocampal system (MSA, FFx) produced a leftward shift (underestimation). Another dissociation was found in the effects of these lesions on working memory for the duration of a stimulus prior to a gap or retention interval interpolated during that stimulus (e.g., Buhusi & Meck, 2000; Meck, Church, & Olton, 1984). Memory of the duration of the event prior to the gap was impaired by FFx lesions, while damage in the frontal system had no effect (Olton, Meck, & Church, 1987; Meck, Church, & Olton, 1984; Meck, Church, Wenk, & Olton, 1987). Lesions of the hippocampal system also produced a total inability of rats to regulate the temporal criterion used on a trial-by-trial basis, whereas lesions of the frontal system did not exhibit such a disruption of temporal regulation (Meck, 1988a; Meck, Church, Wenk, & Olton, 1987).

Simultaneous temporal processing (STP), a variation of the timing procedures described above, includes some trials that require divided attention (Meck, 1987; Meck & Church, 1984; Meck & Williams, 1997b; Olton, Wenk, Church, & Meck, 1988; Pang, Yoder, & Olton, 2001). Divided attention is required when two or more events are presented simultaneously (Malapani, Pillon, Dubois, & Agid, 1992; Meck, 1987; Meck & Church, 1984). The FC has been strongly implicated in temporal processing, attention, and planning (Fuster, 1980; Kolb, 1984; Luria, 1969; Milner, 1982; Milner, Petrides, & Smith, 1985). Consequently, we anticipated another dissociation where lesions in the frontal, but not in the hippocampal, system would impair attention to multiple temporal inputs. Such a result would confirm the importance of the frontal system in temporal discrimination, provide another important dissociation between the functional roles of the frontal and hippocampal systems, and introduce a

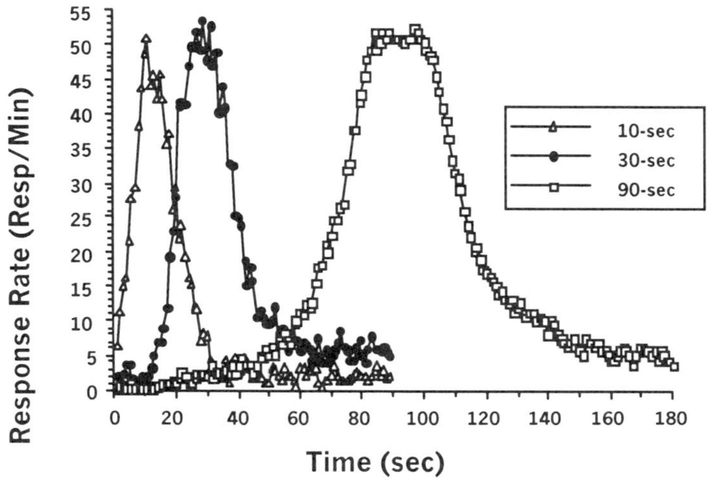

Figure 2. Mean response rate plotted as a function of signal duration for a single rat trained with three different criterion durations (10-s, 30-s, and 90-s) each signalled by the same auditory stimulus but associated with different response levers (Tri-Peak procedure). Data are averaged over 10 sessions.

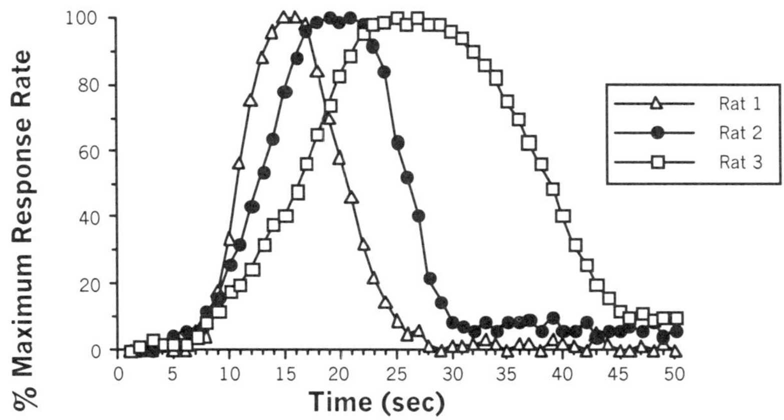

Figure 3. Percent maximum response rate plotted as a function of signal duration for three rats trained with a 20-s criterion duration signalled by an auditory stimulus. Data are averaged over 10 sessions.

useful procedure for the analysis of the neural mechanisms involved in attention in animals.

Our results indicated that lesions of the FC or the NBM produced a severe impairment in the ability to time two stimuli simultaneously. When the second stimulus was presented while the first was still in progress, control rats successfully timed the second stimulus and continued to time the initially presented stimulus as well. These rats gave two peak times, one at the expected time of reinforcement associated with each of the temporal stimuli. In contrast, rats with lesions in the frontal system were unable to time both stimuli simultaneously. These rats correctly timed only the second stimulus. When this second stimulus was presented while the first one was in progress, the rats stopped timing the first stimulus, remembered its value, and started timing the second one. When they finished timing the second stimulus, they recalled the value of the first stimulus and continued to time it. For the first stimulus, however, they displayed a rightward shift of the peak time by an amount equal to the duration of the second, short stimulus (e.g., 10 s). These results support the hypothesis that the FC is important for tasks that require planning and temporal organization of behavior, and attention to different cognitive demands (Milner & Petrides, 1984; Smith & Milner, 1985).

The attempt here has been to conduct studies on animals that combine current anatomical, physiological, and pharmacological techniques with sophisticated behavioral paradigms to study the structural and functional organization of higher brain centers (e.g., frontal and temporal cortices). Ultimately, this approach aims at an IP model of how and where memory storage for temporal information is accomplished by the brain. The results presented here, together with the findings of other timing experiments, demonstrate a sophisticated, integrated action of the FC and HIP in timing. These areas play complementary roles in establishing the correct value in reference memory for the expected duration of an event. Each system plays a separate role in working memory and in attention, with the hippocampal system being critical for normal working memory within a trial, and with the frontal system being critical for paying attention to two or more stimuli simultaneously. In many cases, clinical results have been responsible for the generation of animal models. In the case of the temporal analyses described here, the animal models may have the opportunity to direct the development of appropriate clinical tests (see Malapani et al., 1992). Because of the selective nature of the dissociations observed here it should prove useful to develop similar procedures to evaluate timing and time perception in humans (e.g., Malapani, Rakitin, Meck, Deweer, Dubois, & Gibbon, 1998; Rakitin, Gibbon, Penney, Malapani, Hinton & Meck, 1997). In particular, simultaneous temporal processing procedures may be diagnostic for specific types of brain damage, and informative about the psychological processes that are disrupted in the frontal and temporal syndromes. In this regard, the present chapter proposes to investigate factors that define or produce discrepancies in the content of temporal memory.

INDIVIDUAL DIFFERENCES IN THE CONTENT OF TEMPORAL MEMORY

Temporal memory can vary on two dimensions: clarity and content of the remembered event. Most investigations of memory have dealt with the conditions that affect the clarity of the memory as measured by the amount of recall or recognition. This is the case in most studies of forgetting or amnesia. For example, there is evidence that the clarity of a remembered duration, but not its content, is a function of the length of the retention interval (Church, 1980). With quantitative dimensions such as space, time, or intensity, the content as well as the clarity of a remembered event can be measured. Certain properties of a stimulus other than its duration (e.g., modality, intensity, or frequency) may influence the remembered duration of the stimulus. Various nonspecific factors, such as state of arousal, may affect the clarity of a memory but are unlikely to produce changes in the content of memory. The experiments that follow were designed to investigate the factors that define or produce systematic changes discrepancies in the remembered duration of an event. This type of discrepancy in the content of temporal memory is represented by the difference between the remembered duration of an event and its actual physical duration. Thus, discrepancies in the remembered time of reinforcement can vary in both magnitude and direction.

Although it is frequently observed that timing functions averaged across 10 or more subjects accurately reflect the physical durations of events, there are large and consistent individual differences in the psychophysical functions. The stability and proportionality of these differences suggest that they are the result of a multiplicative transformation of the temporal information that is transferred from the clock to the memory stage. These individual differences are apparently not restricted to a particular species or timing task, because they have been observed in pigeons, rats, and humans in a variety of testing situations (Church & Meck, 1988). For example, we conducted a study in which 10 male rats (Sprague-Dawley strain) were trained on a 50-s PI procedure. We found that the time of the maximum response rate of the mean function relating response rate to time since the trial began was very close to 50 s. However, there were large reliable individual differences in the time of the peak. Although food was available only following the first response after 50 s, some rats appeared to expect food reliably sooner than 50 s (e.g., 40 s) and others reliably later (e.g., 60 s). This bias was not corrected by increased training. We have assumed that the rats do not necessarily store their perception of the physical duration of an event veridically, and that the difference between the measured duration (accumulator value) and the stored duration (memory value) reflects a memory storage constant (K*).

Much of the work described here is based on the interpretation of horizontal shifts in timing functions that represent underestimations (leftward shifts) or overestimations (rightward shifts) of stimulus durations. In order to interpret these changes in timing behavior one must be able to separate clock and memory stages of information processing. Fortunately, a change in clock speed can be distinguished from a change in memory storage speed by the specific behavioral pattern observed. The logic behind the distinction is as follows. If an event regularly occurs at a fixed time, it might subsequently be expected earlier than that time for at least two different reasons: **a)** an internal clock has increased its speed or **b)** the duration has been stored and retrieved as shorter than it was originally perceived to be. Equivalently, it might be expected later than the fixed time for two reasons: **c)** an internal clock has decreased its speed or **d)** the duration has been stored and retrieved as longer than it was originally perceived to be. These two mechanisms can be distinguished behaviorally. The maximal effect of a change in clock speed appears immediately, diminishes with further training until no effect is observed (i.e., normalizes), and reappears immediately to an equal extent and in the opposite direction when the clock speed is returned to normal (rebound effect). In contrast, the maximal effect of a change in the transformation of perceived time (accumulator value) to remembered time (reference memory) does not appear immediately, emerges gradually and once reached is maintained by further training, and gradually returns to baseline with no rebound effect when the conditions are restored to normal. The formal arguments justifying why maintained distortions in the horizontal placement of timing functions must represent a temporal memory dysfunction have been developed in several places (Meck, 1983; Meck & Church, 1987a, b).

CHOLINERGIC FUNCTION AND TEMPORAL MEMORY

An increasing amount of evidence indicates that the cholinergic (Ch) modulation of neural networks in frontal and temporal systems is important for both working and reference memory in a variety of species (e.g., Bartus, Dean, Pontecorvo, & Flicker, 1985; Hasselmo, Anderson, & Bower, 1991). We have previously reported that temporal memory processes involved in time discrimination are selectively sensitive to the destruction of Ch rich areas of the brain such as the MSA, HIP, NBM and FC (Meck, Church, Wenk, & Olton, 1987). It will be noted that the FC receives Ch innervation from the NBM, while the HIP receives Ch innervation from the MSA. Additional evidence has indicated that the pharmacological modification of the Ch nervous system by the administration of Ch receptor blockers (e.g., atropine and scopolamine), anticholinesterases (e.g., physostigmine), and Ch agonists (e.g., oxotremorine) leads to selective changes in the content of temporal memory (Meck, 1983; Meck & Church, 1987a). In general, a decrease in effective Ch function produced by the repeated administration of atropine or related compounds leads to a proportional increase in the remembered times of reinforcement (rightward shift) that are stored in reference memory as well as a total inability to retain trial-specific temporal information stored in working memory. In contrast, the effects of an increase in effective Ch function produced by the repeated administration of physostigmine or related compounds leads to a proportional decrease in the remembered times of reinforcement (leftward shift) that are stored in reference memory with a facilitation of working memory.

AGING, CHOLINE UPTAKE, AND TEMPORAL MEMORY

The integrity of the Ch nervous system decreases with age (Bartus, Dean, Beer, & Lippa, 1982; Blusztajn & Wurtman, 1983). On the basis of the lesion and drug data that has been reported in which the remembered time of reinforcement was inversely related to Ch activity, one would expect the remembered time of reinforcement to increase with age. In a study in which rats were trained on a temporal discrimination task from 10 months to 30 months of age, an age-related increase in the remembered time of reinforcement was demonstrated (Meck, Church, & Wenk, 1986). In addition to the control rats in this aging study, there was a group of rats that received injections of arginine vasopressin (AVP - 0.08 PU/kg, ip) 20 min before 30 training sessions when they were 10-13 months old. AVP administration prevented the age-related increase in the remembered time of reinforcement observed in the control group that had received saline injections. These data suggest that as rats age certain physiological changes occur that lead them to remember the durations of physical events as being proportionally longer than they actually are. Biochemical analysis revealed that in the FC and HIP of control animals, there was an age-related decrease in muscarinic receptor density as measured by [^{3}H]-quinuclidinyl benzilate (QNB) binding, an increase in sodium-dependent high affinity choline uptake (SDHACU), and a modest, but reliable increase in ChAT. AVP administration reduced the age-related increase in SDHACU, but it did not affect the age-related decrease in QNB binding. SDHACU in the FC was significantly correlated with the subjective discrepancy in the remembered time of reinforcement (K*-1) as indicate by the peak time measures. In contrast, the correlation between SDHACU in the HIP and the remembered time of reinforcement was not significant.

These results suggest that certain endogenous peptides exert long-term effects on brain structure and function that are related to temporal memory processes. These effects may be related to the finding that long-term treatment of rats with a neural stimulant, pentylenetetrazole, or with a potent and active analog of the ACTH[4-9] molecule (ORG 2766), can retard the development of both some neuromorphologic and some behavioral correlates of brain aging in rats (Landfield, Baskin, & Pitler, 1981). Of particular importance to the work described here is the observation of a reliable relation between aging processes, SDHACU, and the content of temporal memory. The investigation of these interrelated variables has been advanced during recent years and we have now replicated our earlier results indicating that SDHACU in the FC, but not in the HIP, is positively correlated with the *magnitude* of the discrepancy in the content of temporal memory (Meck, 2001).

One of the major questions remaining is whether the high correlation observed between SDHACU and the remembered time of reinforcement is an artifact of both SDHACU and K*-1 increasing with age or if this correlation holds for any type of discrepancy in K* at any age. To determine this, it is important to obtain SDHACU measures from a relatively large number of experimental subjects that are trained in the PI timing task during EARLY (1-2 months), MIDDLE (10-11 months), and LATE (30-31 months) portions of the lifespan. These data must be compared with data from control subjects of the same ages that do not receive any PI training, but are maintained in an identical manner in all other respects to the experimental subjects. In this way, the relation between the remembered time of reinforcement, age, and SDHACU in the FC and HIP can be further evaluated.

SDHACU has the advantage of being able to reflect the activity of nerve cells that use a specific transmitter substance. Thus, unlike electrophysiological recording techniques, it provides information about a specific transmitter system in the brain. More specifically, SDHACU *in vitro* has been shown to reflect Ch activity *in vivo* (Murrin, DeHaven, & Kuhar, 1977; Samaras & Contrera, 1977; Simon, Atweh, & Kuhar, 1976; Simon & Kuhar, 1975; 1976). Drugs that increase the activity of Ch neurons (e.g., scopolamine) and subsequently increase the release of $[^3H]$-acetylcholine also increase *in vitro* SDHACU (Atweh, Simon, & Kuhar, 1976; Murrin, DeHaven, & Kuhar, 1977). Consequently, SDHACU in the HIP ought to change if behavior alters the firing of MSA Ch neurons, and

SDHACU in the FC ought to change if behavior alters the firing of NBM neurons. This allows for possible dissociations between our standard PI, PI-GAP, and STP procedures that assess reference memory, working memory, and divided attention respectively. Other behavioral training tasks have already been shown to alter SDHACU (Burgel & Rommelspacher, 1978; Raaijmakers, 1982). For example, SDHACU and incorporation of [^{3}H]-choline into [^{3}H]-acetylcholine was higher in the HIP of rats showing good retention in working memory tasks, and in rats that were placed in a novel environment compared to various control conditions.

With special reference to Ch activity in the FC as a function of behavioral training, it is interesting to note that continued training on PI timing procedures leads to an increase in the choline uptake necessary for ACh synthesis in circumstances where uptake may be unaffected by age in the absence of behavioral training. This conclusion is based on the observation that aged rats display a significant positive correlation between choline uptake in the FC (but not the HIP) and the increase in their remembered times of reinforcement. The experimental problem is to determine whether choline uptake in the FC reflects a general monitoring process of temporal expectations regardless of the direction of the discrepancy or whether this positive correlation is artifactually based on aged-related increases in both measures. In order to investigate the significance of these observations we studied the relation between choline uptake and the magnitude and direction of discrepancies in temporal memory at three different ages (1-2 mo, 10-11 mo, and 30-31 mo) during which the effects of age-related changes in cholinergic function would be expected to different markedly. It would be of considerable interest to discover that the activation of choline-uptake in the FC quantitatively reflects a psychological state of "surprise." That is, when the physical time of reinforcement as indexed by a current clock reading is discrepant from the remembered time of previous reinforcement deliveries, memory and/or attentional systems should be activated to an extent that is proportional to the magnitude, and not necessarily the direction, of the discrepancy. Thus, animals that demonstrate consistent horizontal displacements of their timing functions relative to the actual time of reinforcement should be "surprised" by the delivery of reinforcement (i.e., it should be perceived as being too late or too early). The animal should attempt to correct this discrepancy between clock and memory by storing the current time of reinforcement in temporal memory. Unfortunately, if the memory storage process has been corrupted, this becomes a "vicious cycle" that the animal is unable to correct despite the continual activation of the correction process.

Thus, the major overall question is whether choline-uptake in the FC is sensitive to discrepancies in temporal memory in a manner that is at least partially independent of the effects of aging. Three predictions of such a

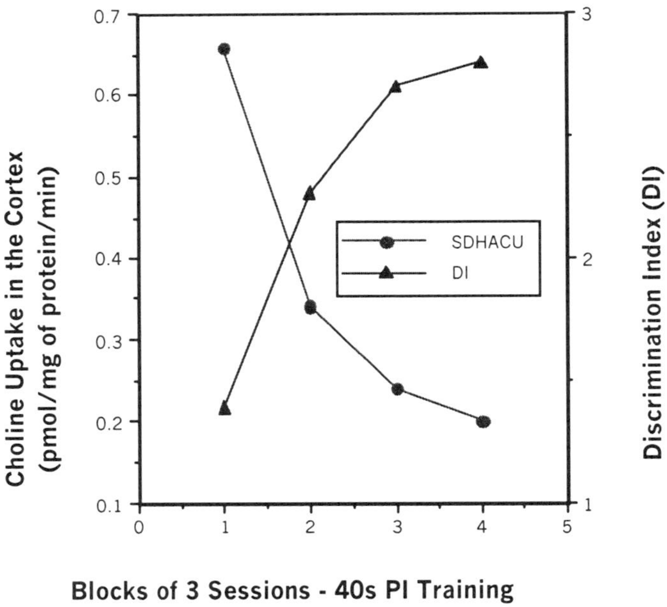

Figure 4. Choline uptake (SDHACU) in the frontal cortex and temporal discrimination index (DI) as a function of peak-interval (PI) training.

hypothesis are that (**a**) rats of all ages trained on the PI timing procedure should initially show high levels of choline-uptake that decrease as a function of their acquisition of an accurate representation of the temporal criterion for reinforcement as measured by the discrimination index (DI = maximum response rate/average response rate), (**b**) rats showing little or no discrepancy in the content of temporal memory at steady-state levels of training should exhibit relatively low levels of choline uptake, (**c**) choline-uptake in the FC should be relatively unaffected by age in the absence of explicit training procedures. Data supporting the first prediction are presented in Figure 4. We are currently exploring the other predictions (cf., Meck, 2001).

We have been concerned that the validity of these experiments may depend on the assumption that the rat's life outside of the testing apparatus is relatively free of time-estimation opportunities – otherwise it would be difficult to separate the effects of aging on choline uptake from the effects due to discrepancies in the content of temporal memory. Given that interval estimation may play a daily role in photoperiodic phenomena it may be virtually impossible to find a situation where aging and interval timing do not coincide. Because of the explicit feedback provided by the delivery of reinforcement in PI timing procedures, it seems unlikely however that the success of these experiments is compromised by the animals' sensitivity (or lack thereof) to the components of the circadian light/dark cycle that they experience in their home cages. It would be of great interest to discover however that the memory translation parameter for the PI timing procedures (K*) is directly correlated with circadian periodicity (τ). Although a number of investigators have proposed such a linkage, to date the data have not supported this type of relation between interval and circadian timing due to the distinctive nature of the feedback and entrainment requirements for each type of timing system (Hinton & Meck, 1997b).

THETA (θ) RHYTHM AND TEMPORAL MEMORY

Recent neuroanatomical and biochemical findings indicate that the MSA, the nuclei of the diagonal band, and the NBM form a complex of subcortical neurons in the basal forebrain that collectively project to most of the limbic, olfactory, and neocortical structures of the telencephalon and account for much of the Ch input to these structures (Bigl, Woolf, & Butcher, 1982). This complex also appears to contain a population of catecholaminergic neurons (Davis & Macrides, 1983) and a variety of peptidergic neurons (Cummings, Elde, Ells, & Lindall, 1983). Several lines of evidence suggest that aging-related changes in this complex, particularly in its Ch component, may contribute to abnormalities of cognition, memory, spatial orientation, and behavior which sometimes occur with normal aging. More severe neurochemical, neuroanatomical, and degenerative aberrations of this complex may account for the debilitating symptoms of Alzheimer's disease and may be generally involved in senile dementia (Coyle, Price, & DeLong, 1983).

Cholinergic and possibly other neurons located in the MSA and predominantly in the vertical limb of the diagonal band are thought to function as pacemakers for hippocampal theta (θ) rhythm (Andersen, 1980; Gaztelu, & Buno, 1982; Mitchell, Rawlins, Steward, & Olton, 1982). Its appearance as 4-12 Hz rhythmic slow-wave activity (RSA) has been associated with attentiveness (Komisaruk, 1977) and with phenomena involving a heightened or focused state of vigilance such as voluntary movement (Vanderwolf, 1971), spatial guidance (O'Keefe & Nadel, 1978), and learning (Landfield, McGaugh, & Tusa, 1972). It has been demonstrated that neurons located throughout the horizontal limb of the diagonal band and in the rodent homolog of the nucleus basalis of Meynert exhibit rhythmic bursting activity in the θ range similar to those neurons in the vertical limb and MSA (McKinney, Coyle, & Hedreen, 1983). Thus, these neuronal groupings may collectively drive θ activity in a variety of olfactory, limbic, and neocortical structures. Furthermore, lesions in this complex which impair learning and memory also reduce or abolish θ activity in the hippocampal formation and related cortical structures (Mitchell, Rawlins, Steward, & Olton, 1982). These various findings suggest that alterations in forebrain θ rhythms may accompany aberrations of the medial septum-diagonal band-nucleus basalis complex and may be important in aging-related impairments of cognitive and learning abilities.

THETA (θ) RHYTHM IN THE HIPPOCAMPUS (HIP) AND FRONTAL CORTEX (FC) AS A INDICATOR OF THE DISCREPANCY IN THE CONTENT OF TEMPORAL MEMORY

Recording Theta (θ) Rhythm in the HIP and FC

As mentioned above, the electrical patterns of the HIP are mainly under the control of subcortical afferents. Of these, the septo-hippocampal path appears essential (Petsche, Stumpf, & Gogolak, 1962; Stewart & Fox, 1990). Cholinergic and probably non-cholinergic neurons of the medial septum and the nucleus of the diagonal band of Broca contain the pattern generator of 'pacemaker' cells which provide a rhythmic input directly, or indirectly via inhibitory interneurons, to the pyramidal and granule cells of the HIP during learning and exploratory behaviors (Buzsaki, 1986). The summed rhythmic membrane currents result in cyclic extracellular field potentials, termed rhythmic slow activity (RSA or theta (θ) rhythm).In the absence of RSA, irregular sharp waves (SPW) of 40-120 ms duration appear in the hippocampal record. Concurrent with SPWs, a number of pyramidal neurons, granule cells and interneurons discharge in concert and then remain virtually silent between SPWs. Several mechanisms assure suppression of SPWs and associated bursting of neurons while RSA associated behaviors are in progress. One of these mechanisms is Ch septal input.

In view of these known physiological electrical patterns, we have investigated the extent to which theta rhythm in the HIP and FC may be related to discrepancies in the remembered time of reinforcement. The major question is whether the magnitude of field potentials in these brain regions is correlated with the degree of irregularity in the content of temporal memory. It has recently been shown that restoration of rhythmic slow activity (θ) in the subcortically denervated HIP by fetal transplants is highly correlated with recovery of memory function (Buzsaki, Gage, Czopf, &

Bjorklund, 1987; Buzsaki, Gage, Kellenyi, & Bjorklund, 1987). This work with fetal transplants has indicated that in some recipient animals brain grafts only partially restore the normal level of rhythmic activity. It is therefore important to establish the quantitative relationship between the level of rhythmic slow activity (θ) in the HIP and the FC and associated temporal memory translation constants (K*) that influence the remembered times of reinforcement.

Driving Theta (θ) Rhythm in the HIP and FC

The major question is whether the remembered time of reinforcement can be modified by driving theta rhythm at a selected rate. Our hypothesis is that there will be a significant negative correlation between theta rhythm and K*. That is to say that the faster theta is, the shorter the remembered time of reinforcement will be. Furthermore, we have been exploring the possibility that there is a direct relation between the magnitude of the change in theta and the magnitude of the change in the content of temporal memory. Our basic goal has been to establish a quantitative relationship between theta and the remembered time of reinforcement, first by recording theta and then by driving theta and thereby possibly influencing the remembered times of reinforcement by directly altering a temporal memory translation constant (K*).

In a recent set of studies we have been able to influence the remembered time of reinforcement stored in reference memory by driving theta rhythm in the FC. If theta is increased by approximately 20% in this region, rats exhibit a proportional leftward shift in their PI timing functions of about 20%. These leftward shifts are maintained by the stimulation and do not normalize with continued testing. These data are compatible with other electrophysiological results that we have obtained and argue in favor of separate clock and memory processing stages for duration discrimination (Meck, 1988b). Our previous electrophysiological work has shown that the pacemaker used to make temporal judgments can be driven at faster rates by implanting electrodes in dopaminergic rich areas such as the medial forebrain bundle (MFB) and the SN. Electrical stimulation in these areas produces leftward shifts in peak times that quickly normalize with continued training and show a rebound effect when stimulation is discontinued. We have shown that these different patterns of results in response to stimulation represent two separate IP stages, one a clock stage, and the other a memory stage (Meck, 1983, 1996). Furthermore, we have shown that the magnitude of the effect given a constant level of electrical stimulation varies as a function of the animal's age. Older rats require increased levels of stimulation in order to obtain the same increase in clock or memory storage speed. This is presumably related to the effects of neuronal degeneration in older animals (Meck, 1988b). Interestingly, in agreement with our data, O'Keefe and Recce (1992) have shown that there is a sufficient range of inter-individual variability in theta in order to establish the importance of phase coding in predicting a variety of spatial behaviors. A future direction for this research will be to determine whether electrical activity in the HIP and FC are differentially related to working memory and reference memory as indexed by the PI-GAP and PI procedures, respectively.

Theta EEG rhythms in temporal and frontal regions may be related to the modifications in the content of temporal memory that have been observed. Interpretation of the magnitude, frequency, and phase information might be problematic, however, given that absolute magnitude is somewhat difficult to analyze in view of differences in electrode placement among animals, etc. In order to address this concern, an experiment was undertaken to demonstrate control of the behavioral and electrophysiological parameters in question so that further inquiries could be justified. This work took advantage of the opposing action of γ-aminobutyric acid (GABA) on Ch activity. Enhancement of cortical and hippocampal ACh transmission can occur by the blockade of GABA receptors which typically supply widespread inhibition on Ch cell activity (Sarter, Bruno, Givens, Moore, McGaughy, & McMahon, 1996). Following the acquisition of the PI timing task, chronic guide cannulae were stereotaxically implanted into the substantia innominata (SI) of both hemispheres, and the rats were retrained. Administration of the GABA-agonist muscimol (25 ng/0.5 μl/hemisphere) or the benzodiazepine receptor (BZR) selective inverse agonist (MDL 26,479 -- 15 ng/0.5 μl/hemisphere) produced opposite effects on both the horizontal placement of PI timing functions for response rate and on theta rhythm power and frequency as shown in Figure 5. Although these data were obtained from different sets of rats, they strongly support the validity of our proposal to conduct more extensive studies aimed at correlating various parameters of theta rhythm with PI timing behavior on a trial-by-trial basis and not simply obtaining session means of each measure with different groups of rats.

In comparison with the effects of muscarinic agonists or cholinesterase inhibitors, trans-synaptic stimulation of forebrain acetylcholine (ACh) release has been discussed as an effective approach for attenuating the behavioral impairments associated with cortical cholinergic hypofunction (e.g., Forster & Lal, 1992). For example, if age is associated with a reduction in basal ACh release, then treatment with BZR inverse agonists might be expected to increase the excitability of cholinergic neurons and attenuate age-related impairments in cholinergic function. Thus, another important aspect of the current data is the suggestion that cortical acetylcholine turnover can be subject to bidirectional modulation by

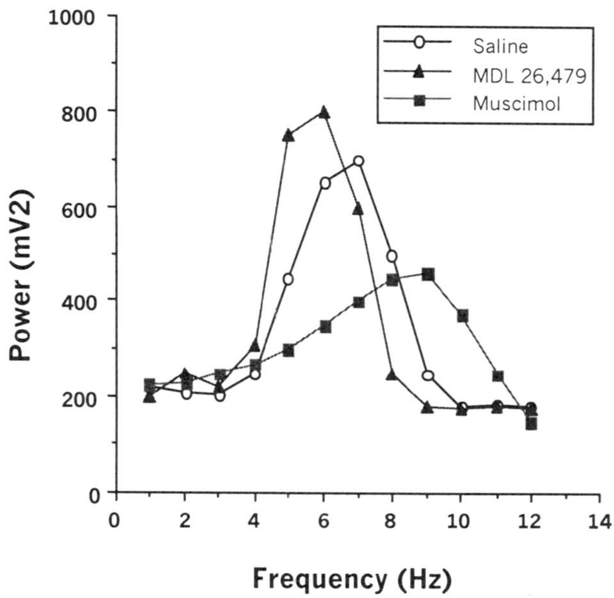

Figure 5. Power spectral analysis of cortical EEG over a 10-min period following microinfusions into the substantia innominata (SI).

manipulating the activity of GABAergic transmission (for reviews see Givens & Olton, 1990; Sarter, Bruno, & Dudchenko, 1990).

This series of experiments dealing with the relation between theta rhythm and the content of temporal memory was designed to complement the SDHACU experiments described above. The major purpose of both sets of experiments was to provide some physiological specification of the two major factors involved in temporal memory discrepancies, i.e., the direction and the magnitude of the discrepancy. Data reported here suggest that theta rhythm in the HIP and FC may only be representative of the direction of the discrepancy in the content of temporal memory. In contrast, as indicated above, SDHACU in the HIP and FC may only be representative of the magnitude of the discrepancy in the content of temporal memory.

CONCLUSIONS

Choline uptake reflects the activity of cholinergic neurons in various brain regions during behavioral activation. Using this measure we have attempted to establish the quantitative relation between SDHACU in the HIP and the

FC and the discrepancies in the remembered times of reinforcement stored in working and/or reference memory as a function of age and behavioral training. Our findings suggest that choline uptake in the FC may serve as an indicator of the *magnitude of the discrepancy* in the content of temporal memory. We also attempted to assess theta rhythm (θ) activity as an indicator of the direction of the discrepancy in the content of temporal memory. Theta rhythm (θ, 4-12 Hz) in the HIP and FC appears to be necessary for certain learning and memory processes. Our first objective was to record theta rhythm in these two brain regions and relate the observed differences in electrical patterns (amplitude and modal frequency) to individual differences in the remembered times of reinforcement stored in working memory and reference memory. A second objective was to attempt to impair or facilitate temporal memory by driving hippocampal rhythmical slow wave activity (RSA or θ) at frequencies that differ significantly from normal patterns of rhythmic excitation. Our findings suggest that θ activity in the FC may serve as an indicator of the *direction of the discrepancy* in the content of temporal memory.

REFERENCES

Andersen, P. (1980). Participating neurones and mechanisms underlying theta activity in unanesthesized rabbits. *Progress in Brain Research*, **54**, 371-380.

Atweh, S., Simon, J. R., & Kuhar, M. J. (1976). Utilization of sodium-dependent high affinity choline uptake *in vitro* as a measure of the activity of cholinergic neurons *in vivo*. *Life Science*, **17**, 1535-1544.

Bartus, R. T., Dean,III, R. L., Beer, B., & Lippa, A. S. (1982). The cholinergic hypothesis of geriatric memory dysfunction. *Science*, **217**, 408-417.

Bartus, R. T., Dean, R. L., Pontecorvo, M. J., & Flicker, C. (1985). The cholinergic hypothesis: A historical overview, current perspective, and future directions. In D. S. Olton, E. Gamzu, & S. Corkin (Eds.), *Annals of The New York Academy of Sciences: Memory Dysfunctions: An Integration of Animal and Human Research From Preclinical and Clinical Perspectives* - Vol. **444**. (pp. 332-358). New York: New York Academy of Sciences.

Bigl, V., Woolf, N. J., & Butcher, L. L. (1982). Cholinergic projections from the basal forebrain to frontal, parietal, temporal, occipital, and cingulate cortices: A combined fluorescent tracer and acetylcholinesterase analysis. *Brain Research Bulletin*, **8**, 727-749.

Blusztajn, J. K., & Wurtman, R. J. (1983). Choline and cholinergic neurons. *Science*, **221**, 614-620.

Buhusi, C. V., & Meck, W. H. (2000). Timing for the absence of a stimulus: The gap paradigm reversed. *Journal of Experimental Psychology: Animal Behavior Processes*, **26**, 305-322.

Bünning, E. (1973). *The physiological clock*, 3rd ed., New York: Springer.

Burgel, P., & Rommelspacher, H. (1978). Changes in high affinity choline uptake in behavioral experiments. *Life Science*, **23**, 2423-2428.

Buzsaki, G. (1986). Generation of hippocampal slow wave patterns. In R. L. Isaacson & K. Pibram (Eds.), *The hippocampus*, Vol. 3, New York: Plenum.

Buzsaki, G., Gage, F. H., Czopf, J., & Bjorklund, A. (1987). Restoration of rhythmic slow activity (θ) in the subcortically denervated hippocampus by fetal CNS transplants. *Brain Research*, **400**, 334-347.

Buzsaki, G., Gage, F. H., Kellenyi, L., & Bjorklund, A. (1987). Behavioral dependence of the electrical activity of intracerebrally transplanted fetal hippocampus. *Brain Research*, **400**, 321-333.

Church, R. M. (1980). Short-term memory for time intervals. *Learning and Motivation*, **11**, 208-219.

Church, R. M. (1984). Properties of the internal clock. In J. Gibbon & L. Allan (Eds.), *Annals of The New York Academy of Sciences: Timing and time perception* (Vol. **423**) (pp. 566-582), New York: New York Academy of Sciences.

Church, R. M., & Meck, W. H. (1988). Biological basis of the remembered time of reinforcement. In R. M. Church, M. L. Commons, J. R. Stellar, & A. R. Wagner (Eds.), *Quantitative analyses of behavior: Biological determinants of reinforcement and memory* (Vol. **7**), (pp. 103-119). Hillsdale, NJ: Erlbaum.

Church, R. M., Meck, W. H., & Gibbon, J. (1993). Application of scalar timing theory to individual trials. *Journal of Experimental Psychology: Animal Behavior Processes*, in press.

Church R. M., Miller, K. D., Meck, W. H., & Gibbon, J. (1991). Symmetrical and asymmetrical sources of variance in temporal generalization. *Animal Learning & Behavior*, **19**, 207–214.

Corkin, S. (1984). Lasting consequences of bilateral medial temporal lobectomy : Clinical course in experimental findings in H. M. *Neurology*, **4**, 249-259.

Coyle, J. T., Price, D. L., & DeLong, M. R. (1983). Alzheimer's disease: A disorder of cortical cholinergic innervation. *Science*, **219**, 1184-1190.

Cummings, S., Elde, R., Ells, J., Lindall, A. (1983). Corticotropin-releasing factor immunoreactivity is widely distributed within the central nervous system of the rat: An immunohistochemical study. *Journal of Neuroscience*, **3**, 1355-1368.

Davis, B. J., & Macrides, F. (1983). Tyrosine hydroxylase immunoreactive neurons and fibers in the olfactory system of the hamster. *Journal of Comparative Neurology*, **214**, 427-440.

Forster, M. J. & Lal, H. (1992). Modulation of learning and memory via benzodiazepine receptors: Potential treatments for age-related dementia. In Meyer, E.M., Simpkins, J.W., Yomamoto, J., & Crews, F.T. (eds), *The Treatment of Dementias: A New Generation of Progress*, pp 25-84. New York: Plenum.

Fraisse, P. (1984). Perception and estimation of time. *Annual Review of Psychology*, **35**, 1-36.

Fuster, J. M. (1980). *The prefrontal cortex: Anatomy, physiology and neuropsychology of the frontal lobe*. New York: Raven Press.

Gaztelu, J. M., & Buno, W. (1982). Septo-hippocampal relationships during EEG theta rhythm. *Electroencephalogy and Clinical Neurophysiology*, **54**, 375-387.

Gibbon, J. (1977). Scalar expectancy theory and Weber's law in animal timing. *Psychological Review*, **84**, 279-325.

Gibbon, J., & Church, R. M. (1984). Sources of variance in an information processing theory of timing. In H. L. Roitblat, T. G. Bever, & H. S. Terrace (Eds.), *Animal cognition* (pp. 465-488). Hillsdale, NJ: Erlbaum.

Gibbon, J. & Church, R. M. (1990). Representation of time. *Cognition*. **37**, 23-54.

Gibbon, J. & Church, R. M. (1992). Comparison of variance and covariance patterns in parallel and serial theories of timing. *Journal of The Experimental Analysis of Behavior*, **57**, 393-406.

Gibbon, J. Church, R. M., & Meck, W. H. (1984). Scalar timing in memory. In J. Gibbon, & L. Allan (Eds.), *Annals of The New York Academy of Sciences: Timing and time perception* (Vol. **423**) (pp. 528-541). New York: New York Academy of Sciences.

Gibbon, J., Malapani, C., Dale, C. L., & Gallistel, C. R. (1997). Toward a neurobiology of temporal cognition; Advances and challnges. *Current Opinion in Neurobiology*, **7**, 170-184.

Givens, B. S., & Olton, D. S. (1990). Cholinergic and GABAergic modulation of medial septal area: Effect on working memory. *Behavioral Neuroscience*, **104**, 849-855.

Gloning, K., & Hoff, H. (1969). Central localization of disorders of higher nervous activity: Clinical, anatomical and physiological aspects. In P. J. Vinken & G. W. Bruyn (Eds.), *Handbook of clinical neurology*. Vol. **3** (pp. 22-47), New York: John Wiley & Sons, Inc.

Goody, W. (1969). Disorders of the time sense. In P. J. Vinken & G. W. Bruyn (Eds.), *Handbook of clinical neurology*.Vol. **3** (pp. 229-250), New York: John Wiley & Sons, Inc.

Hasselmo, M. E., Anderson, B. P., & Bower, J. M. (1991). Cholinergic modulation may enhance cortical associative memory function. In R. P. Lippman, J. Moody, and D. S. Touretzky (Eds.), *Advances in neural information processing systems*, Vol. **3**, (pp. 50-57), San Mateo, CA: Morgan Kaufman.

Hattori, T., & Fibiger, H. C. (1982). On the use of lesions of afferents to localize neurotransmitter receptor sites in the striatum. *Brain Research*, **238**, 245-250.

Hinton, S. H., & Meck, W. H. (1997a). How time flies: Functional and neural mechanisms of interval timing. In C.M. Bradshaw & E. Szabadi (Eds.), *Time and behaviour: Psychological and neurobiological analyses*. (pp. 409-457). New York: Elsevier.

Hinton, S. H., & Meck, W. H. (1997b). The "internal clocks" of circadian and interval timing. *Endeavour*, **21**, 82-87.

Hinton, S. C., Meck, W. H., & MacFall, J. R. (1996). Peak-interval timing in humans activates frontal-striatal loops. *NeuroImage*, **3**, S224.

Johnson, S. W., Mercuri, N. B., & North, R. A. (1992). 5-Hydroxytryptamine$_{1B}$ receptors block the GABA$_B$ synaptic potential in rat dopamine neurons. *The Journal of Neuroscience*, **12**, 2000-2006.

Kolb, B. (1984). Functions of the frontal cortex of the rat: A comparative review. *Brain Research Review*, **8**, 65-98.

Komisaruk, B. R. (1977). The role of rhythmical brain activity in sensorimotor integration. In J. M. Sprague & A. N. Epstein (Eds.), *Progress in Psychobiology and Physiological Psychology* (Vol. **7**) (pp. 55-90), New York: Academic Press.

Landfield, P. W., Baskin, R. K., & Pitler, T. A. (1981). Brain aging correlates: Retardation by hormonal treatments. *Science*, **214**, 581-584.

Landfield, P. W., McGaugh, J. L., & Tusa, R. J. (1972). A temporal correlate of memory storage processes in the rat. *Science*, **175**, 87-89.

Lejeune, H., Ferrara, A., Soffie, M., Bronchart, M., & Wearden, J. H. (1997). Peak procedure performance in young adult and aged rats: Acquisition and adaptation to a changing temporal criterion. *The Quarterly Journal of Experimental Psychology*, in press.

Levin, E. D., Conners, C. K., Sparrow, E., Hinton, S. C., Erhardt, D., Meck, W. H., Rose, J. E., & March, J. (1996). Nicotine effects on adults with attention-deficit/hyperactivity disorder. *Psychopharmacology*, **123**, 55-63.

Levin, E. D., Conners, C. K., Silva, D., Hinton, S. C., Meck, W. H., March, J., & Rose, J. E. (1998). Transdermal nicotine effects on attention. *Psychopharmacology*, **140**, 135-141.

Luria, A. R. (1969). Frontal lobe syndromes. In P. J. Vinken & G. W. Bruyn (Eds.), *Handbook of clinical neurology*. Vol. **2** (pp. 725-757), New York: John Wiley & Sons, Inc.

Lustig, C. A., & Meck, W. H. (1998). Attention mediated temporal processing deficits as a function of age, signal modality, and time of day. *Journal of Cognitive Neuroscience*, **S**, 134.

Lustig, C., & Meck, W. H. (2001). Paying attention to time as one get older. *Psychological Science*, in press.

Malapani, C., Pillon, B., Lo Russo, G., Dubois, B., Munari, C., & Agid, Y. (1992). Damage to striatofrontal loops impairs simultaneous temporal processing (STP) ability. *Neurology*, **42**, 183.

Malapani, C., Rakitin, B., Meck, W. H., Deweer, B., Dubois, B., & Gibbon, J. (1998). Coupled temporal memories in Parkinson's Disease: A dopamine-related dysfunction. *Journal of Cognitive Neuroscience*, **10**, 316-331.

Matell, M. S. & Meck, W. H. (1999). Reinforcement-induced within-trial resetting of an internal clock, *Behavioural Processes*, **45**, 159-171.

Matell, M. S., & Meck, W. H. (2000). Neuropsychological mechanisms of interval timing behaviour. *BioEssays*, **22**, 94-103.

Matell, M. S., Meck, W. H., King, G. R. (1996). Changes in rescaling temporal responding on the tri-peak procedure after modulation of cocaine reactivity. *Society for Neuroscience Abstracts*, **22**, 1130.

McKinney, M., Coyle, J. T., & Hedreen, J. C. (1983). Topographical analysis of the innervation of the rat neocortex and hippocampus by the basal forebrain cholinergic system. *Journal of Comparative Neurology*, **217**, 103-121.

Meck, W. H. (1983). Selective adjustment of the speed of internal clock and memory storage processes. *Journal of Experimental Psychology: Animal Behavior Processes*, **9**, 171-201.

Meck, W. H. (1984). Attentional bias between modalities: Effect on the internal clock, memory, and decision stages used in animal time discrimination. In J. Gibbon, & L. Allan (Eds.), *Annals of The New York Academy of Sciences: Timing and time perception* (Vol. **423**) (pp. 528-541). New York: New York Academy of Sciences.

Meck, W. H. (1986). Affinity for the dopamine D_2 receptor predicts neuroleptic potency in decreasing the speed of an internal clock. *Pharmacology Biochemistry & Behavior*, **25**, 1185-1189.

Meck, W. H. (1987). Vasopressin metabolite neuropeptide facilitates simultaneous temporal processing. *Behavioural Brain Research*, **23**, 147-157.

Meck, W. H. (1988a). Hippocampal function is required for feedback control of an internal clock's criterion. *Behavioral Neuroscience*, **102**, 54-60.

Meck, W. H. (1988b). Internal clock and reward pathways share physiologically similar information-processing stages. In M. L. Commons, R. M. Church, J. R. Stellar, & A. R. Wagner (Eds.) *Quantitative analyses of behavior: Biological determinants of reinforcement*. (Vol. **7**), (pp. 121-138). Hillsdale, NJ: Erlbaum.

Meck, W. H. (1991). Modality-specific circadian rhythmicities influence mechanisms of attention and memory for interval timing. *Learning and Motivation*, **22**, 153-179.

Meck, W. H. (1996). Neuropharmacological properties of timing and time perception. *Cognitive Brain Research*, **9**, 1-16.

Meck, W. H. (2001). [^{3}H]-Choline uptake in the frontal cortex of mature and aged rats is proportional to the absolute error of a temporal memory translation constant. *Learning & Motivation*, in press.

Meck, W. H. & Benson, A. M. (2001). Dissecting the brain's internal clock: How frontal-striatal circuitry keeps time and shifts attention. *Brain & Cognition*, in press.

Meck, W. H., & Church, R. M. (1984). Simultaneous temporal processing. *Journal of Experimental Psychology: Animal Behavior Processes*, **10**, 1-29.

Meck, W. H., & Church, R. M. (1987a). Cholinergic modulation of the content of temporal memory. *Behavioral Neuroscience*, **101**, 457-464.

Meck, W. H., & Church, R. M. (1987b).Nutrients that modify the speed of internal clock and memory storage processes. *Behavioral Neuroscience*, **101**, 465-475.

Meck, W. H., Church, R. M., & Olton, D. S. (1984). Hippocampus, time, and memory. *Behavioral Neuroscience*, **98**, 3-22.

Meck, W. H., Church, R. M., & Wenk, G. L. (1986). Arginine vasopressin inoculates against age-related discrepancies in the content of temporal memory and increases in sodium-dependent high affinity choline uptake. *European Journal of Pharmacology*, **130**, 327-331.

Meck, W. H., Church, R. M., Wenk, G. L., & Olton, D. S. (1987). Nucleus basalis magnocellularis and medial septal area lesions differentially impair temporal memory. *The Journal of Neuroscience*, **7**, 3505-3511.

Meck, W. H., Hinton, S.C., & Matell, M.S. (1998). Coincidence-detection models of interval timing: Evidence from fMRI studies of cortico-striatal circuits. *NeuroImage*, **7**, S281.

Meck, W. H., & Williams, C. L. (1997a). Characterization of the facilitative effects of perinatal choline supplementation on timing and temporal memory. *NeuroReport*, **8**, 2831-2835.

Meck, W. H., & Williams, C. L. (1997b). Simultaneous temporal processing is sensitive to prenatal choline availability in mature and aged rats. *NeuroReport*, **8**, 3045-3051.

Milner, B. (1982). Some cognitive effects of frontal-lobe lesions in man. *Phil. Trans. R. Soc. Lond.*, **298**, 211-226.

Milner, B., & Petrides, M. (1984). Behavioural effects of frontal-lobe lesions in man. *Trends in Neurosciences*, **7**, 403-407.

Milner, B., Petrides, M., & Smith, M. L. (1985). Frontal lobes and the temporal organization of memory. *Human Neurobiology*, **4**, 137-142.

Mitchell, S. J., Rawlins, J. N. P., Steward, O., & Olton, D. S. (1982). Medial septal area lesions disrupt θ rhythm and cholinergic staining in medial entorhinal cortex and produce impaired radial-arm maze behavior in rats. *The Journal of Neuroscience*, **2**, 292-302.

Murrin, L. C., DeHaven, R. N., & Kuhar, M. J. (1977). On the relationship between [^{3}H]-choline uptake activation and [^{3}H]-acetylcholine release. *Journal of Neurochemistry*, **29**, 681-687.

Nichelli, P., Venneri, A., Molinari, M., Tavani, F., & Grafman, J. (1993). Precision and accuracy of subjective time estimation in different memory disorders. *Cognitive Brain Research*, **1**, 87-93.

O'Keefe, J., & Nadel, L. (1978). *The hippocampus as a cognitive map*. Oxford: Claredon Press.

O'Keefe, J., & Recce, M. (1992). Hippocampal complex spike cells systematically shift phase relative to the EEG theta as the rat runs through the place field. *Society for Neuroscience Abstracts*, 297.11.

Olton, D. S., Meck, W. H., & Church, R. M. (1987). Separation of hippocampal and amygdaloid involvement in temporal memory dysfunctions. *Brain Research*, **404**, 180-188.

Olton, D. S., Wenk, G. L., Church, R. M., & Meck, W. H. (1988). Attention and the frontal cortex as examined by simultaneous temporal processing. *Neuropsychologia*, **26**, 307-318.

Pang, K. C. H., Yoder, R. M., & Olton, D. S. (2001). Neurons in the lateral agranular frontal cortex have divided attention correlates in a simultaneous temporal processing task. *Neuroscience*, **103**, 615-628.

Paule, M.G., Meck, W. H., McMillan, D. E., Bateson, M., Popke, E. J., Chelonis, J. J., & Hinton, S. C. (1999). The use of timing behaviors in animals and humans to detect drug and/or toxicant effects. *Neurotoxicology and Teratology*, **21**, 491-502.

Penney, T. B., Gibbon, J., & Meck, W. H. (2000). Differential effects of auditory and visual signals on clock speed and temporal memory. *Journal of Experimental Psychology: Human Perception and Performance*, **26**, 1770-1787.

Penney, T. B., Holder, M. D., & Meck, W. H. (1996). Clonidine-induced antagonism of norepinephrine modulates the attentional processes involved in peak-interval timing. *Experimental and Clinical Psychopharmacology*, 4, 82-92.

Penney, T. B., Meck, W. H., Roberts, S. A., Gibbon, J., & Erlenmeyer-Kimling, L. (2001). Attention mediated temporal processing deficits in subjects at high risk for schizophrenia. *Neuropsychology*, submitted.

Petsche, H., Stumpf, C., & Gogolak, G. (1962). The significance of the rabbit's septum as a relay station between the midbrain and the hippocampus. The control of hippocampal arousal activity by septum cells. *Electroencephalogy Clinical Neurophysiology*, **14**, 202-211.

Raaijmakers, W. G. M. (1982). High-affinity choline uptake in hippocampal synaptosomes and learning in the rat. In C. Ajmone Marsan and H. Matthies (Eds.), *Neuronal plasticity and memory formation*. (pp. 373-385). New York: Raven Press.

Rakitin, B. C., Gibbon, J., Penney, T. B., Malapani, C., Hinton, S. C., & Meck, W. H. (1998). Scalar expectancy theory and peak-interval timing in humans. *Journal of Experimental Psychology: Animal Behavior Processes*, **24**, 15-33.

Rao, S. M., Mayer, A. R., & Harrington, D. L. (2001). The evolution of brain activation during temporal processing. *Nature Neuroscience*, **4**, 317-323.

Richards, W. (1973). Time reproductions by H. M. *Acta Psychologia*, **37**, 279-282.

Roberts, S. (1981). Isolation of an internal clock. *Journal of Experimental Psychology: Animal Behavior Processes*, **7**, 242-268.

Samaras, G. M., & Contrera, J. F. (1977). Choline: High affinity uptake in vivo by rat hippocampus. *Journal of Neurochemistry*, **28**, 1373-1376.

Sarter, M. Bruno, J. P., & Dudchenko, P. (1990). Activating the damaged basal forebrain cholinergic system: Tonic stimulation versus signal amplification. *Psychopharmacology*, **101** 1-17.

Sarter, M. Bruno, J. P., Givens, B., Moore, H., McGaughy, J., & McMahon, K. (1996). Neuroanl mechanisms mediating drug-induced cognition enhancement: Cognitive activity as a necessary intervening variable. *Cognitive Brain Research*, **3**, 329-343.

Simon, J. R., Atweh, S., & Kuhar, M. J. (1976). Sodium-dependent high affinity choline uptake: A regulatory step in the synthesis of acetylcholine. *Journal of Neurochemistry*, **26**, 909-922.

Simon, J. R., & Kuhar, M. J. (1975). Impulse-flow regulation of high affinity choline uptake in brain cholinergic nerve terminal. *Nature*, **255**, 162-163.

Simon, J. R., & Kuhar, M. J. (1976). High affinity choline uptake: Ionic and energy requirements. *Journal of Neurochemistry*, **27**, 93-99.

Smith, M. L., & Milner, B. (1985). Differential effects of frontal-lobe lesions on cognitive estimation and spatial memory. *Neuropsychologia*, **22**, 697-705.

Stewart, M., & Fox, S. E. (1990). Do septal neurons pace the hippocampal theta rhythm? *Trends in Neurosciences*, **13**, 163-168.

Vanderwolf, C. H. (1971). Limbic-diencephalic mechanisms of voluntary movement. *Psychological Review*, **78**, 83-113.

Williams, D. (1969). Temporal lobe syndromes. In P. J. Vinken & G. W. Bruyn (Eds.), *Handbook of clinical neurology*. Vol. **2** (pp. 700-724), New York: John Wiley & Sons, Inc.

9

NEURAL MEDIATION OF MEMORY FOR TIME

Role of the Hippocampus and Medial Prefrontal Cortex

Raymond P. Kesner
University of Utah

INTRODUCTION

The structure and utilization of memory is central to one's knowledge of the past, interpretation of the present, and prediction of the future. Therefore, the understanding of the structural and process components of memory systems at the psychological and neurobiological level is of paramount importance. In recent years, there have been a number of attempts to divide learning and memory into multiple memory systems. Schacter and Tulving (1994) have suggested that one needs to define memory systems in terms of the kind of information to be represented, the processes associated with the operation of each system, and the neurobiological substrates including neural structures and mechanisms that subserve each system. Furthermore, it is likely that within each system there are multiple forms or subsystems associated with each memory system and there are likely to be multiple processes that define the operation of each system. Finally, there are probably multiple neural structures that form the overall substrate of a memory system.

Currently, the most established models of memory can be characterized as dual memory system models with an emphasis on the hippocampus for one component of the model and a composite of other brain structures as the other component. For example, Squire (1994) has proposed that memory can be divided into a hippocampal dependent declarative memory which provides for conscious recollection of facts and events and a non-hippocampal dependent non-declarative memory which provides for memory without conscious access for skills, habits, priming, simple classical conditioning and nonassociative learning. Others have used different terms to reflect the same type of distinction, including a hippocampal dependent explicit memory vs a non-hippocampal dependent implicit memory (Schacter, 1987), and a hippocampal dependent declarative memory based on the representation of relationships among stimuli vs a non-hippocampal dependent procedural memory based on the representation of a single stimulus or configuration of stimuli (Cohen and Eichenbaum, 1993). O'Keefe and Nadel (1978) have suggested that memory is organized as a dual system based on a hippocampal dependent representation of spatial information labeled a locale system and a non-hippocampal dependent representation of non-spatial information labeled a taxon system. Olton (1983) has suggested a different dual memory system in which memory can be divided into a hippocampal dependent working memory defined as memory for the specific, personal and temporal context of a situation and a non-hippocampal dependent reference memory defined as memory for rules and procedures (general knowledge) of specific situations. Different terms have been used to reflect the same distinction including episodic vs semantic memory (Tulving, 1983).

However, because memory involves many neural systems in addition to the hippocampus, Kesner and DiMattia (1987) proposed a neurobiology of a dual system model organized into a data or event-based and knowledge-based memory system which is composed of multiple attributes or forms of memory and subserved by many different neural regions. Based on extensive research aimed at testing this dual system model, the model has been refined and updated (See Kesner, 1998 for an extensive review). In this review, I will briefly present my own comprehensive dual memory model system, followed by a detailed analysis of the temporal form or attribute of memory.

NEUROBIOLOGY OF AN ATTRIBUTE MODEL

Event-Based Memory System

In this comprehensive model, it is assumed that any specific memory is organized within an event-based memory system and a knowledge-based memory system. The event-based memory system provides for temporary representations of incoming data concerning the present, with an emphasis upon facts, data, and events that are usually personal or egocentric and that occur within specific external and internal contexts. The emphasis is upon the processing of new and current information. During initial learning great emphasis is placed on the event-based memory system, which will continue to be of importance even after initial learning in situations where unique or novel trial information needs to be remembered. The event-based memory system is composed of different independently operating forms or attributes of memory. Even though there could be many attributes, the most important attributes include *space, time, response, sensory-perception, and affect*. In humans a *language* attribute is also added. The organization of these attributes within the event-based memory system can take many forms and are probably organized hierarchically and in parallel.

Based on a series of experiments, it can be shown that within the event-based memory system different neural structures and circuits mediate different forms or attributes of memory. The most extensive data set is based on the use of paradigms that measure the short-term or working memory process, such as matching or non-matching to sample, delayed conditional discrimination or continuous recognition memory of single or lists of items. With the use of these memory paradigms it has been shown that a) damage to the hippocampus disrupts spatial and temporal attribute information in rats and humans and linguistic information in humans, b) damage to the caudate affects response attributes in animals and humans, c) damage to the amygdala impairs affect attributes in animals and humans and d) damage to the extra-striate visual cortex interferes with visual object information as an example of a sensory-perceptual attribute in animals and humans (Chiba et al.,1993; Chiba et al.,1994; Heindel et al.,1988; Hopkins et al.,1995b; Horel et al., 1987; Jackson-Smith et al.,1994; Kesner,1990a; Kesner et al.,1993; Kesner and Williams, 1995; McDonald and White, 1993; O'Keefe and Nadel,1978). Furthermore, it can be shown that these neural systems can operate independently of each other. For example, lesions of the hippocampus do not disrupt affect, response, or sensory perceptual (odor or visual object) attributes (Chiba et al.,1993; Kesner et al., 1993; Kesner and Williams,1995; Leonard and Milner,1991; Otto and Eichenbaum, 1992).

Knowledge-Based Memory System

The knowledge-based memory system provides for more permanent representations of previously stored information in long-term memory and can be thought of as one's general knowledge of the world. It can operate in the abstract in the absence of incoming data. The emphasis is on top-down processing. The knowledge-based memory system would tend to be of greater importance after a task has been learned given that the situation is invariant and familiar. In most situations, however, one would expect a contribution of both systems with a varying proportion of involvement of one relative to the other.

The knowledge-based memory system is composed of the same set of different independently operating forms or attributes of memory. These attributes include *space, time, response, sensory-perception, and affect.* In humans a *language* attribute is also added.

The organization of these attributes within the knowledge-based memory system can take many forms and are assumed to be organized as a set of attribute-dependent cognitive maps or neural nets and their interactions that are unique for each memory. It is assumed that long-term representations within cognitive maps are more abstract and less dependent upon specific features.

Based on a series of experiments it can be shown that within the knowledge-based memory system, different neural structures and circuits mediate different forms or attributes of memory. The most extensive data set is based on the use of paradigms that measure the acquisition of new information, discrimination performance, executive functions, strategies and rules to perform in a variety of tasks including skills and the operation of a variety of long-term memory programs.

With the use of these paradigms it has been shown that a) damage to the posterior parietal cortex disrupts spatial attributes (DeRenzi,1982; DiMattia and Kesner, 1988), b) damage to the prefrontal cortex interferes with temporal and response attributes (Fuster, 1995; Kesner and Holbrook, 1987; Kesner et al., 1994; Kolb, 1974; Passingham, 1978), c) damage to the agranular insular and orbitofrontal cortex impairs affect attributes (Butter, 1969; Ragazzino and Kesner, 1996), and d) damage to the inferotemporal cortex interfere with sensory-perceptual attributes (visual object) (Fuster, 1995; Weiskrantz and Sauders, 1984).

It is assumed that interactive patterns of independently operating attributes provide the organizational framework for the existence of each unique memory. At the neurobiological level, it is assumed that each specific brain region code or store the above mentioned attributes as well as their specific interactions. Relative amounts of neural activity within each critical brain region or circuit that represents a critical attribute or a set of attributes provides for the total neural substrate associated with each unique memory representation. Thus, different neural systems are assumed to mediate specific attributes and furthermore, each attribute is processed by different neural systems (event- and knowledge-based memory) depending upon how information is processed. Finally, it is assumed that there is evolutionary continuity between animals and humans not only in terms of mnemonic function, but also in terms of brain-memory function relationships. In this chapter the emphasis will be on the temporal attribute within the context of the neurobiology of attribute model.

TEMPORAL ATTRIBUTE

Within the context of the neurobiology of attribute model, it is possible to divide the temporal attribute into different mnemonic features namely, duration, succession or temporal order, and temporal perspective (Block, 1990). Memory for duration refers to memory for the persistence of an event or the interval between time periods or events, memory for succession refers to the sequential occurrence of events as reflected by memory for relative recency or temporal order, and memory for time perspective is based on memory for the past, present, and future and can be mediated by the use of retrospective and prospective strategies. Because time has no entity of its own, the other attributes of the model, namely space, response, affect, sensory-perception and language provide the content, anchors and time markers for the memory representation of these three different temporal features. A schematic of this model is presented in Figure 1. It is assumed that the hippocampus and interconnected neural circuits in rats and humans mediate the temporal attribute within the event-based memory system and that the medial prefrontal cortex (anterior cingulate) and interconnected neural circuits in the rat and dorsolateral prefrontal cortex and interconnected neural circuits in humans and monkeys mediate the temporal attribute within the knowledge-based memory system. It should be noted that important anatomical and functional links exist between the hippocampus and prefrontal cortex which are likely to aid in the representation and processing of temporal information, but which also will provide for difficulties in dissociating the contributions of the hippocampus and prefrontal cortex in supporting temporal information. In the rat, the neural connections form a circuit starting with a direct hippocampal CA1/subiculum projection to the medial prefrontal cortex, especially the infralimbic/ prelimbic cortex (Swanson, 1981; Ferrino et al., 1987). Interestingly, one can elicit long-term potentiation (LTP) by stimulation of the CA1 hippocampal region and recording from the infralimbic cortex (Laroche et al., 1990). The medial prefrontal cortex is directly connected to

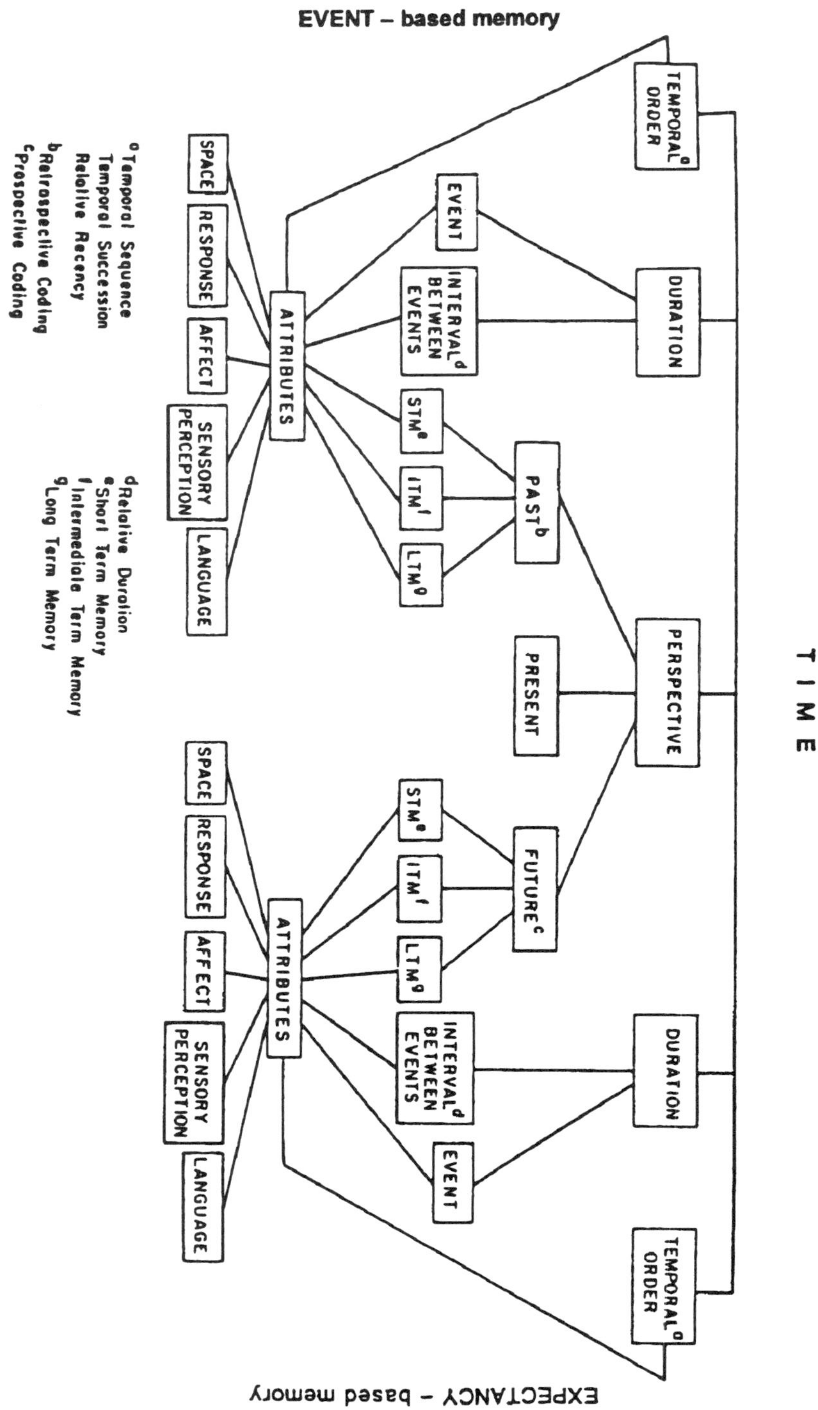
EVENT – based memory
EXPECTANCY – based memory
TIME
TEMPORAL ORDER[a]
DURATION
PERSPECTIVE
PAST[b]
PRESENT
FUTURE[c]
EVENT
INTERVAL BETWEEN EVENTS[d]
STM[e]
ITM[f]
LTM[g]
ATTRIBUTES
SPACE
RESPONSE
AFFECT
SENSORY PERCEPTION
LANGUAGE
[a]Temporal Sequence
Temporal Succession
Relative Recency
[b]Retrospective Coding
[c]Prospective Coding
[d]Relative Duration
[e]Short Term Memory
[f]Intermediate Term Memory
[g]Long Term Memory

Figure 1.

Figure 1 (adjacent page). A schematic of the organization of temporal attribute features within the event-based and knowledge-based memory systems.

the entorhinal cortex, which, in turn, is directly connected to the CA1 and subiculum and thereby completing the circuit (Witter et al., 1989). A similar circuit exists in the monkey in that the hippocampus connects to the subiculum, which, in turn projects directly to the dorsolateral prefrontal cortex. The dorsolateral prefrontal cortex projects to the presubiculum, entorhinal cortex and parahippocampal cortices, which, in turn project to the hippocampus (Goldman-Rakic et al., 1984; Uylings and vanEden, 1990; Witter, 1989).

The purpose of this review is to determine whether some empirical support exists for the above mentioned assumptions. The major questions to be asked are whether each neural system mediates all three of the temporal features associated with the temporal attribute, whether it is possible to dissociate the temporal attribute functions of the hippocampus and prefrontal cortex, and whether there are parallels between animal and humans in terms of the neural basis of memory for the temporal attribute.

Memory for Duration—Animals

In this section I will concentrate on memory for duration rather than the processing of time and time perception as measured by time estimation and time-scale invariance. For a more recent review see Gibbon et al., 1997. Previous research has indicated that fimbria-fornix, but not frontal cortex, lesioned rats are impaired in remembering the duration of a stimulus across a short delay interval, even though there is only a small change in estimating the passage of time (Olton, 1986; Olton et al., 1988; Meck, Church, Olton, 1984). In an attempt to replicate these results using a different paradigm, rats were trained on a short-term memory for duration task using a delayed symbolic conditional discrimination procedure (Jackson-Smith et al, 1994). It had previously been shown that rats acquire high proficiency in short-term memory for duration information (Santi et al., 1995). In the Jackson-Smith et al. (1994), experiment, the rats had to learn that a black rectangle stimulus that was visible for 2 sec would result in a positive (go) reinforcement for one object (a ball) and no reinforcement (no go) for a different object (a bottle). However, when the black rectangle stimulus was visible for 8 sec then there would be no reinforcement for the ball (no go), but a reinforcement for the bottle (go). After rats learned to respond differentially in terms of latency to approach the object, they received large (dorsal and ventral) lesions of the hippocampus, medial prefrontal cortex (anterior cingulate) lesions or lesions of the cortex dorsal to the dorsal hippocampus.

Following recovery from surgery they were retested. The results indicate that there were in contrast to cortical control and medial prefrontal cortex lesions, major impairments following hippocampal lesions as indicated by smaller and statistically nonsignificant latency differences between positive and negative trials on post-surgery tests. In order to ensure that the deficits observed with hippocampal lesions were not due to a discrimination problem, new rats were trained in an object (black rectangle) duration discrimination task. In this situation the rats were reinforced for either a 2 or 10 sec exposure (duration) of the black rectangle. The stimulus was presented and remained visible for either 2 or 10 sec, following which the door was raised and latency to move the stimulus was measured. Half of the animals in each group received a piece of Froot Loop on trials with a short stimulus duration and the other half was reinforced on those trials with a long stimulus duration. After rats learned to respond differentially in terms of latency to approach the object, they received large (dorsal and ventral) lesions of the hippocampus, as well as medial prefrontal cortex lesions for comparison purposes or lesions of cortex dorsal to the dorsal hippocampus. Following recovery from surgery the rats were retested. The results indicate that after hippocampal lesions, there was an initial deficit followed by complete recovery. There were no significant changes for the medial prefrontal or control lesioned animals.

Thus, the hippocampus mediates memory for duration, but does not mediate duration discrimination. The data are consistent with previous research which indicate that fimbria-fornix, but not frontal cortex, lesioned rats are impaired in remembering the duration of a stimulus across a short delay interval, even though there is only a small change in estimating the passage of time (Olton, 1986; Olton et al., 1988; Meck, Church, Olton, 1984). Furthermore, it has been suggested that trace conditioning requires memory for the duration of the conditioned stimulus. Thus, it is of importance to note that rabbits with hippocampal lesions are impaired in acquisition (consolidation) of trace but not delayed eye-blink conditioning (Moyer et al., 1990). In rats, it appears that the hippocampus is actively involved in representing in short-term memory temporal attribute information perhaps based on the use of markers for the beginning and end of the presence (duration) of a stimulus (object). It should be noted that the hippocampus is not directly involved in representing in memory information concerning specific objects (Kesner et al., 1993; Mumby et al., 1992). The medial prefrontal cortex does not appear to play a role in either memory for duration or duration discrimination. It is possible that other neural regions within the prefrontal cortex could mediate memory for duration, but the anterior cingulate does not appear to be critical.

Memory for Duration—Humans

To what extent can one generalize from hippocampal and anterior cingulate function in rats to humans with respect to memory representation of duration as one feature of temporal attribute information. Since there are no data in the literature that have attempted to answer this question, a number of experiments were conducted using humans that have been exposed to hypoxia due to a variety of causes, but primarily carbon monoxide poisoning (Hopkins and Kesner, 1994; Hopkins and Kesner, 1995). These subjects have anterograde amnesia and, based on MRI data, have bilateral damage to the hippocampus, but no detectable damage to entorhinal cortex, parahippocampal gyrus, or temporal cortex. They also show no signs of prefrontal cortex dysfunction based on normal performance on tests of fluency and Wisconsin Card sorting. In addition, patients with damage to the prefrontal cortex were used. The hypoxic and prefrontal cortex damaged subjects and age matched controls were tested for short-term memory for duration of a visual object. Subjects were presented with a single object (square, circle, etc) on a computer screen for a duration of 1 or 3 sec. They were instructed to remember the duration of presentation of the object. After a delay of 1, 4, 8, 12, 16, or 20 sec, the same object appeared for the same or different duration. The subjects were asked to indicate whether the duration was the same or different from the duration shown in the study phase. The results indicate that the hypoxic subjects were impaired relative to control and prefrontal cortex damaged subjects in short-term memory for duration for all but the shortest delay (Kesner and Hopkins, 2001). There were no deficits for the prefrontal cortex damaged subjects. In order to determine whether the deficits may have been due to impaired memory for the objects per se, a control task was administered to the same subjects. They were presented with a single object for a duration of 1 or 3 sec and were asked to remember the object. After a delay of 1, 4, 8, 12, 16, or 20 sec either the identical or a different object appeared on the screen. The subjects were asked if it was the same or a different object. The results indicate that there were no or minimal differences between the hypoxic, prefrontal cortex, and control subjects (Kesner and Hopkins, 2001). The impairment could not be due to an inability to estimate time accurately, because in an additional experiment with objects the subjects were asked to estimate the time elapsed before each of the 1, 4, 8, 12, 16, 20 sec delay intervals. The results indicate that for hypoxic subjects time estimates were accurate up to 8 sec followed by some underestimation with longer delays, so that short-term memory for the duration of 1 or 3 sec stimulus exposure could not be due to difficulty in estimating time (Kesner and Hopkins, 2001). The process of estimating time may not require active participation of short-term memory and may, therefore, appear to be independent of short-term memory for the duration of exposure of a stimulus. Similarly, for the prefrontal cortex damaged subjects time estimates were accurate up to 8 sec followed by a slight underestimation with longer delays.

The data are consistent with previous research which indicate that humans with hypoxia resulting in bilateral hippocampal damage are impaired in acquisition (consolidation) of trace but not delayed eye-blink conditioning (Disterhoft et al., 1996). Thus, the results suggest that like rodents, humans with hippocampal damage have difficulty in representing short-term memory for duration of an object, but not short-term memory for a single object. In contrast, both rodents and humans with prefrontal cortex damage do not appear to have any difficulty in short-term memory for duration of an object, suggesting a dissociation between hippocampal and prefrontal cortex representation of duration information.

Memory for Temporal Order—Animals

In order to test whether the hippocampus or medial prefrontal cortex subserve memory for temporal order, rats were tested for order memory for a list of items (places on a maze) (Kesner and Holbrook, 1987; Kesner and Novak, 1982). Rats were trained on an eight-arm radial maze for Froot Loop reinforcement. After training each animal was allowed to visit eight arms on each trial (one per day) in an order that was randomly selected for that trial (study phase). The sequencing of the eight arms was accomplished by sequentially opening Plexiglas doors (one at a time) located at the entrance of each arm. Immediately after the animal had received reinforcement from the last of the eight arms, the test phase began. During the test phase for order memory either the first and second, fourth and fifth, or seventh and eighth door that occurred in the sequence was opened. The rule to be learned leading to an additional reinforcement was to choose the arm that occurred earlier in the sequence. There were 24 tests with only one test per trial.

Following training, animals performed better than chance for the first (primacy) and last (recency) order position. The animals then received medial prefrontal cortex or dorsal hippocampal lesions. After recovery from surgery animals were given an additional twenty-four tests. Results indicate that medial prefrontal cortex lesioned animals had an order memory deficit for all spatial locations. In additional tests, it was shown that this order deficit appeared even when the animals were allowed to self-order the spatial locations during the study phase, were presented with a fixed sequence of four spatial locations or when the list length was only two spatial locations (Kesner and Holbrook, 1987). In contrast, dorsal hippocampal lesioned rats were impaired only for the first and middle choice orders, but they performed without difficulty on the last choice order,

i. e. they showed a primacy deficit, but an intact recency effect. The normal recency effect could have been due to small lesions of the hippocampus. Larger lesions (dorsal and ventral) of the hippocampus also impair the recency effect (unpublished observations). These data suggest that both neural regions play a role in memory for temporal order.

The possibility exists, however, that order memory was impaired in the Kesner and Holbrook (1987) and Kesner and Novak (1982) experiments, because only temporally adjacent items were selected. This is especially a problem because Estes (1986) has summarized data in humans demonstrating that order or sequential information is remembered better when more items (lag) occur between any two items to be tested. Thus, an experiment was designed to test order memory for items that were further apart temporally in the to-be-remembered sequence (Chiba et al., 1992; Chiba et al., 1994). In this experiment, rats are required to remember an event (e.g. spatial location, visual object) dependent upon the temporal order of occurrence of events. More specifically, on an eight arm maze during the study phase of each trial rats were allowed to visit each of eight arms once in an order that is randomly selected for that trial. The test phase required the rats to choose which of two arms occurred earlier in the sequence of arms visited during the study phase. The arms selected as test arms varied according to temporal lag or distance (0-6) or the number of arms that occurred between the two test arms in the study phase. After the rats reached a criterion of 75% or better performance on all the distances but zero, the rats received large (dorsal and ventral), small (dorsal) hippocampus, medial prefrontal cortex, or cortical control lesions dorsal to the dorsal hippocampus. Following recovery from surgery, the rats were retested. The results indicate that for post-surgery tests, the control rats performed at chance at a temporal distance of zero, but their performance was excellent for the remaining temporal distances. In contrast, on post-surgery tests rats with dorsal hippocampal lesions did not perform well for temporal distances of 0 and 2, but performance was excellent for the longest temporal distances of 4 and 6. Furthermore, on post-surgery tests rats with large (dorsal plus ventral) hippocampal and medial prefrontal cortex lesions had a marked deficit for all temporal distances with a slight improvement for the longest temporal distance. In this task it is necessary to separate one event from another, a process which has been labeled temporal pattern separation. Hippocampal or medial prefrontal cortex lesioned rats cannot separate events across time, perhaps because of an inability to inhibit interference that is likely to accompany sequential occurring events. The resultant increase in temporal interference impairs the rat's ability to remember the order of specific events. It appears that the larger the damage to the hippocampus, the greater the temporal interference.

In the previous experiment the selected sequence to be remembered varied on each trial emphasizing the importance of new information which is likely to be processed by the event-based memory system. An alternative procedure is based on the presentation of a single fixed or constant sequence to be remembered. After learning the temporal order of this sequence, there is high likelihood that the knowledge-based system rather than the event-based system would mediate memory for this constant order. Based on the hypothesis that the medial prefrontal cortex rather than the hippocampus mediates temporal order within the knowledge-based memory system, it should be possible to determine whether the medial prefrontal cortex plays a more important role than the hippocampus in the constant temporal distance task. To test this idea rats were presented with a constant sequence of eight spatial locations followed by temporal distance tests. The results indicate that relative to controls small and large hippocampal lesions do not result in any deficits, but medial prefrontal cortex lesions do produce an impairment (Chiba et al., 1994; Chiba et al., 1992). Thus, these data support the importance of medial prefrontal cortex for maintaining well-learned temporal sequences in long-term memory. It should be noted that the events do not have to be based on only spatial location information. Similar deficits have been observed with lists of visual objects or odors in rats (unpublished observations).

In general, these data are consistent with previous descriptions of deficits following medial prefrontal cortex or hippocampal lesions in tasks in which rats emit a specific sequence of behavioral responses requiring temporal organization (Barker, 1967; Slotnick, 1967; Stamm, 1955). In monkeys, dorsolateral prefrontal cortex lesions result in deficits in temporal ordering of events comparable to what has been described in rats. This temporal ordering deficit is evidenced by deficits in self-ordering of a sequence of responses (Petrides and Milner, 1982). In another experiment by Pribram and Tubbs (1967) monkeys with frontal cortex lesions could not perform a 5 sec right-left delayed alternation task (order deficit), but with a 15 sec delay between a right-left couplet of responses, no deficit was found. In this latter condition the animal could chunk, or categorize, each couplet as an item of information and would not need to remember the order of a right-left response. In a more recent study Funahashi et al. (1997) showed that neurons within the prefrontal cortex (peri-principal and superior arcuate areas) of primates participated in retaining the temporal order of stimulus presentation during the delay period prior to a test for temporal order memory. Based upon the above mentioned findings, it is suggested that the prefrontal cortex is primarily involved in temporal structuring of information in short-term memory (Fuster, 1980).

A somewhat different function based on temporal order relates to memory for frequency of occurrence of events. In order to test the role of the hippocampus and medial prefrontal cortex in memory for frequency, rats were tested on an eight-arm radial maze (Kesner, 1990b). On each trial (one per day) each animal was allowed to visit four arms to receive reinforcement in an order that was randomly selected for that trial. One of the arms was repeated with a lag of 1, 2, or 3 arms in between the repetition. The sequencing of the four arms and the repetition of one of the arms is accomplished by sequentially opening Plexiglas doors to each arm (one at a time). This constituted the *study phase*. Within 20 seconds after entry into the last (fifth) arm of the study phase, the door in front of the repeated arm and the door in front of a nonrepeated arm were opened simultaneously. This constituted the *test phase*. The rule to be learned leading to an additional Froot Loop reinforcement was to choose the arm that had been *repeated* in the study phase sequence.

After training and based on at least 8 observations per lag, animals showed excellent memory for the repetition with a lag of 3 arms between repetitions, but performed poorly for a lag of 1 or 2 arms. The effect is not due to the variable number of spatial locations that occur between the repeated spatial locations and the test, because analysis of only the repeated spatial locations that occur at the end or the beginning of the list revealed the same repetition lag affect. This repetition lag effect (i.e., better retention with more items between a repetition) has also been reported for humans (Madigan, 1969). The rats then received medial prefrontal cortex or hippocampus lesions or served as sham-operated controls. After recovery from surgery animals were given an additional 24 tests with 8 tests for each lag condition. Sham-operated animals showed no significant changes in performance. In contrast, animals with medial prefrontal cortex or hippocampus lesions displayed a deficit for all lag conditions. Thus, both lesions disrupted the repetition lag effect further supporting a role for the hippocampus and medial prefrontal cortex in mediating the temporal order feature of the temporal attribute.

Memory for Temporal Order—Humans

Previous research with frontal lobe damaged patients indicate that these patients cannot remember the order in which information was experienced, plan and create a complex set of motor movements, or program a temporally ordered set of activities. For example, it has been shown that frontal cortex damaged patients cannot discriminate the more from the less recent, even though they can remember that certain words or pictures have been presented (Milner, 1971; Lewinsohn, et al., 1972). In addition, frontal cortex-damaged patients are impaired in a short-term memory task in which two stimuli had to be remembered for a 60-s time interval (paired-comparison task) (Milner, 1964). This task requires short-term temporal memory for two events, implying that frontal cortex-damaged patients cannot remember the order of stimulus presentation. In another experiment, frontal cortex damaged patients were impaired in their ability to self-order a sequence of stimuli presented one at a time (Petrides and Milner, 1982). Additional evidence comes from the finding that frontal-lobe damaged patients are very poor at copying a series of facial movements. These patients have difficulty in ordering the various components of the sequence. The components are remembered correctly, but are expressed in an incorrect order (Kolb and Milner, 1981).

Previous research also indicates that amnesic patients also have difficulty in processing temporal order information (Huppert and Piercey, 1976; Hirst and Volpe, 1982; Squire et al, 1981). Only one study, Sagar et al. (1990) found no deficit for patient H.M. who had a bilateral temporal lobectomy. However, Sagar used a different testing procedure for assessing temporal order memory. Reanalyzing Sagar's data using a temporal distance measure as described below resulted in the same pattern of results as described above. Even though the tasks used in previous studies are not analogous to what has been used in rat studies, is it possible to generalize from hippocampal and medial prefrontal cortex (anterior cingulate) function in rats to humans with respect to memory representation of temporal succession as one feature of temporal attribute information? In order to answer this question, tasks were selected that were more analogous to previous research with rats, namely short-term order memory for a six item list of X's that appear in specific locations. Furthermore, to compare the right and left side in humans a comparable task using words was also used. In this study patients with right or left temporal lobe resection, hypoxics with damage to the hippocampus, bilateral, left, and right prefrontal cortex, epileptic controls or age-matched controls were tested for order memory. More specifically for the spatial task, during the study phase a series of random (novel) sequences of six Xs appeared on the screen, for a period of 5 sec each. Subjects were instructed to pay attention to the locations of the Xs as well as to the order in which they occurred. In the test phase subjects were presented with two Xs and were asked to determine which one occurred earlier in the study phase. For the verbal task during the study phase a series of random (novel) sequences of six words appeared on the screen, for a period of 5 sec each. Subjects were instructed to pay attention to the words as well as to the order in which they occurred. In the test phase subjects were presented with words and were asked to determine which one occurred earlier in the study phase. The procedures used for the order memory tests were first developed for the assessment of list learning for spatial location information in rats and were then adapted for use with brain damaged patients. Because in rats only one test was given for each serial list, the same procedure was adopted for humans. This procedure has the advantage of minimizing response interference. Because the set size is only 6 items, normal subjects perform well on both order memory tests. The results indicate that relative to age-matched controls for spatial location temporal order memory the right temporal lobe resected patients including hippocampus, right prefrontal cortex damaged patients and the hypoxics patients are markedly impaired. For word temporal order memory relative to age-matched controls the left temporal lobe resected patients including hippocampus, left and bilateral prefrontal cortex damaged patients and the hypoxics patients are markedly impaired. Similar results were obtained for temporal order for abstract pictures and motor responses. Also, for frontal cortex damaged subjects there were in contrast to order memory deficits no deficits for memory for items per se, but for hippocampal damaged subjects there were deficits not only for order but for item information as well (Kesner and Holbrook, 1987; Kesner et al., 1994; Kesner, 1990a; Hopkins et al., 1995a).

Based on the demonstration that in humans order or sequential information is remembered better when more items (lag) occur between any two items to be tested an additional experiment was designed to test order or distance memory for items that were further apart temporally in the to-be-remembered sequence. Patients with right temporal lobe resection, hypoxics with damage to the hippocampus, right prefrontal cortex damage or controls were presented with a list of eight spatial locations (Xs) on a grid on a Macintosh computer and tested for memory for temporal distances (Chiba et al.,submitted; Hopkins et al.,1995b).

In this task during the study phase a series of random (novel) sequences of eight Xs appeared on the screen, for a period of 5 sec each. Subjects were instructed to pay attention to the locations of the Xs as well as to the order in which they occurred. In the test phase subjects were presented with two Xs and were asked to determine which one occurred earlier in the study phase. Unlimited time was allowed for the subjects to make their choices. Temporal distances of 0, 2, 4, and 6 were assessed with eight observations for each distance for each type of task. Temporal distance is determined by the number of items in the study phase that occur between the two test items. For each study phase, four tests were given, one for each temporal distance. The results indicate that relative to controls, the hypoxics and right temporal lobe resected subjects showed an impairment for all temporal distances, but with some improvement for the longest distances, whereas the right prefrontal cortex damaged patients showed a deficit for all temporal distances. These results are similar to what was found in rats and suggest that patients with right or bilateral hippocampal dysfunction have difficulty in encoding or remembering temporal contexts for spatial information perhaps due to increased interference in the temporal resolution of specific spatial locations. In the previous experiment the selected sequence to be remembered varied on each trial emphasizing the importance of new information which is likely to be processed by the event-based memory system. An alternative procedure is based on the presentation of meaningful geometric patterns to emphasize the contribution of the knowledge-based memory system. Based on the hypothesis that the medial prefrontal cortex rather than the hippocampus mediates temporal order within the knowledge-based memory system, it should be possible to determine whether the medial prefrontal cortex plays a more important role than the hippocampus in a structured spatial temporal distance task. To test this idea further the same subjects were given sequences where the Xs appeared in a meaningful geometric pattern. An example of a meaningful and familiar pattern would be the presentation of the locations in the pattern of a large X. The first X would be presented in the top left corner of the grid and then in each subsequent square down the diagonal ending in the bottom right corner. The next X would be presented in the bottom left corner of the grid and in each subsequent square up the diagonal ending in the top right corner. The results indicate that relative to controls there were no deficits for the right temporal lobe resected and hypoxic patients, whereas there was a profound deficit for the right frontal cortex damaged subjects for all but the longest temporal distance.

In general, the above mentioned data support the idea that the hippocampus and prefrontal cortex in rats and humans mediate new temporal order information, but when the information is well known or a structured rule can be applied only the prefrontal cortex in rats and humans continues to mediate temporal order information. These data suggest that whenever knowledge-based information can be used to remember temporal order of specific sequences of events, there is a greater involvement of the prefrontal cortex in comparison with the hippocampus. It is also important to point to significant parallels in mediating temporal order information between animals and humans.

Since temporal organization is required for frequency memory (repetition), it is not surprising that patients with prefrontal cortex damage and to some extent temporal lobe resected patients with hippocampal damage are impaired in judgment for frequency of occurrence of abstract designs within a list (Milner, et al., 1985). In a more recent study Jurado et al. (1997) showed that frontal lobe damaged patients had a similar impairment in estimating the frequency of occurrence of words within a list, yet item recognition for those same words remained intact. It has been suggested that frequency judgements require a search through memory for temporal information (Hintzman, et al., 1981; Hintzman, et al., 1982; Milner et al., 1985), i.e., one needs to search one's memory for each specific instance of a specific item. In summary, it appears that the prefrontal cortex in humans and medial prefrontal cortex in the rat mediate memory for frequency, further supporting the importance of prefrontal cortex and hippocampus in mediating temporal order information.

Time Perspective—Memory for the Past and Future—Animals, Humans

It has been suggested that the hippocampus mediates memory for past events within a relatively short time frame (short-term, intermediate term or working memory) (Olton, 1983; Kesner et al., 1990a). Thus, as a consequence of damage to the hippocampus, one would expect severe difficulty in remembering relatively new spatio-temporal information that would, in turn, create a problem in using memory for the future based on memories necessary to plan or anticipate future events. In contrast, it has been suggested that the prefrontal cortex mediates memory for future events, given that memory for the past within either a short or long time frame is not based on temporal order.

In order to test these ideas in rats, sham-operated rats or animals with medial prefrontal cortex or hippocampus lesions were tested in a task that provided an opportunity for rats to utilize retrospective (past as time perspective) and prospective (future as time perspective) memory codes, while remembering items (spatial locations) within short or long lists (Kesner, 1989; Kametani and Kesner, 1989). More specifically, on any one trial a rat was presented with 2, 4, 6, 8, or 10 items (spatial locations) on a 12-arm radial maze followed 15 minutes later by two win-shift tests comprising a choice between a place previously visited and a novel place. Each animal was given a total of 20 trials with 8 tests for each point of interpolation or each list length (2, 4, 6, 8, or 10). During learning, rats show an increase in errors as the number of places to be remembered is increased from 2 to 6 to 8 reflecting the use of a retrospective memory code (the ability to remember the spatial locations previously visited). These rats also show a decrease in errors as the number of places is increased from 8 to 10 reflecting the use of a prospective memory code (the ability to anticipate the spatial locations that have not yet been visited). Results indicate that sham-operated animals display an increase in errors as a function of point of interpolation or set size (2 to 8 items) followed by a decrease in errors with a set size of 10 items suggesting the use of both retrospective and prospective memory codes. In contrast, animals with medial prefrontal cortex lesions made few errors for short list lengths, but a large number of errors for the longer list lengths, reflecting an inability to shift from a retrospective to a prospective memory code, whereas animals with hippocampus lesions made errors across all list length, reflecting a deficit in utilizing boh retrospective and prospective strategies. It is likely that the hippocampus lesioned rats were not able to use a prospective strategy, because they were not able to remember spatial information needed to initiate a prospective strategy.

Additional support for the idea that the hippocampus mediates the past comes from the findings that post-trial disruption of normal hippocampal function with, for example, electrical brain stimulation, results in time-dependent memory impairments (Kesner and Wilburn, 1974). These effects reveal that the hippocampus is involved in short-term consolidation processes, because the gradients are usually short within minutes to a few hours. In a more recent study (Kim et al., 1995), it was shown that trace eye-blink conditioning is impaired when hippocampal lesions are made 1 day, but not 30 days, after acquisition, suggesting that trace conditioning as a measure of duration information reveals the involvement of the hippocampus in representing the past via short-term consolidation processes. Long-term temporally graded functions have also been observed for previously learned spatial discriminations prior to surgery in rats and mice, but these long-term gradients (2-4 weeks) are observed primarily following entorhinal cortex rather than hippocampal lesions (Cho et al., 1993; Cho et al., 1995; Cho and Kesner, 1996). A long term gradient following hippocampal lesions has been reported following contextual fear conditioning (Kim and Fanselow, 1992), but recent results have questioned whether these gradients can be reliably measured (Maren et al., 1996; Weisend et al.,1996). Thus, it is possible that short-term consolidation gradients derive from hippocampal dysfunction, whereas entorhinal cortex dysfunction is necessary to produce long-term retrograde amnesia consolidation gradients. Very long-term memory losses often observed in patients with damage to the hippocampus and interconnected neural circuitry tend to result in flat retrograde amnesia gradients, suggesting a role for the hippocampus in retrieving very long term information. For a review see (Nadel and Moscovitch, 1997). Whether the hippocampus promotes the transfer of information to the knowledge-based system, promotes the consolidation of information already processed in the knowledge-based system or is intimately involved in retrieval of past information still needs to be resolved.

The results of the Kametani and Kesner (1989) study suggest that the medial prefrontal cortex mediates prospective memories in an item recognition task that utilizes spatio-temporal information, suggesting an important role for the medial prefrontal cortex in mediating a future time perspective within the knowledge- based memory system. Fuster (1985) has also suggested that dorsolateral prefrontal cortex of monkeys and humans mediates prospective functions of temporal ordering of information. Support for this possible prospective function comes from the finding that some cells in the dorsolateral frontal cortex of monkeys show a gradual increase in their firing rate during the delay in apparent anticipation of the test phase of a delayed matching-to-sample, delayed response or delayed alternation task (Fuster, et al., 1982; Niki, 1974a,b; Kojima, et al., 1981). Furthermore, Shallice (1982) and Owen et al. (1990) have shown that left frontal-lobe damaged patients have a difficult time solving the Tower of London problem. This problem requires temporal ordering of simple moves and thus requires planning and the utilization of prospective codes. Also, during learning of the Tower of London task and based on SPECT or PET scans an increased cerebral blood flow was measured in the left dorsolateral prefrontal cortex, but it should be noted that other cortical areas are also activated including parietal cortex and other frontal cortical areas (Morris et al., 1993; Baker et al., 1996). Finally, in another experiment it has been shown that frontal cortex damaged subjects are impaired in their ability to self-order a sequence of stimuli presented one at a time. It is assumed that to perform well in this task that the subject has to initiate a plan to follow and initiate the order of responding (Petrides and Milner, 1982). Finally, it should be noted that there is some support that the prefrontal cortex also plays a role in representing the past in that damage to especially the lateroventral prefrontal cortex produces a long-term retrograde amnesia gradient, which has been interpreted to reflect a loss or an impairment in the use of strategic retrieval processes (Markowitsch, 1995; Nadel and Moscovitch, 1997).

SUMMARY

Memory for the temporal attribute within the context of the neurobiology of attribute model is composed of at least three features including memory for duration, memory for succession or temporal order, and memory for past and future time perspective within a dual-based (event and knowledge) memory system. Research aimed at testing the assumption that the hippocampus and interconnected neural circuits mediate the temporal attribute within the even-based memory system and the prefrontal cortex and interconnected neural circuits mediate the temporal attribute within the knowledge-based memory system in animals and humans is reviewed. The research indicates that a) memory for the duration feature of the temporal attribute is mediated by the hippocampus, but not prefrontal cortex in both animals and humans, b) memory for the temporal order feature of the temporal attribute based on new information is subserved by both the hippocampus and prefrontal cortex, but based on prior knowledge or the ability to use prior knowledge is supported only by prefrontal cortex, but not the hippocampus, in both animals and humans, c) memory for the past (time perspective) feature of the temporal attribute is mediated by the hippocampus, whereas memory for the future (time perspective) feature of the temporal attribute is supported by the prefrontal cortex in both animals and humans. There is a clear parallel between animals and humans in terms of hippocampal and prefrontal cortex mediation of the temporal attribute supporting the assumption of evolutionary continuity. There is support for a greater involvement of the hippocampus in comparison with the prefrontal cortex in mediating temporal attribute information within the event-based memory system. Conversely, there is support for a greater involvement of the prefrontal cortex in comparison with the hippocampus in mediating temporal attribute information within the knowledge-based memory system. Future research needs to concentrate on the development of new paradigms to measure memory for different temporal features and to uncover the critical neural circuits that subserve these temporal features.

REFERENCES

Baker, S.C., Rogers, R.D., Owen, A.M., Frith, C.D., Dolan, R.J., Frackowiak, R.S.J., & Robbins, T.W. (1996). Neural systems engaged by planning: a PET study of the Tower of London task. *Neuropsychologia*, **34**, 515-526.

Barker, D.J. (1967). Alterations in sequential behavior of rats following ablation of midline limbic cortex. *Journal of Comparative and Physiological Psychology*, **3**, 453-604.

Block, R.A. (1990). Models of psychological time. In R.A. Block (Ed.), *Cognitive models of psychological time* (pp. 1-36). Hillsdale, NJ: Lawrence Erlbaum.

Chiba, A.A., Johnson, D.L., & Kesner, R.P. (1992). The effects of lesions of the dorsal hippocampus or the ventral hippocampus on performance of a spatial location order recognition task. *Society for Neuroscience Abstracts*, **18**, 1422.

Chiba, A.A., Kesner, R.P., Matsuo, F., & Heilbrun, M.P. (1993). A dissociation between affect and recognition following unilateral temporal lobectomy including the amygdala. *Society for Neuroscience Abstracts*, **19**, 792.

Chiba, A.A., Kesner, R.P., Matsuo, F., Heilbrun, M.P., & Plumb, S. (Submitted). A double dissociation between the right and left hippocampus in processing the temporal order of spatial and verbal information.

Chiba, A.A., Kesner, R.P., & Reynolds, A.M. (1994). Memory for spatial location as a function of temporal lag in rats: Role of hippocampus and medial prefrontal cortex. *Behavioral and Neural Biology*, **61**, 123-131.

Cho, Y.H., Beracochea, D., & Jaffard, R. (1993). Extended temporal gradient for retrograde and anterograde amnesia produced by ibotenate entorhinal cortex lesions in mice. *Journal of Neuroscience*, **13**, 1759-1766.

Cho, Y.H., & Kesner, R.P. (1996). Involvement of entorhinal cortex or parietal cortex in long-term spatial discrimination memory in rats: Retrograde amnesia. *Behavioral Neuroscience*, **110**, 436-442.

Cho, Y.H., Kesner, R.P., & Brodale, S. (1995). Retrograde and anterograde amnesia for spatial discrimination in rats: Role of hippocampus, entorhinal cortex and parietal cortex. *Psychobiology*, **23**, 185-194.

Cohen, N.J. & Eichenbaum, H.B. (1993). *Memory, Amnesia, and Hippocampal Function*. Cambridge, MA: The MIT Press.

DeRenzi, E. (1982). *Disorders of space exploration and cognition*. New York: Wiley.

DiMattia, B.V., & Kesner, R.P. (1988). Spatial cognitive maps: Differential role of parietal cortex and hippocampal formation. *Behavioral Neuroscience*, **102**, 471-480.

Disterhoft, J.F., Carrillo, M.C., Hopkins, R.O., Gabrieli, J.D.E., & Kesner, R.P. (1996). Impaired trace eyeblink conditioning in severe medial temporal lobe amnesics. *Society for Neuroscience Abstracts*, **22**, 1866.

Estes, W.K. (1986). Memory for temporal information. In J.A. Michon & J.L. Jackson (Eds.), *Time, mind and behavior* (pp.151-168). New York: Springer-Verlag.

Funahashi, S., Inoue, M., & Kubota, K. (1997). Delay-period activity in the primate prefrontal cortex encoding multiple spatial positions and their order of presentation. *Behavioural Brain Research*, **84**, 203-223.

Fuster, J.M. (1980). *The Prefrontal Cortex: Anatomy, Physiology, and Neuropsychology of the Frontal Lobe*. New York: Raven.

Fuster, J.M. (1985). The prefrontal cortex, mediator of cross-temporal contingencies. *Human Neurobiology*, **4**, 169-179.

Fuster, J.M. (1995). *Memory in the cerebral cortex: An empirical approach to neural networks in the human and nonhuman primate*. Cambridge, MA: The MIT Press.

Fuster, J.M., Bauer, R.H., & Jervey, J.P. (1982). Cellular discharge in the dorsolateral prefrontal cortex of the monkey in cognitive tasks. *Experimental Neurology*, **77**, 679-694.

Gibbon, J., Malapani, C., Dale, C.L., & Gallistel, C.R. (1997). Toward a neurobiology of temporal cognition: Advances and challenges. *Current Opinion in Neurobiology*, **7**, 170-184.

Heindel, W., Butters, N., & Salmon, D. (1988). Impaired learning of a motor skill in patients with Huntington's disease. *Behavioral Neuroscience*, **102**, 141-147.

Hintzman, D.L., Grandy, C.A., & Gold, E. (1981). Memory for frequency: A comparison of two multiple trace theories. *Journal of Experimental Psychology: Human Learning and Memory*, **7**, 231-240.

Hintzman, D.L., Nozawa, G., & Irmscher, M. (1982). Frequency as a nonpropositional attribute of memory. *Journal of Verbal Learning and Verbal Behavior*, **21**, 127-141.

Hirst, W., & Volpe, B. (1982). Temporal order judgments with amnesia. *Brain and Language*, **1**, 294-306.

Hopkins, R.O., & Kesner, R.P. (1994). Short-term memory for duration in hypoxic subjects. *Society for Neuroscience Abstracts*, **20**, 1075.

Hopkins, R.O., & Kesner, R.P. (1995). Item and order recognition memory in subjects with hypoxic brain injury. *Brain and Cognition*, **27**, 180-201.

Hopkins, R.O., Kesner, R.P., & Goldstein, M. (1995a). Item and order recognition memory for words, pictures, abstract pictures, spatial locations, and motor responses in subjects with hypoxic brain injury. *Brain and Cognition*, **27**, 180-201.

Hopkins, R.O., Kesner, R.P., & Goldstein, M. (1995b). Memory for novel and familiar spatial and linguistic temporal distance information in hypoxic subjects. *Journal of the International Neuropsychological Society*, **1**, 454-468.

Horel, J.A., Pytko-Joiner, D.E., Voytko, M.L., & Salsbury, K. (1987). The performance of visual tasks while segments of the inferotemporal cortex are suppressed by cold. *Behavioral Brain Research*, **23**, 29-42.

Huppert, F., & Piercey, M. (1976). Recognition memory in amnesic patients: Effects of temporal context and familiarity of material. *Cortex*, **4**, 3-28.

Jackson-Smith, P., Kesner, R.P., & Amann, K. (1994). Effects of hippocampal and medial prefrontal lesions on discrimination of duration in rats. *Society for Neuroscience Abstracts*, **20**, 1210.

Jurado, M.A., Junque, C., Pujol, J., Oliver, B., & Vendrell, P. (1997). Impaired estimation of word occurrence frequency in frontal lobe patients. *Neuropsychologia*, **35**, 635-641.

Kametani, H., & Kesner, R.P. (1989). Retrospective and prospective coding of information: Dissociation of parietal cortex and hippocampal formation. *Behavioral Neuroscience*, **103**, 84-89.

Kesner, R.P. (1989). Retrospective and prospective coding of information: Role of the medial prefrontal cortex. *Journal of Experimental Brain Research*, **74**, 163-167.

Kesner, R.P. (1990a). Learning and memory in rats with an emphasis on the role of the hippocampal formation. In Kesner, R.P. & Olton, D.S. (Eds.), *Neurobiology of comparative cognition* (pp. 179-204). Hillsdale, NJ: Erlbaum.

Kesner, R.P. (1990b). Memory for frequency in rats: Role of the hippocampus and medial prefrontal cortex. *Behavioral and Neural Biology*, **53**, 402-410.

Kesner, R. P. (1998). Neurobiological views of memory. In J. L. Martinez & R. P. Kesner (Eds.), *The Neurobiology of Learning and Memory*. Academic Press: San Diego, CA, 361-416.

Kesner, R.P., Bolland, B.L., & Dakis, M. (1993). Memory for spatial locations, motor responses, and objects: triple dissociation among the hippocampus, caudate nucleus, and extrastriate visual cortex. *Experimental Brain Research*, **93**, 462-470.

Kesner, R.P., & DiMattia, B.V. (1987). Neurobiology of an attribute model of memory. *Progress in Psychobiology and Physiological Psychology*. New York: Academic Press.

Kesner, R.P., & Holbrook, T. (1987). Dissociation of item and order spatial memory in rats following medial prefrontal cortex lesions. *Neuropsychologia*, **25**, 653-664.

Kesner, R.P., Hopkins, R.O., & Fineman, B. (1994). Item and order dissociation in humans with prefrontal cortex damage. *Neuropsychologia*, **32**, 881-891.

Kesner, R.P., & Novak, J. (1982). Serial position curve in rats: Role of the dorsal hippocampus. *Science*, **218**, 173-174.

Kesner, R.P., & Wilburn, M.W. (1974). A review of electrical stimulation of the brain in context of learning and retention. *Behavioral Biology*, **10**, 259-293.

Kesner, R.P., & Williams, J.M. (1995). Memory for magnitude of reinforcement: Dissociation between the amygdala and hippocampus. *Neurobiology of Learning and Memory*, **64**, 237-244.

Kesner, R. P., & Hopkins, R. O. (2001). Short-term memory for duration and distance in humans: Role of the hippocampus. *Neuropsychology*, **15**, 58-68.

Kim, J.J., Clark, R.E., & Thompson, R.F. (1995). Hippocampectomy impairs the memory of recently, but not remotely, acquired trace eyeblink conditioned responses. *Behavioral Neuroscience*, **109**, 195-203.

Kim, J.J., & Fanselow, M.S. (1992). Modality-specific retrograde amnesia of fear. *Science*, **256**, 675-677.

Kojima, S., Matsumura, M., & Kubota, K. (1981). Prefrontal neuron activity during delayed-response performance without imperative GO signals in the monkey. *Experimental Neurology*, **74**, 396-407.

Kolb, B. (1974). Social behavior of rats with chronic prefrontal lesions. *Physiological Psychology*, **87**, 466-474.

Kolb, B., & Milner, B. (1981). Performance of complex arm and facial movements after focal brain lesions. *Neuropsychologia*, **19**, 491-504.

Leonard, G., & Milner, B. (1991). Contribution of the right frontal lobe to the encoding and recall of kinesthetic distance information. *Neuropsychologia*, **29**, 47-58.

Lewinsohn, P.M., Zieler, J.L., Libet, J., Eyeberg, S., & Nielson, G. (1972). Short-term memory: A comparison between frontal and nonfrontal right- and left-hemisphere brain-damaged patients. *Journal of Comparative and Physiological Psychology*, **81**, 248-255.

Madigan, S.A. (1969). Intraserial repetition and coding processes in free recall. *Journal of Verbal Learning and Verbal Behavior*, **8**, 828-835.

Maren, S., Aharonov, G., & Fanselow, M.S. (1996). Excitotoxic dorsal hippocampus lesions and Pavlovian fear conditioning in rats. *Society for Neuroscience Abstracts*, **22**, 1379.

Markowitsch, H.J. (1995). Which brain regions are critically involved in the retrieval of old episodic memory? *Brain Research Reviews*, **21**, 117-127.

McDonald, R.J., & White, N.M. (1993). A triple dissociation of systems: hippocampus, amygdala, and dorsal striatum. *Behavioral Neuroscience*, **107**, 3-22.

Meck, W.H., Church, R.M., & Olton, D.S. (1984). Hippocampus, time and memory. *Behavioral Neuroscience*, **98**, 3-22.

Milner, B. (1964). Some effects of frontal lobectomy in man. In J.M. Warren, & K. Akert (Eds.), *The Frontal Granular Cortex and Behavior* (pp. 313-334). New York: McGraw-Hill.

Milner, B. (1971). Interhemispheric differences in the localization of psychological processes in man. *British Medical Bulletin*, **27**, 272-277.

Milner, B., Petrides, M., & Smith, M.L. (1985). Frontal lobes and the temporal organization of memory. *Human Neurobiology*, **4**, 137-142.

Morris, R.G., Ahmed, S., Syed, G.M., & Toone, B.K. (1993). Neural correlates of planning ability: Frontal lobe activation during the Tower of London test. *Neurophsychologia*, **31**, 1367-1378.

Moyer, J.R. Jr., Deyo, R.A., & Disterhoft, J.F. (1990). Hippocampectomy disrupts trace eye-blink conditioning in rabbits. *Behavioral Neuroscience*, **104**, 243-252.

Mumby, D.G., Wood, E.R., & Pinel, J.P.J. (1992). Object recognition memory is only mildly impaired in rats with lesions of the hippocampus and amygdala. *Psychobiology*, **20**, 18-27.

Nadel, L. & Moscovitch, M. (1997). Memory consolidation, retrograde amnesia and the hippocampal complex. *Current Opinion in Neurobiology*, **7**, 217-227.

Niki, H. (1974a). Prefrontal unit activity during delayed alternation in the monkey. I. Relation to direction of response. *Brain Research*, **68**, 185-196.

Niki, H. (1974b). Prefrontal unit activity during delayed alternation in the monkey. II. Relation to absolute versus relative direction of response. *Brain Research*, **68**, 185-196.

O'Keefe, J., & Nadel, L. (1978). *The hippocampus as a cognitive map*. Oxford: Clarendon Press.

Olton, D.S. (1983). Memory functions and the hippocampus. In W. Seifert (Ed.), *Neurobiology of the Hippocampus*. New York: Academic Press.

Olton, D.S. (1986). Hippocampal function and memory for temporal context. In R.L. Isaacson & K.H. Pribram (Eds.), *The Hippocampus*, Vol.**3**. New York : Plenum Press.

Olton, D.S., Wenk, G.L., Church, R.M., & Meck, W.H. (1988). Attention and the frontal cortex as examined by simultaneous temporal processing. *Neuropsychologia*, **26**, 307-318.

Otto T., & Eichenbaum, H. (1992). Complementary roles of the orbital prefrontal cortex and the perirhinal-entorhinal cortices in an odor-guided delayed-nonmatching-to-sample task. *Behavioral Neuroscience*, **106**, 762-775.

Owen, A.M., Downes, J.J., Sahakian, B.J., Polkey, C.E., & Robbins, T.W. (1990). Planning and spatial working memory following frontal lobe lesions in man. *Neuropsychologia*, **28**, 1021-1034.

Passingham, R. (1978). Information about movements in monkeys (Macaca mulatta) with lesions of dorsal prefrontal cortex. *Brain Research*, **152**, 313-328.

Petrides, M., & Milner, B. (1982). Deficits on subject-ordered task after frontal- and temporal-lobe lesions in man. *Neuropsychologia*, **20**, 249-262.

Pribram, K.H., & Tubbs, W.E. (1967). Short-term memory, parsing, and the primate frontal cortex. *Science*, **156**, 1765-1767.

Sagar, H.J., Gabrieli, J.D.E., Sullivan, E.V., & Corkin, S. (1990). Recency and frequency discrimination in the amnesic patient H.M. *Brain*, **113**, 581-602.

Santi, A., Weise, L., & Kuiper, D. (1995). Memory for event duration in rats. *Learning and Motivation*, **26**, 83-100.

Schacter, D.L. (1987). Implicit memory: History and current status. *Journal of Experimental Psychology: Learning, Memory, and Cognition*, **13**, 501-518.

Schacter, D.L., & Tulving, E. (1994). *Memory Systems 1994*. Cambridge, MA: The MIT Press.

Shallice, T. (1982). Specific impairments of planning. *Philosophical Transactions of the Royal Society of London B*, **298**, 199-209.

Slotnick, B.M. (1967). Disturbances of maternal behavior in the rat following lesions of the cingulate cortex. *Behavior*, **29**, 204-236.

Squire, L.R. (1994). Declarative and nondeclarative memory: Multiple brain systems supporting learning and memory. In D.L. Schacter & E. Tulving (Eds.), *Memory Systems 1994* (pp. 203-231). Cambridge, MA: The MIT Press.

Squire, L., Nadel, L., & Slater, P. (1981). Anterograde amnesia and memory for temporal order. *Neuropsychologia*, **19**, 141-146.

Stamm, J.S. (1955). The function of the median cerebral cortex in maternal behavior of rats. *Journal of Comparative and Physiological Psychology*, **48**, 347-356.

Tulving, E. (1983). *Elements of Episodic Memory*. Oxford: Clarendon Press.

Weisend, M.P., Astur, R.S., & Sutherland, R.J. (1996). The specificity and temporal characteristics of retrograde amnesia after hippocampal lesions. *Society for Neuroscience Abstracts*, **22**, 1118.

Part 4: Cognitive Structure

10

CONSERVATION OF A HIPPOCAMPAL ROLE IN REPRESENTATIONAL FLEXIBILITY

Michael Bunsey
Kent State University

A critical distinction in the study of human memory is that between declarative and nondeclarative memory. This distinction is biologically privileged, in that the hippocampal system appears to subserve declarative but not nondeclarative memory. Thus, humans with hippocampal damage are impaired in their ability to recollect and verbalize information about events in their lives, but demonstrate intact memory in many cases in which conscious recollection is not required, for example showing normal levels of priming, i.e. the facilitated processing of previously experienced material (Cohen, 1984; Squire, 1992).

It is a matter of considerable controversy whether an analogous distinction obtains in nonhumans (Jarrard, 1993; Cohen & Eichenbaum, 1993; Eichenbaum, Otto & Cohen, 1992). At first glance, it would seem unlikely: The emphasis on verbal processing associated with the concept of declarative memory appears to preclude its application to nonhumans. It is perhaps not surprising, then, that many of the proposed distinctions between hippocampal-dependent and hippocampal-independent memory in the animal literature have differed dramatically from that in the human literature. For example, one of the most powerful current conceptions holds that the critical dichotomy in animals is between spatial and nonspatial memory, with the hippocampus being dedicated to spatial memory processing (O'Keefe & Nadel, 1978; Jarrard, 1993). This hypothesis derives from the observation that animals with hippocampal lesions are generally impaired in mazes and other spatial tasks, while only rarely demonstrating impairments on nonspatial tasks. Other theories, based on different pools of data, have outlined different distinctions, e.g. between working and reference memory (Olton, Becker & Handlemann, 1979), contextual encoding versus learning along the performance line (Hirsch, 1974), configural learning versus simple association (Sutherland and Rudy, 1989), and representational versus dispositional memory (Thomas and Gash, 1986). In none of these cases is there an immediately apparent connection with the declarative-nondeclarative framework of the human literature.

A recently advanced alternative view is that a distinction analogous to the declarative-nondeclarative one does, in fact, hold in nonhumans (Cohen & Eichenbaum, 1993; Eichenbaum, Otto, & Cohen, 1992). According to this view, the conservation of hippocampal function across species has been obscured by the emphasis in the human literature on the verbal aspects of declarative memory. Cohen and Eichenbaum (1993) have argued that declarative memory must first be redefined in terms that are applicable across species before we can hope to assess whether hippocampal function has been conserved. In an attempt to do so, they posit that the fundamental features of declarative or "relational" memory -- features which can be assessed across species -- consist of the formation of associations among independent items or events that can be accessed flexibly to guide memory expression in various and even novel situations. In humans the ability to flexibly access memory associations in novel conditions would include the ability to access memories with language systems. Thus, the ability to verbally recollect and declare a memory would, by this view, be merely one reflection of a more fundamental process.

This conceptualization of declarative memory, although derived from work on human amnesia, can be applied across species. Importantly, it can also be applied across the disparate data pools which have led to the various formulations of hippocampal-dependent memory enumerated above. This is in contrast to many of these formulations which have focused on limited domains of data. Redefining declarative memory thus offers the possibility of an account of hippocampal function which is comprehensive, relevant both across species and across research approaches.

To illustrate using the most prominent example, we can consider how this generalized framework accounts for the prevalence of spatial memory deficits in hippocampal rats. Eichenbaum and colleagues hold that this imbalance reflects the fact that spatial tasks represent the only animal memory tasks which consistently load on relational processing. Spatial tasks usually demand the formation of associations among distal visual cues into maps which can be used flexibly to guide navigation. In the Morris water maze, for example, rats are trained to find a hidden goal from varying start positions (Morris, 1981). Remembering goal location requires memory for the relationships among multiple distal visual cues, while the varied start positions force subjects to infer goal location using novel views of the controlling cues. These processes epitomize declarative or relational memory. The nonspatial tasks used with rats, in contrast, do not require declarative processing, and generally involve only the gradual development of biases concerning individual stimuli (Eichenbaum, Otto & Cohen, 1992). In a simple visual discrimination, for example, learning is indexed by a gradual increase in responding to an individual cue over the course of trials. It is precisely this type of implicit or "nonrelational" memory which is consistently spared in human amnesics (Cohen, 1984; Schacter, Chiu, & Ochsner, 1993). According to the framework followed here, then, the notion that the hippocampus is dedicated to spatial processing is a consequence of a confounding between the extent to which animal memory tasks are spatial and the extent to which they require relational processing. In our view, spatial memory is just one example of declarative memory, albeit a very good one.

To address these issues, our recent research has worked towards two aims: 1. To develop tasks which assess the nonspatial declarative memory capacities of rats, and 2. To use these tasks to assess the hippocampal-dependence of these declarative processes. This approach is designed both to test the competing spatial and declarative memory theories and, more fundamentally, to better characterize the critical features of hippocampal-dependent memory. Our results have indicated that rats are capable of declarative-type memory, and have supported the notion that the hippocampus subserves this memory in rats as in humans. The data thus imply conservation of hippocampal function.

Along the way, this work has also led to somewhat surprising dissociations between the roles of the different components of the hippocampal system in declarative memory formation. These dissociations, in combination with similar findings from other labs, have been providing new insight into how hippocampal components interact both with each other and with the cortex during the composition of declarative memory.

PAIRED ASSOCIATE LEARNING AND THE HIPPOCAMPAL SYSTEM

Our first step was to develop a paired associate task for use in rats. The verbal paired associate task is a classic test of human memory and has been a valuable tool in the study of the cognitive aspects of human associative learning and in the study of amnesia (Winocur & Weiskrantz, 1976). In this task, subjects are presented with a list of word pairs, such as dog-table, and are then given one of the words of each pair and are required to recall its mate. Paired associate learning requires the rapid formation of associations among multiple, independent sensory cues, hallmarks of declarative memory. It is perhaps not surprising, then, that this task is hippocampal-dependent; humans with hippocampal amnesia are severely impaired on paired associate learning (Winocur & Weiskrantz, 1976; Zola-Morgan, Squire, & Amaral, 1986). We reasoned that the development of an analogous version of this nonspatial task for use in rats could aid progress in the study of the mnemonic role of the rodent hippocampus.

We designed a paired associate task for rats using odors as cues and a recognition format. Rats were trained in an automated apparatus to poke their noses into a conical odor port at the beginning of a trial. Once a rat initiated a nose-poke, a pair of odors was presented through the odor port. The two odors were briefly emitted in succession with a short "blank" interval between them. We arbitrarily designated four pairs of odors (A-B, C-D, E-F, G-H) as paired associates. When these pairings were presented, a subject could obtain a sweetened water reward by responding to a water port located directly above the odor port. If any other combination of these eight odors was presented (e.g. A-D, C-H) no reward was available, and rats were required to inhibit water port responses. It should be noted that these "mispairings" were comprised of the same odors as the paired associates, only in different combinations. Thus, the only way for an animal to distinguish between paired associates and mispairings was to form an association between the members of a pair. Learning about individual odors did not aid performance, as any single odor was either rewarded or unrewarded depending on the odor it was presented with on a given trial. In contrast, a third type of trial, nonrelational sequences, could be learned on the basis of individual odors. Nonrelational sequences were comprised of one of the eight odors from a paired associate in combination with another odor from a separate, never rewarded pool (W, X, Y, Z). Learning to distinguish paired associates from nonrelational sequences did not require the formation of odor-odor associations, but rather could be solved solely on the basis of the presence of an odor from the never rewarded pool.

In the following studies, rats were trained with each of the three trial types intermixed, for sessions of 120 trials, until they reached a learning criterion. Our pilot work indicated that normal rats learn to distinguish paired associates from the two types of foil fairly rapidly. In our first experiment, we used this task to test rats with relatively large lesions of the hippocampal system (Bunsey & Eichenbaum, 1993a). Specifically, we removed perirhinal and entorhinal cortices, i.e., PRER, through aspiration. This surgery ablates the parahippocampal regions which funnel information into and out of the hippocampus, thereby functionally isolating the

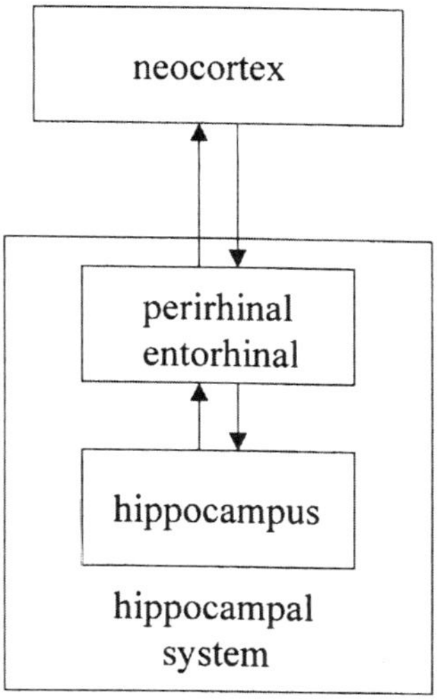

Figure 1. Wiring diagram illustrating schematic connections within the hippocampal system.

hippocampus itself (Figure 1). Because perirhinal and entorhinal cortices feed back into other cortical areas, this surgery also directly removes any functions of this parahippocampal region which are independent of the hippocampus. We expected this surgery to cause severe impairments in paired associate learning, given that it removes areas comparable to those ablated in the famous amnesic patient H.M. (Scoville & Milner, 1957).

Following recovery from surgery, PRER rats and sham-operated controls were trained on the paired associate task. Both groups learned to distinguish paired associates from nonrelational sequences at rapid and equivalent rates. In addition, controls gradually learned to distinguish paired associates from mispairings. PRER rats, in contrast, failed to learn this latter distinction. Indeed, most of these subjects had not reached criterion at the termination of testing, having received nearly twice as many trials as controls (Figure 2). Despite the similarities between the nonrelational and mispair trials -- e.g. both involved attending as two odors were presented and both required inhibition of responses -- PRER rats were only impaired on the latter trial type, in which they were required to form an association between two odors. The finding that PRER rats were selectively impaired when required to associate independent nonspatial sensory stimuli supports the hypothesized role of the hippocampal system in declarative-type or relational memory.

Simultaneous with this work, other labs have demonstrated similar results. Murray, Gaffan, & Mishkin (1993) found that lesions of parahippocampal regions in monkeys impaired visual-visual association learning, while Sutherland & Rudy (1989) and Davidson, McKernan &

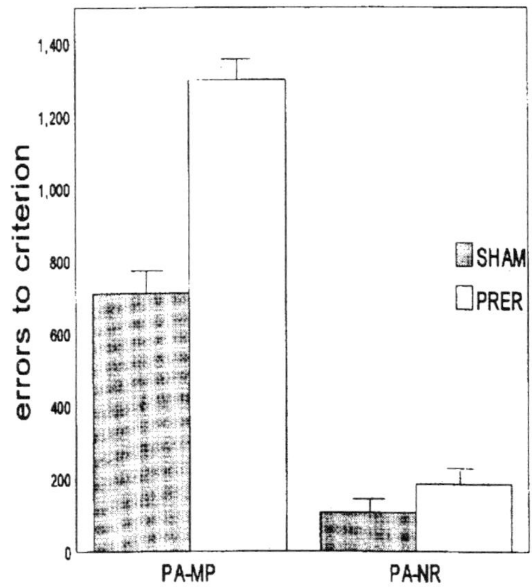

Figure 2. Errors to criterion for rats with perirhinal/entorhinal lesions (PRER) and sham controls on the discriminations between paired associates and mispairs (PA-MP) and between paired associates and nonrelational sequences (PA-NR).

Jarrard (1993) demonstrated that large lesions of hippocampal system in rats caused deficits in configural learning, a form of learning functionally similar to paired associate learning. In a configural learning task two cues may predict reinforcement when presented together but predict no reinforcement when presented separately. In combination with the human paired associate data, these results provide converging evidence across species that the hippocampal system is necessary for the formation of associations among independent sensory stimuli.

PAIRED ASSOCIATE LEARNING AND THE HIPPOCAMPUS ITSELF

The above results indicated that large lesions of the hippocampal system which ablate PRER and isolate the hippocampus impair relational memory. In the next study we examined the role of the hippocampus itself, i.e., dentate gyrus and CA1-CA3 (Bunsey and Eichenbaum, 1993b). Selectively removing the hippocampus is now possible due to Jarrard's (1989) development of a neurotoxic technique to ablate hippocampal cell bodies while sparing neighboring parahippocampal areas and fibers of passage. Using this technique, Jarrard (1993) has shown that many of the behavioral effects consequent to traditional hippocampal lesions are not due to specific

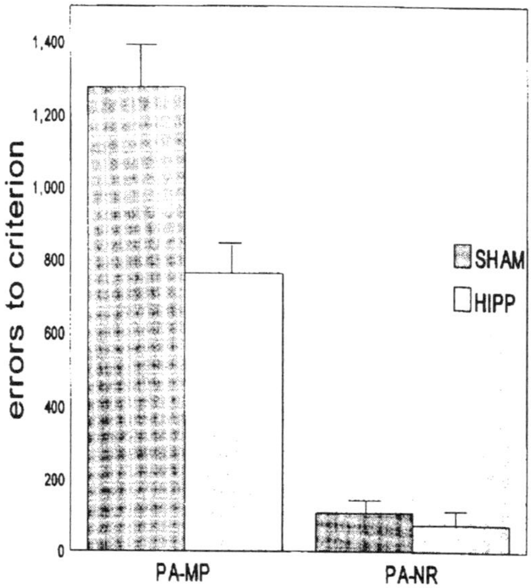

Figure 3. Errors to criterion for rats with ibotenate lesions of hippocampus (HIPP) and sham controls on the discriminations between paired associates and mispairs (PA-MP) and between paired associates and nonrelational sequences (PA-NR).

hippocampal damage, but may instead be attributable to damage to other parts of the hippocampal system, such as hippocampal cortex. It thus seemed important to test animals with these selective ibotenic acid-induced lesions using the protocol described above.

Our results indicated that hippocampal rats were able to learn to distinguish paired associates from nonrelational sequences at a normal rate. However, in marked contrast to the PRER rats, hippocampal subjects were facilitated in learning the paired associate-mispair distinction (Figure 3). This unexpected result suggested that the hippocampus itself and the parahippocampal region serve different and, in fact, complementary functions in paired associate learning.

Although these data were surprising, there is now converging evidence from several labs using several different preparations, demonstrating comparable results. Both hippocampal rats and hippocampal monkeys have been shown capable of normal paired associate acquisition (Cho & Kesner, 1993; Saunders and Weiskrantz, 1989; Gallagher & Holland, 1992). More strikingly, at least two other studies have demonstrated a dissociation across lesion types comparable to that seen in our data. Thus, although configural learning in rats was impaired following large hippocampal system lesions, it was intact following ibotenic acid lesions restricted to the hippocampus (Davidson et al., 1993). Likewise, although monkeys with parahippocampal lesions were impaired in the acquisition of visual paired associates, monkeys with more selective lesions of hippocampus were not (Murray et al., 1993).

Of course, all of these results seem at odds with the data from human amnesics. We originally chose a paired associate task in part because its acquisition is sensitive to hippocampal damage -- even specific CA1 damage -- in humans (Zola-Morgan et al., 1986). The normal or facilitated acquisition of paired associates by hippocampal animals implies important differences in hippocampal function across species. Recent work with amnesics, however, suggests reconciliation. Specifically, although they are consistently impaired on the standard paired associate task, amnesics learn pairs at normal rates if an implicit measure of memory is taken. For example, Musen and Squire (1993a) demonstrated that when word pairs were repeatedly presented to subjects, amnesics developed gradual changes in reading speed of previously seen pairs, and this facilitation occurred at normal rates. These test conditions are in fact more analogous to those used with animal subjects than standard verbal paired associate conditions. In its standard form, paired associate learning requires not only the formation of arbitrary associations but also the expression of these associations under novel conditions. That is, subjects learn word pairs by reading the pairs, but at testing are presented only one word of each pair and must retrieve its mate. This difference at the time of study and the time of test necessitates a flexible representation of the pairings to guide appropriate performance. Such representational flexibility -- one of the key components of declarative memory -- is not required in the versions of paired associates used with animals nor in the human implicit paired associate tasks.

On the basis of these data, we have speculated that paired associate learning may be achieved in either of two ways, each subserved by a distinct hippocampal component (Eichenbaum & Bunsey, 1995). We suggest that the parahippocampal area is sufficient to associate independent sensory stimuli by acting to "fuse" together items which co-occur, while the hippocampus itself encodes relationships among items which maintain the compositionality, or independent identity, of the items. It is this latter type of association which makes representational flexibility possible (It is noteworthy that William James (1890) proposed a similar functional subdivision between fusion and association over 100 years ago.)

According to this framework, animals with PRER lesions are unable to learn paired associate tasks because both of these forms of stimulus-stimulus association are disrupted. Animals with specific hippocampal lesions, on the other hand, can learn paired associates via stimulus-stimulus fusion, mediated by the intact parahippocampal region. This fusion is sufficient to guide accurate performance under the standard animal testing conditions, though it results in an abnormal representation. That is, in the absence of the hippocampus, compositionality of the elements is lost, resulting in what is called in the human literature a unitized structure (Schacter, 1985). It is interesting to note that Gluck & Myers (1995) have recently developed computational models of these hippocampal areas which seem to react to "lesions" in a similar fashion.

REPRESENTATIONAL FLEXIBILITY AND THE HIPPOCAMPUS

Our interpretation of the above results is that although hippocampal rats are able to learn paired associates at rapid rates, they do so in a manner qualitatively different than controls, resulting in an abnormal fused, inflexible representation of the pairings. It follows that hippocampal rats, despite rapid acquisition rate, should be impaired under conditions in which they have to use their associations flexibly or when the pairings have to be "de-composed" into their elements.

To test this prediction we developed a novel task, using somewhat unusual testing conditions in which paired associate learning would be particularly rapid and robust (Bunsey & Eichenbaum, 1996). Specifically, rats were trained to dig through cups filled with a sand substrate for highly palatable sweetened cereal rewards. Discriminations can be tested in this paradigm by scenting these cups with different spices and associating reward with only certain scents. Rats rapidly develop digging behavior under these conditions and show remarkable rates of discrimination learning. In this particular set of experiments we trained rats on two overlapping paired associate problems. Schematically, rats were given a rewarded sample cup scented with either odor A or odor X, followed immediately by a pair of choice cups scented with odors B and Y. Subjects had to learn to pair A with B by digging in cup B on those choices following sample A. Only cup B was baited on these trials. Likewise, subjects had to learn to dig in cup Y to receive reward on those trials in which X was the sample. Following learning of this paired associate task, subjects were trained on another paired associate problem in which odors B and Y were now the sample odors, and two novel odors, C and Z, were the choice odors. Subjects were required to learn the pairings of B-C and Y-Z. The experiment was thus designed such that subjects were exposed to two groupings of odors, A-B-C and X-Y-Z.

Acquisition of these paired associates does not require a flexible memory representation. Learning can presumably occur through a gradual fusion of pair members in response to their repeated presentation, as in the previous experiments. Learning of the pairs was thus not expected to critically depend on the hippocampus. Representational flexibility was only

Figure 4 (adjacent page). Performance on the three task components. **A.** Errors to criterion for both sets of paired associates **B.** Preference index for the unrewarded transitivity trials. This index was calculated as (X-Y)/(X+Y) where X and Y were the digging times in the transitive and alternate choices, respectively. Positive scores reflect a preference for the transitive choice **C.** Preference index for symmetry trials. This index was calculated as the transitivity measure but using first choice rather than digging time as dependent measure.

required in the next two phases of the experiment, in which transitivity and symmetry were measured. Transitivity requires an inference among items which have never been experienced together. In the present case, transitivity was assessed by asking whether rats would pair odor A with C and odor X with Z on the basis of their linked training history (i.e. A-B, B-C and X-Y, Y-Z). Transitivity requires subjects to form an integrated representation of overlapping stimulus associations and to spontaneously access this organization when challenged with items not paired before. As such, transitivity requires representational flexibility. Using different paradigms, Davis (1992) and Roberts & Phelps (1994) have both demonstrated that normal rats are capable of a similar form of transitivity.

Symmetry requires a subject to express a bidirectional association among items which have only been trained in a unidirectional fashion (Schusterman & Gisiner, 1988; Devany, Hayes & Nelson, 1986). In the present case, subjects were originally trained to choose element B following sample A and element Y following sample X: The subsequent symmetry test probed whether subjects would spontaneously choose A following sample B and X following sample Y. An animal which formed a declarative-type representation during original training would hypothetically do so. Like transitivity, symmetry requires a subject to spontaneously access an organized representation of associations to answer a novel challenge, in this case a reversed presentation. Symmetry has not previously been demonstrated in rats.

Rats with selective lesions of the hippocampus and controls were tested using this protocol. Normal animals learned paired associates at a rapid rate and showed significant levels of both transitivity and symmetry. In displaying this representational flexibility, then, these rats demonstrated nonspatial declarative-type memory (Cohen, 1984; Cohen & Eichenbaum, 1993). Animals with hippocampal damage learned the paired associates at the same rate as controls, as expected (Figure 4a). However, in contrast to controls, hippocampals showed no evidence of transitivity (Figure 4b) or symmetry (Figure 4c), performing at chance levels on both of these measures.

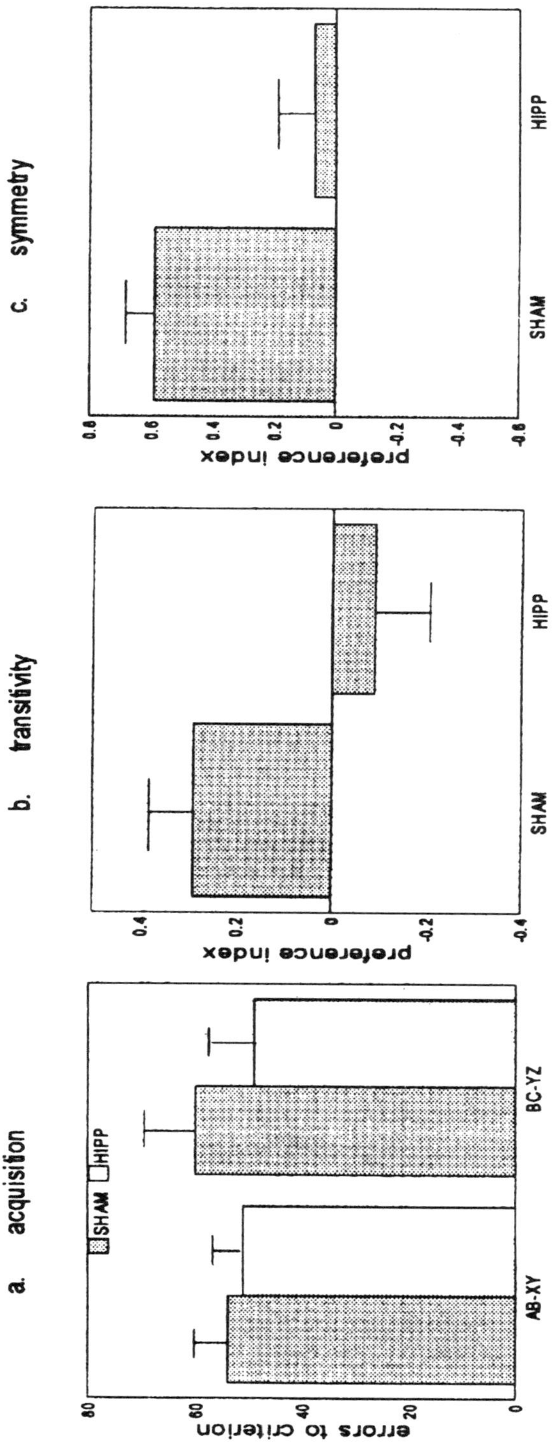
a. acquisition
SHAM
HIPP
errors to criterion
80
60
40
20
0
AB-XY
BC-YZ
b. transitivity
preference index
0.4
0.2
0
-0.2
-0.4
SHAM
HIPP
c. symmetry
preference index
0.8
0.6
0.4
0.2
0
-0.2
-0.4
-0.6
SHAM
HIPP

Figure 4.

These data provide striking evidence that a fundamental deficit in hippocampal rats is in the ability to create memory representations which can be used flexibly. This deficit is present regardless of whether or not the relevant information is spatial and is seen despite intact stimulus-stimulus association. Hippocampals learn odor paired associates at normal or above normal rates, but fail when asked to use these pairings in novel situations, either when challenged to make inferences across pairs or when required to decompose pairs and recognize them during reversed presentation. This is analogous to the situation with human subjects, in which amnesics can implicitly learn paired associates if pairs are repeatedly presented but fail if presented with a different situation at the time of testing (i.e. if given only one of the two words of a pair). These results are also compatible with earlier findings by Saunders & Weiskrantz (1989). They found that monkeys with hippocampal lesions were able to learn an object paired associate task, but were impaired on probe trials in which the spatio-temporal ordering of items was changed. Together, these data support the notion that hippocampal function has been conserved across species.

It should be noted this emphasis on representational flexibility is also compatible with performance of hippocampals in spatial memory tasks. Eichenbaum, Stewart & Morris (1990) demonstrated that rats with fornix lesions could learn to navigate to a goal in the Morris water maze if the same start position was used on every trial, but were severely impaired relative to controls if they were then started from a new position. As noted earlier, the ability to find the goal from a new start position requires a subject to use a map flexibly. These data indicate that even within the spatial domain, a deficit may only be seen to the extent that representational flexibility is required.

HIPPOCAMPAL ROLE IN ACQUISITION OF SOCIAL MEMORY

One troubling aspect of these data is the consistently rapid learning rate demonstrated by hippocampal subjects. One explanation for the rapid learning of hippocampals is that our particular testing procedures may favor them. Specifically, these relatively unnatural learning conditions which involve arbitrary associations among completely novel odors may bias subjects to fuse items. We have suggested an analogy (Eichenbaum & Bunsey, 1995) in which English-speaking humans are required to learn associations between unfamiliar German words, e.g. heer and tisch; in this situation subjects might be prone to fuse the words into a compound, heertisch, in order to remember. This bias would not be expected under conditions in which long-standing representations of the elements preexist.

Thus, we would not expect fusion when the subjects were given a pair of familiar English words (dog and table). Rather, in these latter conditions, subjects would form associations among items which maintain their compositionality. There is, in fact, evidence that human amnesics are successful in implicit pair learning when stimuli are readily fused, e.g. when associations must be formed between words and the colors they are printed in (Musen & Squire, 1993b) or when words to be paired are the elements of an idiom, e.g. sour-grapes (Schacter, 1985).

Based on this reasoning, we predicted that hippocampal rats would be impaired in acquiring pairings involving non-arbitrary cues, which are less likely to be fused. To test this prediction, we employed an ethologically relevant form of learning which involves the pairing of a biologically important odor with a second odor, i.e. the social transmission of food preferences paradigm (Galef & Wigmore, 1983; Strupp and Levitsky, 1984). Social transmission involves the learning of which foods are safe to eat. Rats which are briefly housed with a conspecific which has eaten a distinctively flavored diet will later demonstrate a preference for that diet when given a choice of novel foods. Galef et al. (1988) have shown that the mechanism underlying this learning involves an association between two odors, the diet odor and carbon disulfide, a constituent of rat breath.

We expected retention of this form of learning to be dependent on the hippocampus. We posited that the biological distinctiveness of one of the odors, carbon disulfide, would render it less likely to be fused with the second odor. Indeed, in this form of prepared learning fusion would seem maladaptive, given that only the diet odor is present at the time of "testing." This task differs from the previously described tasks in other respects as well. In social transmission, learning occurs in a single trial. Such rapid acquisition is characteristic (though not defining) of declarative memory (Cohen, 1984). More cogently, there is direct evidence that implicit stimulus-stimulus learning does not occur in a single trial, instead requiring multiple trials to develop (Musen & Squire, 1993a). This social task also differs from the previous tasks in that memory is tested under conditions different from those of learning. Together these features indicated to us that this task would be hippocampal-dependent.

In fact, evidence to date had already suggested that the hippocampus is involved in this form of learning. Winocur (1990) showed that electrolytic lesions of the hippocampal system impaired memory for social transmission. Unfortunately, as noted earlier, electrolytic lesions disrupt multiple components of the hippocampal system and it is thus difficult to attribute any deficit specifically to hippocampal damage (Jarrard, 1993). We examined whether lesions restricted to the hippocampus would impair retention of this form of learning. Using the ibotenic acid technique employed above, we found that hippocampal lesions did, indeed, impair the

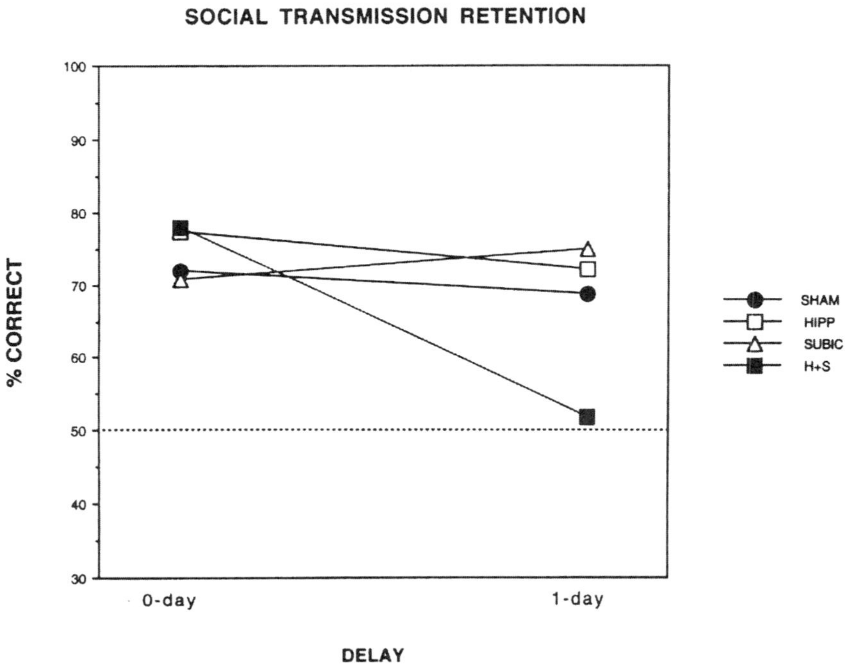

Figure 5. Retention for the two delay conditions for controls (SHAM), and rats with lesions of hippocampus (HIPP), subiculum (SUBIC), or hippocampus plus subiculum (H+S). Different groups were tested at each delay.

memory for social transmission (Bunsey & Eichenbaum, 1995). Moreover, this deficit was delay-dependent, as hippocampal rats were impaired after a 24 hour retention interval, but not when given an immediate test (Figure 5). Delay-dependence provides strong evidence for a specific memory effect as the only difference between the impaired and unimpaired groups is the length of retention interval. However, in contrast to the results with transitivity and symmetry, the lesion effect was only observed if both hippocampus and subiculum were removed -- ablation of either structure alone had no effect. We have argued that this pattern indicates that these hippocampal components may be considered a functional unit (Bunsey & Eichenbaum, 1995). Thus, damage to either one of the components would properly be considered a partial lesion, and would be expected to produce only a partial deficit at most. A close analysis of the patterns of results seen with lesions of these components and performance in spatial tasks is consistent with this framework (Morris et al., 1990). Nonetheless, given that transitivity and symmetry were impaired with the more limited lesions, further work will be needed to clarify the conditions under which deficits are to be expected. It is possible that the social memory task is more robust and/or less demanding than transitivity and symmetry, making it more easily mediated by the circuitry of one of these structures following removal of the other.

Regardless, this experiment indicated that retention of a natural form of rapidly learned pairing in which fusion is unlikely to occur was severely impaired by hippocampal region lesions. These results add further support to the hypothesized role of hippocampus in declarative memory. Equally important, these results also highlight the value of ethologically relevant measures of memory in the study of hippocampal function. There is evidence that non-declarative processes are more likely to come into play the more training a subject undergoes (Musen & Squire, 1993). Indeed, extended training is an integral part of the definition of hippocampal-independent habit memory (Mishkin, Malamut, & Bachevalier, 1984). Ethologically relevant tasks, in being acquired rapidly and automatically, are less likely to recruit nondeclarative memory processes, and thus may be more appropriate and sensitive tests of declarative memory deficits. It is interesting that Alvarez & Squire (1995) recently reported that the aspect of the delayed non-match to sample task which is most sensitive to limited hippocampal damage is the first block of trials. It is during this early training phase that normal monkeys show spontaneous nonmatching, reflecting their preference for novelty. After extended training, hippocampals and controls nonmatch at equivalent rates. Perhaps the nature of the task changes after extended training, with skill learning assuming a larger role, despite the fact that each trial involves unique stimuli. Regardless, as in our work the most sensitive measure of hippocampal-dependent memory involved an untrained behavior which occurs "naturally".

CONCLUSIONS

This ongoing research aims at two intertwined goals: First to explore the ability of rats to form declarative-type memory in the nonspatial domain, and second to assess the hippocampal-dependence of this memory. Our results have supported the notion that rats are capable of declarative memory and indicate that this memory is indeed dependent on hippocampal function. They are thus consistent with the hypothesis that hippocampal function has been conserved across species. The data help elucidate what the fundamental features of declarative memory are, characterizing them in nonverbal terms which can be applied across species, and suggest ways in which different hippocampal subcomponents contribute to this memory. Specifically, the parahippocampal region is sufficient to associate, via "fusion", items which occur simultaneously or in sequence, whereas the hippocampus itself acts to form relational representations; organized associations among items which maintain their compositionality and are capable of being accessed flexibly. How these two regions act together to form declarative memory in intact organisms is a question for future work.

REFERENCES

Alvarez, P, Zola-Morgan, S., & Squire, L.R. (1995) Damage limited to the hippocampal region produces long-lasting memory impairment in monkeys. *Journal of Neuroscience*, **15**, 3796-3807.

Bunsey, M. & Eichenbaum, H. (1993a). Paired associate learning in rats: Critical involvement of the parahippocampal region. *Behavioral Neuroscience*, **107**, 740-747.

Bunsey, M. & Eichenbaum, H. (1993b). Selective hippocampal lesions facilitate performance in a paired associate task in humans. *Abstracts of the Society for Neuroscience*, **19**, 358.

Bunsey, M. & Eichenbaum, H. (1995). Selective damage to the hippocampal region blocks long-term retention of a natural and nonspatial stimulus-stimulus association. *Hippocampus*, **5**, 546-556.

Bunsey, M., Eichenbaum, H. (1996). Conservation of hippocampal memory function in rats and humans. *Nature*, **379**, 255-257.

Cho, Y.H. & Kesner, R.P. (1995). Relational object association learning in rats with hippocampal lesions. *Behavioural Brain Research*, **67**, 91-98.

Cohen, N.J. (1984). Preserved learning capacity in amnesia: Evidence for multiple memory systems. In: The Neuropsychology of Memory. N. Butters & L.R. Squire (ed.), Guilford Press, New York.

Cohen, N.J., Eichenbaum, H. (1993). Memory, Amnesia and the Hippocampal System. MIT Press, Cambridge, MA.

Davidson, T.L., McKernan, M.G., & Jarrard, L.E. (1993). Hippocampal lesions do not impair negative patterning: A challenge to configural association theory. *Behavioral Neuroscience*, **107**, 227-234.

Davis, H. (1992). Transitive inference in rats. *Journal of Comparative Psychology*, **106**, 342-349.

Devany, J.M., Hayes, S.C., & Nelson, R.O. (1986). Equivalence class formation in language-able and language-disabled children. *Journal of the Experimental Analysis of Behavior*, **46**, 243-257.

Eichenbaum, H. & Bunsey, M. (1995). On the binding of associations in memory: Clues from studies on the role of the hippocampal region in paired-associate learning. *Current Directions in Psychological Science*, **4**, 19-23.

Eichenbaum, H.E., Otto, T. & Cohen, N.J. (1992). The hippocampus -- What does it do? *Behavioral and Neural Biology*, **57**, 2-36.

Eichenbaum, H.E., Otto, T. & Cohen, N.J. (1994). Two functional components of the hippocampal system. *Behavioral and Brain Sciences*, **17**, 449-518.

Eichenbaum, H.E., Stewart, C., Morris, R.G.M. (1990). Hippocampal representation in spatial learning. *Journal of Neuroscience*, **10**, 331-339.

Galef, B.G., Mason, J.R., Preti, G., & Bean, N.J. (1988). Carbon disulfide: A semiochemical mediating socially induced diet choice in rats. *Physiology and Behavior*, **42**, 119-124.

Galef, B.G. & Wigmore, S.W. (1983). Transfer of information concerning distant foods: A laboratory investigation of the "information centre" hypothesis. *Animal Behavior*, **31**, 748-758.

Gallagher, M. & Holland, P.C. (1992). Preserved configural learning and spatial learning impairment in rats with hippocampal damage. *Hippocampus*, **2**, 81-88.

Gluck, M.A. & Myers, C.E. (1995). Representation and association in memory: A neurocomputational view of hippocampal function. *Current Directions in Psychological Science*, **4**, 23-29.

Hirsch, R. (1974). The hippocampus and contextual retrieval of information from memory: A theory. *Behavioral and Neural Biology*, **12**, 421-444.

James, W. (1918). The Principles of Psychology. Holt, New York (Original work published 1890)

Jarrard, L.E. (1989). On the use of ibotenic acid to lesion selectively different components of the hippocampal system. *Journal of Neuroscience Methods*, **29**, 251-259.

Jarrard, L.E. (1993). On the role of the hippocampus in learning and memory in the rat. *Behavioral and Neural Biology*, **60**, 9-26.

Mishkin, M., Malamut, B., & Bachevalier, J. (1984). Memories and habits: Two neural systems. In: The Neurobiology of Learning and Memory. J.L. McGaugh, G. Lynch, & N.M. Weinberger (ed.), Guilford Press, New York.

Morris, R.G.M. (1981). Spatial localization does not require the presence of local cues. *Learning and Motivation*, **12**, 239-260.

Morris, R.G.M., Schenk, F., Tweedie, F., & Jarrard, L.E. (1990). Ibotenate lesions of hippocampaus and/or subiculum: Dissociation components of allocentric spatial learning. *European Journal of Neuroscience*, **2**, 1016-1028.

Murray, E.A., Gaffan, D., Mishkin, M. (1993). Neural substrates of visual stimulus-stimulus association in rhesus monkeys. *Journal of Neuroscience*, **13**, 4549-4561.

Musen, G. & Squire, L.R. (1993a). On the implicit learning of novel associations by amnesic patients and controls. *Neuropsychology*, **7**, 119-135.

Musen, G. & Squire, L.R. (1993b). Implicit learning of color word associations using a Stroop paradigm. *Journal of Experimental Psychology: Learning, Memory, & Cognition*, **19**, 789-798.

O'Keefe, J. & Nadel, L. (1978). The Hippocampus as a Cognitive Map. Clarendon Press, Oxford.

Olton, D.S., Becker, J.T., & Handlemann, G.E. (1979). Hippocampus, space, and memory. *Behavioral and Brain Sciences*, **2**, 313-365.

Roberts, W.A. & Phelps, M.T. (1994). Transitive inference in rats: A test of the spatial coding hypothesis. *Psychological Science*, **5**, 368-374.

Saunders, R.C. & Weiskrantz, L. (1989). The effects of fornix transection and combined fornix transection, mammilary body lesions and hippocampal ablations on object-pair association memory in the rhesus monkey. *Behavioural Brain Research*, **35**, 85-94.

Schacter, D.L. (1985). Multiple forms of memory in humans and animals. In: Memory Systems in the Brain. Weinberger, N.M., J.L. McGaugh, & G. Lynch (ed.). Guilford Press, New York.

Schacter, D.L., Chiu, C.P., Ochsner, K.N. (1993). Implicit memory: A selective review. *Annual Review Neuroscience*, **16**, 159-182.

Schusterman, R.J. & Kastak, D. (1993). A California sea lion (Zalophus californianus) is capable of forming equivalence relationships. *The Psychological Record*, **43**, 823-839.

Scoville, W.B. & Milner, B. (1957). Loss of recent memory after bilateral hippocampal lesions. *Journal of Neurology, Neurosurgery, and Psychiatry*, **20**, 11-21.

Squire, L.R. (1992). Memory and the hippocampus: A synthesis of the findings with rats, humans, and monkeys. *Psychological Review*, **99**, 195-231.

Strupp, B.J. & Levitsky, D.A. (1984). Social transmission of food preferences in adult hooded rats (Rattus Norvegicus). *Journal of Comparative Psychology*, **98**, 257-266.

Sutherland, R.J. & Rudy, J.W. (1989). Configural association theory: The role of the hippocampal formation in learning, memory, and amnesia. *Psychobiology*, **17**, 129-144.

Thomas, G.J. & Gash, D.M. (1986). Differential effects of posterior septal lesions on dispositional and representational memory. *Behavioral Neuroscience*, **100**, 712-719.

Winocur, G. (1990). Anterograde and retrograde amnesia in rats with dorsal hippocampal or dorsomedial thalamic lesions, *Behavioural Brain Research*, **38**, 145-154.

Zola-Morgan, S., Squire, L.R., & Amaral, D.G. (1986). Human amnesia and the medial temporal region: Enduring memory impairments following a bilateral lesion limited to field CA1 of the hippocampus. *Journal of Neuroscience*, **6**, 2950-2967.

11

THE STRUCTURE OF EVENTS

Herbert L. Roitblat
University of Hawaii

Until recently the paradigm representation for animals was a reflex and a copy. A simple stimulus is detected and a simple response results. Learning, it would follow, would correspond to the formation of connections between these stimuli and responses. Internal changes resulting from experience were thought to mirror the corresponding physical events about which the animal was learning (Bolles, 1975). In such a paradigm it is difficult to conceive of how an animal would represent a complex or structured event. Considerable research effort was dedicated to discovering the simple stimulus properties that mediated performances that otherwise appeared to involve complex events (see Roitblat, 1987, for a review). The goal was to explain these apparently complex events by discovering the simple properties to which the animal was actually responding. The investigator could thereby reduce the apparently complex event to a simple one.

HIERARCHICAL OBJECT STRUCTURE

For example, when Herrnstein, Loveland, and Cable (1976) found that pigeons could apparently recognize natural categories of objects, such as trees, a substantial effort was directed at identifying the specific visual cues that the pigeons were using as a surrogate for trees. The concept of tree was generally considered by the investigators to be outside the realm of possible representational elements. Instead, it was thought that the pigeons would have to be using something more perceptually primitive, such as specific colors or a specific geometric shape that would allow them to identify photographic slides with trees and reject slides without trees. Despite this effort, no one has yet been able to identify the specific features of these slides that would provide the necessary and sufficient cues for classification.

From an alternative perspective of holistic representations, one might argue, as a colleague once did, that discriminating the slides with trees in them is easy because they look like trees. These slides did, indeed, look like they contained trees, but the holistic explanation leaves unstated just what it means to look like a tree. We may not be able to identify the necessary and sufficient cues to looking like a tree (either for a human or for a pigeon), but we may be able to explore in more detail the characteristic features that trees have and the means by which they are recognized. Somewhere between the assertion that the pigeon's were discriminating nothing but simple visual characteristics and the assertion that they were discriminating holistic trees that looked like trees is an appropriate level of analysis for categorical representation. Rather than take the reflex as the paradigm representational primitive, it might be possible to take a hierarchical view of object representation that would permit, for example, pigeons to use a variety of cues to recognize trees because these cues are each indicators of trees as objects rather than of slides as configurations of lines and colors.

ALTERNATIVES TO COPY THEORIES OF LEARNING

Similarly, our dominant model of learning is little changed since the early part of the 20th century. The essential idea is that learning is a process by which experience is stamped into the brain. Thorndike (1911) is most famous for this view. His theory of reinforcement argued that reward stamped in the stimulus-response connections that preceded its delivery. Another way of saying this same thing is that reward strengthens a neural copy of the stimulus-response connection that preceded it. This copy view is not limited to comparative psychologists, but is actually pervasive throughout psychology, even among investigators of human cognitive function.

Recent developments in connectionist (McClelland & Rumelhart, 1986; Rumelhart & McClelland, 1986) and dynamic systems approaches to cognition provide an alternative approach to the copy theory of representation and learning (Smith & Thelen, 1993; Thelen & Smith, 1994). On this view, learning does not stamp experience into the brain, rather it provides guidance for changing the dynamic properties of the system. The task for psychology then is to identify the nature of these changes rather to identify the particular experience that was stamped in.

THE PROBLEM OF TIME

Another type of problem that presents difficulties for the traditional reflex-paradigm account is the set of tasks involving temporal structure. Time is particularly problematical as a stimulus because there is no obvious sensory system that detects time in the same way that the eyes detect light. No single observable stimulus corresponds to the passage of time. As a result, explaining behavior on the basis of the passage of time was an assertion that the animal was responding to something that could not be observed, but had to be inferred from changing experience. Such inferences are incompatible with an account of behavior that takes the reflex as its fundamental unit.

Learning has traditionally been viewed as requiring synchronous pairings between events. Delayed conditioning, for example, was somewhat problematic, but was explained by the persistence of neural events for a short time following environmental events. Without an obvious stimulus and without temporally contiguous pairings, learning about temporally structured events provided some difficulties. Nevertheless, temporal conditioning (Pavlov, 1927/1960) and other examples of timing in animals made very clear that animals do indeed respond to the passage of time. The discovery of circadian rhythms and the brain structures that mediate them provided a means by which internal brain events could be used to structure daily activity and to anticipate daily and seasonal events.

CIRCADIAN STRUCTURE

Just as scenes have hierarchical structure, events extended in time also have hierarchical structure at various time scales. Circadian (daily) and circannual (yearly) rhythms structure events for animals. They can presumably deal with these events by relating them to the various points in the daily or annual cycle. These rhythms represent regularities of the day/night and annual cycles. Animals that respond in anticipation of daily and seasonal events are not reacting directly to external stimuli, rather they are using internal clocks to structure the time and the sequence of events. The evidence for this internal as opposed to external structuring of events comes from many sources, but among the most compelling is how the animals continue to respond when the direct physical stimulation (e.g., temperature variation, light-level changes, food abundance, and so forth) are removed. Animals continue to show rhythmic behavior patterns even in the

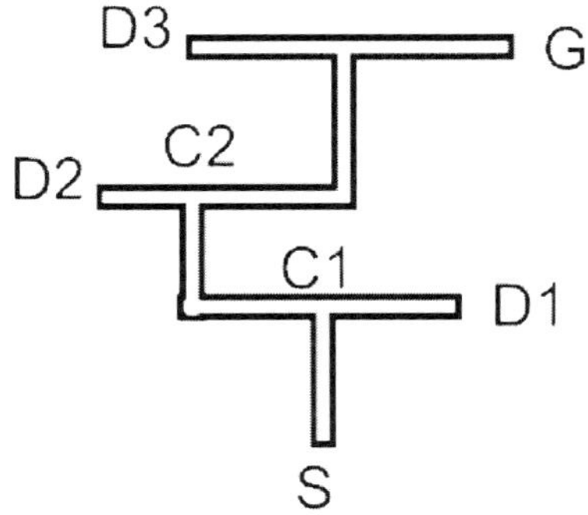

Figure 1. A multiunit T-maze. C1, C2, & C3 are choice points, D1, D2, & D3 are dead ends.

absence of these cues (e.g., Rusak & Zucker, 1975; Silver & Bittman, 1984).

Circadian rhythms showed how to account for behavior by reference to some kind of internal representation of the passage of time. The fact that such representations were posited by biologists rather than flaky cognitivists lent credibility to their existence. Psychologists could treat the output of circadian pacemakers as a stimulus like a visual stimulus and so incorporate it into the reflex paradigm.

An important feature of circadian rhythms is that they are entrained by environmental events, they are not merely responses to environmental events. The Zeitgeiber "tuned", but did not cause the circadian rhythm. The same rhythm could serve as the input for a large number of behaviors.

SERIAL STRUCTURE

Events over shorter durations also have structure that animals can utilize by relating the elements of the sequence to internal representations. A caveat is in order here. Not all serially structured events are necessarily represented hierarchically. For example, when a rat goes through a maze (Figure 1) it must execute a sequence of responses corresponding to each of the choice points in the maze. Successful traversal of the maze depends on the correct sequencing of these events, but the rat need have no knowledge of their sequence. In fact, such knowledge of a sequence could actually interfere with the animal's performance if the maze is constructed so that the rat can reach some of the choice points by more than one route. If the animal were compelled to execute the sequence in a fixed order, and it were somehow blocked, displaced, or otherwise caused to travel by a different route to get to a particular choice point it would then fail to complete the maze, because the same sequence of choices is no longer available. The rat in the maze need not know about the sequences of responses that brought it to a choice point, it needs only to know what choice to make at that point (Bever, Straub, Terrace, & Townsend, 1980). The structure of the maze structures the sequence. The rat need not represent that structure explicitly. A relatively more simple representation pairing choice points and responses would be adequate.

Serial anticipation learning, on the other hand, does seem to imply that the animal learns and uses a structured representation of a temporally extended event. In the serial anticipation task, rats are trained to run down a straight alley for a food reward that changes in magnitude from one run to the next in a predictable order. Hulse and Campbell (1975) and Hulse and Dorsky (1977) found that rats were better able to learn a sequence when it was strongly monotonic than when it either weakly monotonic or nonmonotonic. For example, the rats learned in fewer trials to anticipate, that is to run slowly to, the final element in a sequence such as 14-7-3-1-0 than a sequence such as 14-1-3-7-0 (where each digit represents the number of food pellets received as a reward for the run). The rats in the latter sequence ran at the same high speed to each of the elements in the sequence, apparently not anticipating the final 0 element. Apparently, the redundancy in the structure of the monotonic sequence allowed the rats to represent the sequence more easily. It may not be clear from these results, however, whether the animals represented the sequence in a hierarchical fashion as they represent circadian events, or whether they represented it in a linear fashion as they appear to represent the serial structure of a maze.

As Bever and his colleagues (1980) pointed out, there is serial structure inherent in the maze that the animal could use to structure its performance. One way this can be done is with a Markov process. The animal's representation of the event is characterized in terms of certain kinds of internal states and simple rules for moving from one state to another. In the case of the maze, the states might correspond to each choice point, and the movements in one direction or another from that choice point would then be the transitions between states. Each state is uniquely characterized by the stimulus configuration at a choice point and each transition that follows that state is learned as a function of the success of the animal in traversing the maze. In analyzing such processes we describe the states and their transitions in terms of a transition matrix.

In the transition matrix shown Table 1, the rat's current position or "state" is given by the row in the table. The possible positions or states to which it can move are given by the nonzero entries in the row. So, for example, the rat faces its first choice at choice-point 1. Before training, it may be equally likely to turn either to choice-point 2 or to dead-end 1, so the probabilities of these two transitions are both shown as 0.5. It cannot reach any other states or positions directly from choice-point 1, so the cells

Table 1. A Hypothetical Transition Matrix for Traversing the Maze in Figure 1

	State						
State	C1	D1	C2	D2	C3	D3	G
C1	0.0	0.5	0.5	0.0	0.0	0.0	0.0
D1	1.0	0.0	0.0	0.0	0.0	0.0	0.0
C2	0.33	0.0	0.0	0.33	0.33	0.0	0.0
D2	0.0	0.0	1.0	0.0	0.0	0.0	0.0
C3	0.0	0.0	0.33	0.0	0.0	0.33	0.33
D3	0.0	0.0	0.0	0.0	1.0	0.0	0.0

corresponding to these transitions have a value of 0.0. These reachable alternatives are represented with equal probability in this matrix, but they do not need to start out with equal probability. As the animal learns the maze, these probabilities will change to values approaching 0.0 for backtracking and dead ends, and 1.0 for correct traversal of the maze.

The order in which the rows appear in the table is merely a presentational convenience for readability. There is no need to organize the rows in any particular order. The transitions are determined by the state of the animal (i.e., its position) at any point in time and the set of states that are reachable from that state, and these could be presented in any arbitrary order. The order in which the rat moves through the maze is not determined by the order of the states in the rat's memory, but by the structure of the maze. Only certain positions are reachable from any given choice point. The rat merely has to know which direction to turn at each choice point (the transition probabilities for the corresponding row), it does not need to know the order in which the choice-points appear (the order of rows in the table).

The essential feature of a Markov model is that it has "no memory." In a Markovian process, the history of the preceding sequence of states is not contained in the representation. Instead, the representation consists of a set of states and rules for traversing from one state to the next. The only way information can affect the performance of the system is if the information changes the state. The history of the system is contained only in the its current state. If there is more than one path by which the system can arrive at a particular state, the path by which it did actually arrive at that state is not known.

In the case of serial anticipation learning, we could construct a similar representation to describe what the rat knows as it determines its running speed in the runway. Instead of describing choice points, the states

Table 2. Hypothetical Transition Matrix for Serial Anticipation Learning

Belief State	Belief State				
	14	7	3	1	0
Start	1.0	0	0	0	0
14	0	1.0	0	0	0
7	0	0	1.0	0	0
3	0	0	0	1.0	0
1	0	0	0	0	1.0

of this representation are "beliefs" or "expectations" about the number of food pellets to be found at the end of the maze. At this point I do not want to attach too much meaning to the term belief. I intend it merely to capture the notion that the rat regulates its running speed roughly consistent with the number of pellets that are likely to be present at the end of the runway. A model of this sort has been described by Capaldi and his associates (e.g., Capaldi, Verry, & Davidson, 1980). Such a Markov representation might resemble that in Table 2. According to this representation, at the start of a trial, the rat anticipates that there will be 14 pellets (more precisely, the rat anticipates that there will be a lot of pellets, the precise number is probably not known) at the end of the runway. Having received 14 pellets, the rat then anticipates that the next run will be rewarded with 7 pellets, and so forth. According to the matrix in Table 2, the rat anticipates perfectly the size of the reward on each run. This is surely an oversimplification. Imperfect anticipation would be represented by values in the matrix that are less than 1.0 and more than 0.0.

An alternative, but related, hypothesis argues that the rats employ a rule, of the form, $item_{i+1} < item_i$. Hulse and Dorsky (1977, 1979) proposed a model of this type. In both of these models, the rat's knowledge of where it is in the sequence (its current state) is exclusively determined by the number of pellets it has received. Neither model includes information about the sequence per se, only what to expect following specific events. The association-generalization model (Capaldi, Verry, & Davidson, 1980) attributes the animal's response to learned associations between successive items, and the rule-based model (Hulse & Dorsky, 1977, 1979) attributes it to the use of a rule. A third possibility is that the rats employ a different kind of representation that is not essentially Markovian in structure.

For example, the rats could employ a hierarchical representation in which the sequence is represented as a whole and as a set of elements that constitute that whole. In one experiment (Roitblat, Pologe, & Scopatz, 1983) rats were trained to anticipate the sequence 14-7-3-1-0. They were then tested under a regimen that allowed us to discriminate between a

Markovian and a hierarchical representation. The rats were tested with occasional probe trials in which one of the elements of the sequence was replaced with the 0-pellet element. For example, they were tested with the probe sequences 14-7-3-**0**-0, 14-7-**0**-1-0, 14-**0**-3-1-0, or **0**-7-3-1-0. The running speeds to the final 0-pellet element was indistinguishable on the probe trials from that obtained on the standard sequence trials. Changing the elements by replacing them with a 0-pellet element presumably changed the state of the animal, but it did not change their ability to anticipate the subsequent elements in the sequence.

Clearly, the rats were using more information than the immediately preceding element to predict the amount of food to be delivered following the next run. Animals appear to form coherent hierarchical representations of this kind of sequence that contain information about the sequence that is more global and more abstract than a simple Markov type model would predict. These representations are not disrupted when a novel element occasionally replaces one of the elements in the sequence.

SEQUENCE DISCRIMINATION

Another sequentially structured task on which animals have been trained is sequence discrimination. In this task, an animal, usually a pigeon, must discriminate one sequence of events from other sequences of the same events. Explicit feedback is delayed until the end of the trial. For example, Terrace (1986) trained pigeons to peck one key if the sequence ABC (e.g., the colors red, green, blue projected on a pecking key) was presented and to peck a different key if any other combination of A, B, and C were presented. The choice keys were located in fixed locations, one to the left and one to the right of the key on which the sequence was presented. Although the keys were not illuminated until after the sequence had been completed, Terrace found that the pigeons were responding to the locations of these two keys before the end of the sequence on a substantial number of trials. These data suggested that the pigeons were deciding the classification of the sequence (i.e., whether it was positive or negative) during its presentation.

A Markov type model of this task might be similar to that presented in Table 3. According to this kind of model, the pigeon starts off in a memory state that corresponds to the start of a trial. State transitions depend on the combination of the pigeon's current memory state and the stimulus that was presented. If the first element in the sequence is presented while the pigeon is in memory state, *start*, then its memory changes into a state reflecting that the first element has been presented. I have labeled this *State A*. The system changes to *State A* only if it is currently in memory state *start* and the first element of the sequence is presented. If it is in memory

Table 3. Hypothetical Transition Matrix for Discriminating 3-Event Sequences

Memory State	Stimulus	Memory State: A	B	C	Peck Yes	Peck No
Start	A	1.0	0.0	0.0	0.0	0.0
Start	B	0.0	0.0	0.0	0.0	1.0
Start	C	0.0	0.0	0.0	0.0	1.0
A	A	0.0	0.0	0.0	0.0	1.0
A	B	0.0	1.0	0.0	0.0	0.0
A	C	0.0	0.0	0.0	0.0	1.0
B	A	0.0	0.0	0.0	0.0	1.0
B	B	0.0	0.0	0.0	0.0	1.0
B	C	0.0	0.0	0.0	1.0	0.0

state *Start* and any of the other stimuli are presented, the system changes to state *Peck-No*, and, according to this simple model, does not process further any subsequent elements in the current trial. Similarly, if it is in State *A*, and the second element of the sequence is presented, the memory state changes to *State B*. All other stimuli cause the system to change to state *Peck-No*. Such a model has no memory, because the state transitions depend only on the combination of memory state and stimulus, not on any of the preceding states the system may have occupied. Errors would presumably result from transitions to states other than the correct state. Combinations of state and stimulus other than the last one could, for example, have values above 0.0 in the Peck-Yes column. Once the system is state *Peck-Yes*, the pigeon will respond positively, whatever the actual sequence that had been presented.

In contrast to the prediction of this Markov system, Roitblat, Scopatz, & Bever (1987) found evidence that pigeons were employing a hierarchical representation of such sequences. We trained pigeons to discriminate 3-item sequences of colored lights. ABC was the positive sequence and the other 26 possible combinations of A, B, and C were the negative sequences. The same white test stimulus followed each sequence and the bird's task was to peck at this white stimulus if the positive sequence had been presented. Pecks to the test stimulus following one of the negative sequences were not reinforced.

We used a mathematical model to analyze the patterns by which the elements of the sequence gained control over responding. We identified seven possible factors that the pigeon could use to discriminate the positive sequence from each of the negative sequences. These elements were the three stimuli alone (A, B, & C), the stimuli in pairs, (AB, AC, & BC), and

Table 4. The Values Used in the Hierarchical Model of Sequence Discrimination

Presented Stimuli	Value in the Model						
	A	B	C	AB	A_C	BC	ABC
AAA	+	-	-	-	-	-	-
AAB	+	-	-	-	-	-	-
AAC	+	-	+	+	-	-	-
ABA	+	+	-	+	-	-	-
ABB	+	+	-	-	-	-	-
ABC	+	+	+	+	+	+	+
ACA	+	-	-	-	-	-	-
ACB	+	-	-	-	-	-	-
ACC	+	-	-	-	+	-	-
BAA	-	-	-	-	-	-	-
BAB	-	-	-	-	-	-	-
BAC	-	-	+	-	-	-	-
BBA	-	-	-	-	-	-	-
BBB	-	+	-	-	-	-	-
BBC	-	+	+	-	-	+	-
BCA	-	-	-	-	-	-	-
BCB	-	-	-	-	-	-	-
BCC	-	-	+	-	-	-	-
CAA	-	-	-	-	-	-	-
CAB	-	-	-	-	-	-	-
CAC	-	-	+	-	-	-	-
CBA	-	+	-	-	-	-	-
CBB	-	+	-	-	-	-	-
CBC	-	-	+	-	-	+	-
CCA	-	-	-	-	-	-	-
CCB	-	-	-	-	-	-	-
CCC	-	-	+	-	-	-	-

the triplet of the three stimuli considered as a single unit (ABC). Each of these seven units was assigned a positive value in the model if the corresponding stimuli were presented in the sequence. These values are shown in Table 4.

During the early stages of acquisition the pigeon's performance was controlled largely by the final element in the sequence. If the final element was C, then the bird pecked at the test stimulus, otherwise the bird did not peck. Later in training, the birds' performance was controlled by higher order combination elements, for example, by the combination of the first and third elements in the sequence (A_C). If the first element in the sequence was stimulus A and the last was stimulus C, then the bird pecked. Control by higher-order combination units indicates that the pigeons were representing the sequences in a hierarchy consisting of the stimulus elements at the bottom of the hierarchy and the combinations at the higher levels. They were basing their decisions not just on individual elements, but on combinations of elements. They responded to the combination A_C even when an element other than B was in the middle position, indicating that they did not simply process the stimuli one by one and terminate processing when the first negative stimulus was perceived. These data are inconsistent with a simple linear model of the sort described in Table 3 because the response to the sequence depends on memory for nonadjacent items in the sequence. Consistent with this interpretation, Terrace (1986) found that sequences ending with stimulus C were the most difficult for the pigeon to discriminate (consistent with control by the final element found in this experiment), but 24 of the 26 negative sequences would be rejected even before reaching the final element (because either A would be absent from the first position or B would be absent from the second position).

TEMPORAL STRUCTURE OF ACOUSTIC EVENTS

Temporal structure also occurs over much smaller time scales, especially in acoustic events. The structure of acoustic events consists in the pattern by which the spectrum of an acoustic signal changes over time. Speech, for example, depends on the pattern by which the distribution of energy at various frequencies changes over time. Figure 2 shows an example of digitized speech. There is a complex relationship between frequency and time as shown by the changing shapes of the frequency bands. These variations in the acoustic signal are so well ingrained (either well learned or innately structured) that it is difficult even to recognize that they are structured. As Figure 2 shows, however, there is actually a complex structure to speech signals. We have evolved mechanisms in our ears and brains that make interpretation of these signals automatic and effortless.

Even finer grain temporal structure is apparent in animal biosonar signals. Several species are known to use biological sonar to obtain information about the world around them. Among these, bats and dolphins have received the most study (See Simmons, et al. 1995, for a review of bat biosonar).

All of the toothed cetaceans (as opposed to the baleen whales) that have been studied show evidence of biological sonar. Of these, we know most about the bottlenose dolphin, *Tursiops truncatus*. Dolphins emit

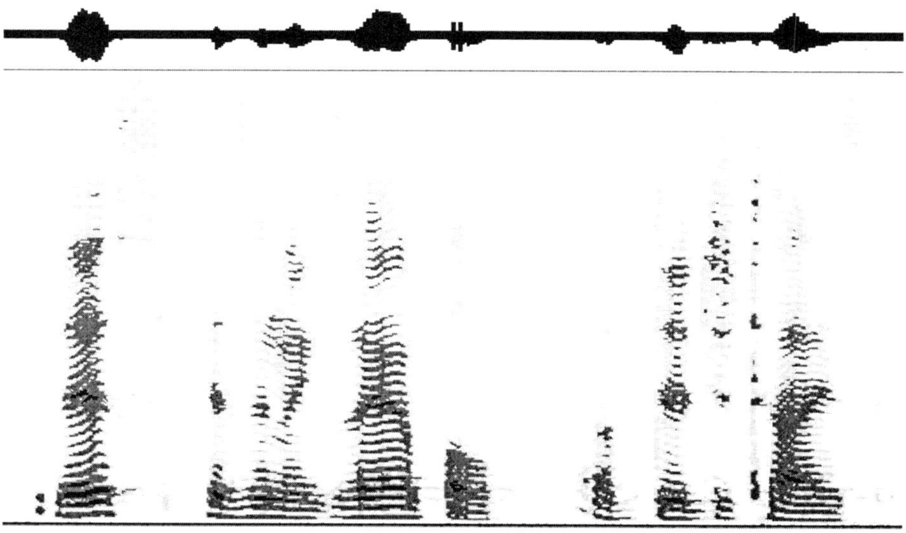

Figure 2. A speech spectrogram for the phrase, "This is an example of digitized speech." The top of the figure shows the waveform and the lower part of the figure shows the spectrogram of the signal. Frequency is shown along the vertical axis, and time is along the horizontal axis. The darkness of a point indicates the intensity of the signal at that time and frequency. The duration is about 2.5 s and the frequency ranges from 0 to 5.5 kHz.

biosonar signals as brief clicks generated in their nasal apparatus within their heads, below the blow-hole. These signals are transmitted through the fatty tissue in the forehead or melon. The sound travels through the water and reflects off nearby objects. The structure of the returning echo provides information about the structure and material composition of the reflecting object. Bottlenose dolphins have extremely sensitive, broad range hearing, spanning over 7 octaves, and can detect frequencies as high as 150 kHz (Johnson, 1967). Peak sensitivity is in the range of 15 to 110 kHz. Hearing is directionally sensitive, with maximum sensitivity directly ahead of the dolphin's rostrum. Sensitivity drops in a frequency-sensitive pattern to either side (see Au, 1993).

Bottlenose dolphins can identify through echolocation many characteristics of submerged objects, including size, structure, shape, and material composition (Au, 1993; Bel'kovich & Dubrovskiy, 1976; Nachtigall, 1980). For example, they can discriminate between circles, squares, and triangular targets covered with neoprene (Barta, 1969). They can also discriminate a 16.1-cm from a 15.2-cm diameter disk at a distance of 0.7 m (Barta, 1969). Dolphins have also been found to discriminate aluminum cylinders differing in wall thickness by as little as 0.23 mm,

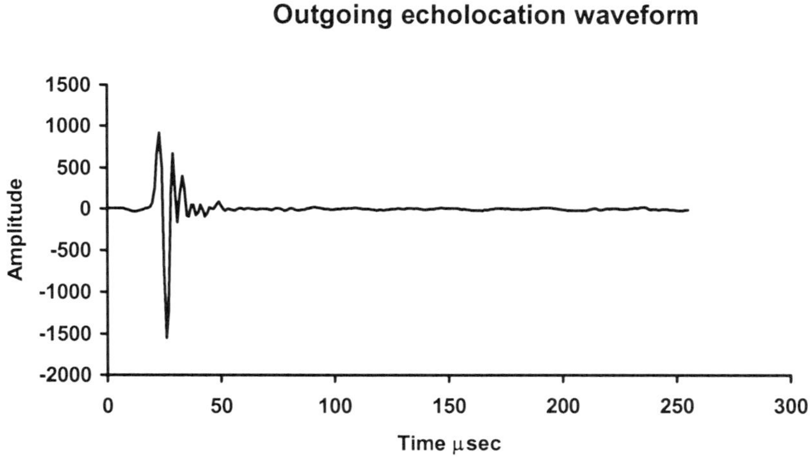

Figure 3. The waveform for an example outgoing echolocation click. The click was recorded from a hydrophone placed 1 m in front of the dolphin.

presented at a distance of 8 m (Au & Pawloski, 1992), as well as freely rotating three dimensional geometric objects, such as cubes, rectangular prisms and pyramids (Helweg, Roitblat, Nachtigall, & Hautus, 1996).

Dolphin echolocation clicks typically have peak energy at frequencies ranging from 40 to 130 kHz and with recorded source levels over 220 dB re: 1 μPa at 1 m, (Au, 1980; Moore & Pawloski, 1990). Figure 3 shows an example click recorded about 1 meter in front of a dolphin. Figure 4 shows the spectrum of the same signal. Notice that the signal is

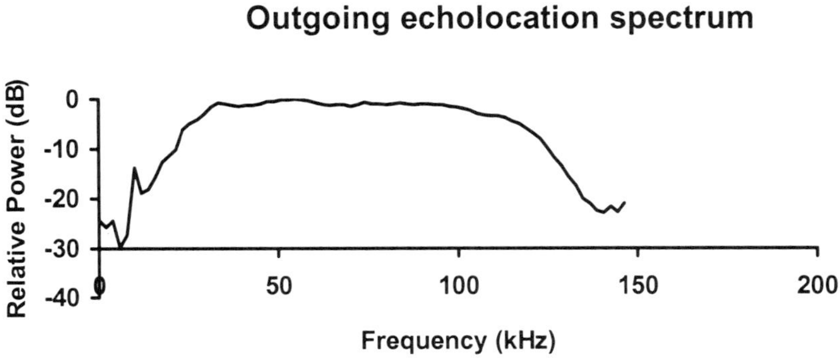

Figure 4. The spectrum of the outgoing echolocation click shown in Figure 3.

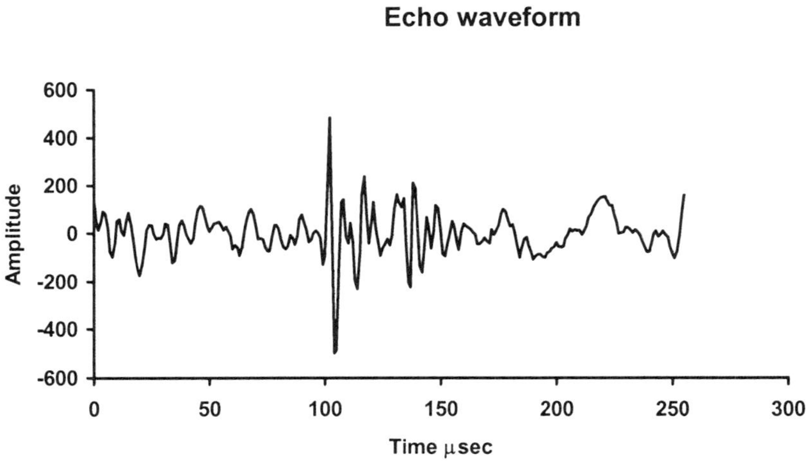

Figure 5. The waveform of the returning echo from the cylindrical object ensonified by the click in Figure 3.

very brief, lasting about 50 μsec. The resulting spectrum is extremely broadband, spanning about 100 kHz. By comparison human hearing tends to be limited to the band between 18 Hz and 20 kHz. The echo returning from a cylindrical object located about 5 m in front of the animal is shown in Figure 5. The temporal structure of this echo corresponds to the structure and composition of the object. The corresponding spectrum of this echo is shown in Figure 6.

Dolphins can also discriminate between aluminum, copper, and brass circular targets (Evans & Powell, 1967. In a number of experiments, dolphins were also found to be able to discriminate hollow aluminum cylinders (e.g., 3.81 cm OD) from the same size hollow cylinder made of stainless steel or a solid cylinder formed of coral rock pebbles embedded in degassed epoxy resin (e.g., Hammer & Au, 1980).

All of the information the dolphin has about the object is contained in the one-dimensional time series of the echo. The dolphin must extract information about the identity and shape of the object from the pattern by which the signal changes over time. This temporal structure occurs at two levels. At the grosser level, dolphins tend to emit multiple clicks to the same object before they respond to it. A dolphin performing a familiar task with familiar objects might emit as many as 35 or more clicks to each of three objects. It appears that the dolphin integrates over multiple samples of the object to build up a more complete or less ambiguous representation (Roitblat, Penner, & Nachtigall, 1990). The dolphin also utilizes the fine-grain temporal structure of the echo to recognize the features of the object

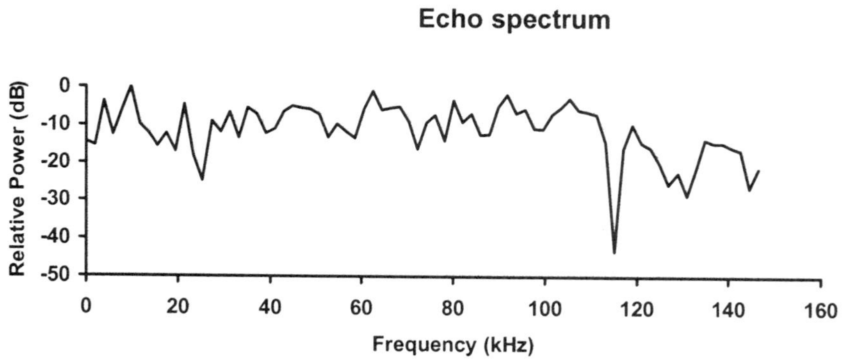

Figure 6. The spectrum of the echo shown in Figure 5.

that returned the echo. I want to focus here on the finer-grain temporal structure of the echolocation task. Models of the apparent interclick integration are discussed elsewhere (Moore, Roitblat, Penner, & Nachtigall, 1991; Roitblat & Moore, 1991; Roitblat, Au, Nachtigall, Shizumura, & Moons, 1995).

The physics of objects determines the relationship between the signal that strikes the object and the echo that returns from it. The basic element of the echo is a copy of the outgoing signal simply reflected from the front face of the object. There may also be other reflections from the internal structure of the object, which may include sound traveling via a number of different paths through and around the object. The material of the object can also have resonance characteristics which will be excited by the echolocation signal. In addition, there may be other objects in the vicinity (clutter) that also return complex echoes. All of these "features" add together and form a complex time series (echo waveform) when it is received by the dolphin. The dolphin filters through this mass of data to derive the characteristics of the object to which it is attending.

The fact that the information-carrying part of the echo time series lasts a fraction of a millisecond (often about 200 μsec or less) does not diminish the fact that there is no single stimulus that corresponds to the object, rather it is the complex progression of the time series that provides the information about the object's structure and composition. The dolphin has evolved mechanisms for transforming this time-extended representation of the object's characteristics into a time-independent framework that it can use for processing further the information about the object. Signal processing engineers approximate this mechanism when they obtain the Fourier transform of the signal. The Fourier transform characterizes the complex wave in terms of its spectrum, containing the distribution of energy at each of the frequencies contained in the signal (see Figures 4 and 6). This transform changes the time series into a stable spectral representation, but at the cost of some temporal uncertainty. In contrast, bats (Simmons et al., 1995) and dolphins (Au & Pawloski, 1992) show evidence of temporal resolution on the order of less than 2 μsec. Hence, their representations include temporal resolution beyond that contained in a single spectrum of the whole echo.

Mammalian ears perform a transformation similar to a Fourier transform. The structure of the cochlea, especially the thickness and stiffness of the basilar membrane causes different portions of the cochlea to be differentially sensitive to frequencies. The base of the basilar membrane is differentially responsive to high frequency signals and the apex is differentially sensitive to low frequencies and the areas in between are proportionally distributed (Ketten, 1992). The dolphin cochlea can be well described as a bank of frequency filters (Au, 1994; Roitblat, Moore, Helweg, & Nachtigall, 1993).

The transformation performed by the ear is mechanical and neural rather than mathematical. It does not require any cognitive or general purpose computational ability because the transformation comes directly from the physical structure of the cochlea and the neurons that connect to it. Nature does not care about labels. Evolution keeps those mechanisms that solve problems effectively, by whatever means such solutions are accomplished. The case of the cochlea shows how the brain and sensory system of the dolphin has evolved to be able to deal with transforming temporally structured events into atemporal configurations. Part of this process is to evolve mechanisms that compare portions of incoming signals to prespecified internal representations. In the cochlea this comparison is performed by the frequency-response characteristics (filters) of the basilar membrane and the hair-cell and ganglion-cell structures that connect to it. I would suggest that this kind of process could serve as the paradigm by which animals represent other kinds of temporally extended event information.

Circadian rhythms, for example, appear to be represented by brain pacemakers. Temporally structured daily events can be represented atemporally relative to the state of these pacemakers. Serial anticipation learning and sequence discrimination may employ similar kinds of representations by comparing the elements of the sequence to a more global representation of sequential relations. The temporal sequence of events is compared to an atemporal representation of these events. Such representations are possible if allow ourselves to consider kinds of representations beyond those inherent in copying and reflexes.

In the case of the circadian and auditory representational mechanisms the comparison comes from the interaction of perceptual and identifiable physical processes. The structure of the cochlea and the physiological function of the pacemakers provide the basis for the representation. Serial anticipation and sequence discrimination may similarly operate on the basis of preexisting structures, for example, short-term oscillators, but such mechanisms have not been identified. Nevertheless, interactions between perceptual mechanisms and various structures can be a very effective means of transforming a potentially difficult problem into a simpler one. For example, when landing a commercial jetliner, there are strict requirements for the exact speed of the plane at different altitudes, and wing-flap configurations (Hutchins, 1995). The speed must be closely monitored during landing to avoid crashing. Such a monitoring task is made simpler, however, by a feature of the air-speed indicator that was not planned to have this function. The rules call for the pilot to take appropriate action if the speed of the plane differs by more than 5 miles per hour from the prescribed speed. The air-speed indicator is a circular meter with the speeds marked off in 10 mph intervals.

The prescribed speed is marked on the air-speed indicator by a movable marker called a speed bug. The speed bug was designed to be large enough to be easily visible, but small enough that it would not obscure more than one hash mark on the air speed indicator. Its width, therefore, matches the spacing of the 10 mph intervals on the air-speed indicator. The problem of monitoring the airspeed within the prescribed limit is, thereby, transformed to the problem of making sure that the airspeed indicator needle stays within the arc of the speed bug. This property of the speedbug is an inadvertent byproduct of other design decisions, but it is still extremely useful to help pilots manage the cognitive load inherent in landing a commercial jetliner.

Transformations such as the speed bug or the frequency analysis of the cochlea do not make these tasks any less cognitive. Instead, they provide a mechanism by which the organism is executing its cognition. Discovering the mechanism by which a psychological phenomenon occurs does not render it nonpsychological, merely comprehensible. These transformations suggest the possibility that organisms employ other transformations that help to simplify the cognitive properties of a task. Such mechanisms merit investigation.

REFERENCES

Au, W. W. L. (1980). Echolocation signals of the Atlantic bottlenose dolphin (*Tursiops truncatus*) in open waters. In R. G. Busnel & J. F. Fish (Eds.), *Animal sonar systems* (pp. 251-282). New York: Plenum Press.

Au, W. W. L. (1993). *The sonar of dolphins*. New York: Springer Verlag.

Au, W. W. L. (1994). Comparison of sonar discrimination: Dolphin and an artificial neural network. *Journal of the Acoustical Society of America*, **95**, 2728-2735.

Au, W. W. L., & Pawloski, D. A. (1992). Cylinder wall thickness discrimination by an echolocating dolphin. *Journal of Comparative Physiology A.*, **172**, 41-47.

Barta, R. E. (1969). Acoustical pattern discrimination by an Atlantic bottlenosed dolphin. Unpublished manuscript, Naval Undersea Center, San Diego, CA.

Bel'kovich, V. M. & Dubrovskiy, N. A. (1976). *Sensory basis of cetacean orientation*, Leningrad: Nauka.

Bever, T. G., Straub, R. 0., Terrace, H. S., & Townsend, D. J. (1980). Study of serial behavior in humans and nonhumans. In P. Jucsyk & D. Klein (Eds.), *The nature of thought: Essays in honor of D. 0. Hebb*. (pp. 51-93). Mawah, NJ: Lawrence Erlbaum Associates.

Bolles, R. C. (1975). Learning, motivation and cognition. In W. K. Estes (Ed.), *Handbook of learning and cognition, vol. 1*. (pp. 249-280). Mawah, NJ: Lawrence Erlbaum Associates.

Capaldi, E. J., Verry, D. R., & Davidson, T. L. (1980). Memory, serial anticipation pattern learning and transfer in rats. *Animal Learning and Behavior,* **8**, 575-585

Evans, W. E. & Powell, B. A. (1967). Discrimination of different metallic plates by an echolocating delphinid. In R. G. Busnel (Ed.), *Animal sonar systems, biology and bionics*, Vol. 1. (pp. 363-383). Jouy-en-Josas, France: Laboratoire de Physiologie Acoustique.

Hammer, C. E., Jr., & Au, W. W. L. (1980). Porpoise echo-recognition: An analysis of controlling target characteristics. *Journal of the Acoustical Society of America*, **68**, 1285-1293.

Helweg, D. A., Roitblat, H. L., Nachtigall, P. E., & Hautus, M. J. (1996). Recognition of aspect-dependent three-dimensional objects by an echolocating Atlantic bottlenose dolphin. *Journal of Experimental Psychology: Animal Behavior Processes,* **22**, 19-31.

Herrnstein, R. J., Loveland, D. H. & Cable, C. (1976). Natural concepts in pigeons. *Journal of Experimental Psychology: Animal Behavior Processes*, **2**, 285-302.

Hulse, S. H., & Campbell, C. E. (1975). "Thinking ahead" in rat discrimination learning. *Animal Learning and Behavior,* **3**, 305-311.

Hulse, S. H., & Dorsky, N. P. (1977). Structural complexity as a determinant serial pattern learning. *Learning and Motivation,* **8**, 488-506.

Hulse, S. H. & Dorsky, N. P. (1979). Serial pattern learning by rats: Transfer of a formally defined stimuli relationship and the significance of nonreinforcement. *Animal Learning and Behavior,* **7**, 211-220.

Hutchins, E. (1995). How a cockpit remembers its speeds. *Cognitive Science*, **19**, 265-288.

Johnson, C. S. (1967). Sound detection thresholds in marine mammals. In W. Tavolga (Ed.), *Marine bioacoustics*. (pp. 247-260). New York: Pergamon Press.

Ketten, D. R. (1992). The marine mammal ear: Specializations for aquatic audition and echolocation. In D. R. Webster, R. R. Fay, & A. N. Popper (Eds.) *The evolutionary biology of hearing*. (pp. 717-750). New York: Springer Verlag.

McClelland, J. L. & Rumelhart, D. E. 1986. *Parallel distributed processing: Explorations in the microstructure of cognition, Vol. 2. Psychological and biological models*. Cambridge, MA: MIT Press.

Moore P. W. B. & Pawloski, D. (1990). Investigations on the control of echolocation pulses in the dolphin (*Tursiops truncatus*). In J. Thomas & R. Kastelein (Eds.) *Sensory abilities of cetaceans*. (pp. 305-316). New York: Plenum Press.

Moore, P. W. B., Roitblat, H. L., Penner, R. H., & Nachtigall, P. E. (1991). Recognizing successive dolphin echoes with an integrator gateway network. *Neural Networks*, **4**, 701-709.

Nachtigall, P. E. (1980). Odontocete echolocation performance on object size, shape, and material, In R. G. Busnel & J. F. Fish (Eds.), *Animal Sonar Systems* (pp. 71-95). New York: Plenum Press.

Pavlov, I. P. (1927/1960). *Conditioned reflexes*. Oxford: Oxford University Press. Reprinted 1960. New York: Dover.

Roitblat, H. L., Au, W. W. L., Nachtigall, P. E., Shizumura, R. & Moons, G. (1995). Sonar recognition of targets embedded in sediment. *Neural Networks*, **8**, 1263-1273.

Roitblat, H. L., Moore, P. W. B., Helweg, D. A., & Nachtigall, P. E. (1993). Representation and processing of acoustic information in a biomimetic neural network. In J.-A. Meyer, H. L. Roitblat, & S. W. Wilson (Eds.) *From animals to animats 2: Proceedings of the second international conference on simulation of adaptive behavior* (pp. 90-99). Cambridge, MA: MIT Press.

Roitblat, H. L., Moore, P. W. B., Nachtigall, P. E. & Penner, R. H., (1991). Natural Dolphin echo recognition using an integrator gateway network. In D. S. Touretsky, & R. Lippman (Eds.), *Advances in neural information processing systems* 3. (pp. 273-281). San Mateo, CA: Morgan Kaufmann.

Roitblat, H. L., Penner, R. H., & Nachtigall, P. E. (1990). Matching-to-sample by an echolocating dolphin. *Journal of Experimental Psychology: Animal Behavior Processes*, **16**, 85-95.

Roitblat, H. L., Pologe, B., & Scopatz, R. A. (1983). The representation of items in serial position. *Animal Learning and Behavior*, **11**, 489-498.

Roitblat, H. L., Scopatz, R. A., & Bever, T. G. (1987). The hierarchical representation of three-item sequences. *Animal Learning and Behavior*, **15**, 179-192.

Rumelhart, D. E. & McClelland, J. L. 1986. *Parallel distributed processing: Explorations in the microstructure of cognition, Vol. 1. Foundations*. Cambridge, MA: MIT Press.

Rusak, B. & Jucker, I. (1975). Biological rhythms and animal behavior. *Annual Review of Psychology*, **26**, 137-171.

Silver, R. & Bittman, E. L. (1984). Reproductive mechanisms: Interaction of circadian and interval timing. In J. Gibbon & L. Allen (Eds.), *Timing and time perception*. (pp. 488-514). New York: Annals of the New York Academy of Sciences, vol. 423.

Simmons, J. A., Saillant, P. A., Wotton, J. M., Haresign, T., Ferragamo, M. J., & Moss, C. F. (1995). Composition of biosonar images for target recognition by echolocating bats. *Neural Networks*, **8**, 1239-1261.

Smith, L. B. & Thelen, E. (1993; Eds.) A dynamic systems approach to development. Cambridge, MA: MIT Press.

Terrace, H. S. (1986). Positive transfer from sequence production to sequence discrimination in a nonverbal organism. *Journal of Experimental Psychology: Animal Behavior Processes*, **12**, 215-234.

Thelen, E. & Smith, L. B. (1994). A dynamic systems approach to the development of cognition and action. Cambridge, MA: MIT Press.

Thorndike, E. L. (1911). *Animal intelligence: Experimental studies*. New York: Macmillan.

12

THE GROWTH OF COGNITIVE STRUCTURE IN MONKEYS AND MEN

Brendan McGonigle and Margaret Chalmers
University of Edinburgh

BACKGROUND: AN UNBRIDGEABLE GAP IN THE STUDY OF COGNITIVE FUNCTIONING

There is a widespread view that the sorts of animal learning mechanisms most frequently studied in the laboratory are inductively too weak and unproductive to generate the kinds of behaviours expressed in higher order forms of human cognitive and linguistic adaptation (Chomsky, 1980; Fodor & Pylyshyn, 1988; Piaget, 1971). One reason for this (Harlow, 1949) is that investigations are rarely followed through from one learning episode to another to assess the cumulative benefits (if any) as a function of the agent's task and life history. Yet the course of human development is protracted, and even sophisticated adult subjects frequently show dramatic changes in strategy when confronted with many problems of the same type, detecting pattern and structural invariance in some (e.g., Wood, 1978), using analogies to bridge problems of a different surface structure (Gentner, 1983), and devising progressively economic, data reducing procedures to secure success with the least investment in resource (Anderson, 1990; McGonigle & Chalmers, 1996, 1998; McGonigle & Chalmers, 2001). Whilst Harlow's pioneering work on the learning set (LS) using primates is an exception, indicating the vast potential for accelerated learning, his generative claim that the LS leaves the organism free to attack problems of a new hierarchy of difficulty has never been properly realised due to the fact that all problems in conventional LS studies are of the same (simple binary) type and level of difficulty (but see Terrace this volume). Worse still, LS tasks conventionally involve arbitrarily chosen pairings of stimuli so that it is impossible in these circumstances for subjects to devise any rules of relation or stimulus classifier system which would connect one learning episode with another, despite the fact that relational codification and learning arguably constitutes a quite different type of adaptation with a different inductive and generative profile (Bryant, 1974; Bunsey & Eichenbaum, 1996; Dusek & Eichenbaum, 1997; Gazzaniga, Ivry, & Mangun, 1998; McGonigle & Chalmers, 1996; McGonigle & Jones, 1978; Wills & Mackintosh, 1998).

In the domain of human cognition, on the other hand, experimental paradigms heavily emphasise language based tasks often as their only means to access (or so it is perceived) complex language-like abstractions which can be operated only by symbolically competent agents (Russell, 1983). Thus thinking and reasoning tasks attempt to probe cognitive representation by giving subjects texts or verbal inference-drawing tasks from input such as "All of the bee-keepers are artists; some of the chemists are bee-keepers" (Johnson-Laird, 1983). Children have not been spared deep linguistically based probing either as in Piaget's celebrated tests of transitive reasoning, for example; "Edith is fairer than Lilli; Edith is darker than Suzanne, who is the fairest/darkest" and in his assessment of class inclusion competences in such tests as "are there more flowers or more daffodils ?" (Inhelder & Piaget, 1964). Beyond success, furthermore, justification of the answer is often a part of assessment as well (Karmiloff-Smith, 1992; Piaget, 1928), if only to help determine how explicitly the subject was aware of the **necessity** of the conclusion at an explicit propositional level of comprehension, yielding further insights into whether the answer was driven by the demands of a modal (logical) understanding rather than by empirical knowledge alone (Smith, 1993).

As a consequence, human cognitive psychology with its emphasis on abstract representational devices has left psychological and cognitive neuroscience with a huge credibility gap between bottom-up, knee-jerk portrayals of intelligence, and the sorts of organisation which support the highest achievements of man as portrayed in the laboratories of cognitive psychologists. When characterised as abstract, symbolic, and domain independent (see 'proof' based theories of representation (Fodor, 1983)), such architectures of the mind seem often to have no credible roots in earlier phylogenetic or ontogenetic adaptations. Whether concerned with representational format (see e.g. the analogue *versus* propositional issue (Pylyshyn, 1981)), structural organisation or problem solving (Anderson, 1990; Johnson-Laird, 1983), the currency of explanation starts from beyond the point at which most behaviour based learning accounts leave off, i.e. the currency of symbol manipulation.

The consequences for both cognitive theory and application are far-reaching. For if language is the only serious instrument of probing 'higher' intellectual functioning, where does this leave the psychology of the non-human, the developing human and the cognitively impaired? If impotent without "semiotic instruments and the like ", as Piaget (1971) puts it, to assess representational factors, a first casualty would be the investigation of cognitive functioning in non-humans and with it any prospect of a neurosciences model based on non-human research which could provide useful insights into high level deficits in executive memory and related subsystems—the dementias, for example. Such in-principle limitations would also apply to developmental approaches—especially those which are themselves informed by non-human comparisons (Chalmers & McGonigle, 1997; McGonigle & Chalmers, 1996, 1998; Terrace & McGonigle, 1994) and those focusing on transitions in human cognitive growth from its earliest stages (Fischer & Canfield, 1986). Finally, the clinician attempting to evaluate subtle cognitive deficits where language disorder itself may be a confounding factor; for example a syndrome such as Fragile X would be left 'high and dry' without adequate substitutes for language (and other symbol) based evaluations (Chalmers, 1998)

In this chapter, we provide a characterisation of cognitive systems from a comparative and developmental perspective, reviewing research that has led to our current focus on the dynamics of self organisation and the self-selection of procedures which provide the agent with progressively enhanced adaptive power with greater economy of resource (Anderson, 1990). Embracing many of the core phenomena such as linear and hierarchical organisation perceived by many as important indicants of high level brain organisation our aim has been to help establish a common currency of tasks and measures beyond the scope of knee jerk level intelligence, yet without pre-supposing linguistic competence.

In particular, we shall review a corpus of work on transitivity, seriation, and classification designed to evaluate both the power and the primacy of relational competences, seen initially by James (1891) as central to an evolutionary shift from 'habit of mind' mechanisms to the association of ideas by similarity (or the perception of relations (Boakes, 1984)). Whilst James saw the shift as a basic difference between human minds and those of 'brutes', Herrnstein (1989) provides a more accurate contemporary view:

> "abstract relations differentiate most sharply among animals. It has been noted that we see the largest gaps in comparative performance at the level of abstract relations... once a species has a foothold at the level of abstract relations, the possibilities are unbounded".

This switch of emphasis is also timely. For many now doubt that traditional learning approaches based on association principles are either the whole story or indeed any story at all (Gallistel, 1995). As Gazzaniga et al put it, perhaps more inclusively, ".... when one is trying to understand how the brain enables learning, one must realise that there may be several mechanisms, not just one." (Gazzaniga et al., 1998), p. 521.

Given the context of the forum which motivated this volume, furthermore, we shall also highlight some of the more immediate neurosciences implications and applications which now follow from our stance.

COGNITIVE STRUCTURE: RELATIONAL MECHANISMS AND THEIR ORIGINS

Relationally based organisation is undoubtedly a key player in supporting the structural properties which lie at the core of human cognitive organisation. As Hummel and Holyoak (2001) point out, "human thinking is structure sensitive.... in the sense that we can represent and reason about abstract relationships, appreciating the similarities and differences between the idea that "John loves Mary" and the idea that "Mary loves John" (see Fodor and Pylyshyn, 1988)". In addition, a formidable weapon which makes human cognition both powerful and economic, is the human brain's ability to exploit analogy (Gentner, 1983; Luger, 1994) using a common relational structure to bind otherwise related objects and events.

However, switching emphasis to relational rather than associative competences confronts the investigator of non humans with some serious, historically entrenched problems both conceptual and procedural, all of which have had a blighting influence on decades of comparative research. First is the long-standing assumption already noted that it is only at the abstract level of representation that such systematic, productive aspects of relational comprehension operate. Linked to this has been the notion that language may provide not only an essential window on cognition but is a key casual instrument in the determination of core cognitive matters. Indeed, Fodor and Pylyshyn (1988) claim that the systematicity of thought *follows from* the systematicity of language. Certainly, the most palpable evidence of relational competence in humans is found in the vocabulary of relational connectives such as comparatives and scalars (we **point** to objects; we **declare** relationships!). So it is perhaps unsurprising that relational differences between objects expressed by terms such as 'bigger' and 'smaller' once attracted the view that language itself provides the bridge or the mediating device which make such achievements possible.

Two factors reinforced this view. The first was that many demonstrations designed to demonstrate relational competence in non-humans, and based on a one-step transposition paradigm (Reese, 1968), were also interpretable in simple stimulus generalisation terms (Spence, 1937). The second factor inhibiting the development of relationally based learning theories was the emergence of language based 'mediation' theories to account for the development of such expertise in human cognitive development. Even within a behaviourist stance, some appeal to linguistic factors was made necessary because older children and adult subjects transposed relationally on what were known as 'far' tests i.e. to stimulus values outside the critical transfer range of Spence's generalisation model. Advocated by Kuenne (Kuenne, 1946) and others, the idea was that a relevant linguistic label could come to act as an internal conditioned response (R_m) to the stimulus, itself then acting as mediated stimulus (S_m) for the final observable choice. The final chain is no longer S - R, but S - R_m - S_m - R. Kuenne argued that there were 'levels' of verbal behaviour itself in the context of discrimination learning. Only at later levels, she argued, do children's verbal responses come to control their choice behaviour, performing a role more like dimensional abstraction, even though, as Spence's student, she did not actually invoke such an overtly 'mentalistic' interpretation.

In discussing Kuenne's position, Bryant (Bryant, 1974) drew attention to the tautology noted by Lashley (1929). "The main trouble with the hypothesis that children begin to take in and use relations to help them solve problems because they learn the appropriate comparative terms like 'larger' is that it leaves unanswered the very awkward question of how they learned the meaning of these words in the first place." A further problem for the mediation hypotheses was in determining whether or not language *per se* changes the 'level' of stimulus analysis. As late as 1972, Fein observed that there were at least four types of possible stimulus analysis which transposition research had failed to differentiate. A stimulus field, she argued, could be analysed as a succession of e.g. small and large objects, or as an array of objects, whose absolute codes are contextually coded, e.g. the large in the training pair, the small in the test pair. Higher order 'relational' types of analysis would be scalar in nature, i.e. a series of objects, in which e.g. A is bigger than B; B bigger than C. In a further refinement, such scalars may carry only ordinal information without representing the magnitude of differences between items thus scaled. A more advanced form would carry metrical information of the amount by which the items differed. As for 'levels', the logic of Kuenne's mediational analysis **had** to suggest that the controlling function of language was based on a grasp of the **scalar** nature of the dimension, necessary to give contrastive meaning to verbal labels such as 'big' and 'small'. But if achieved by behavioural rather than linguistic regulation, the determiners of such relational achievements remain submerged under the dominance of associative learning stances. Yet if behavioural, it should be possible to show how changes in levels of stimulus analysis proceeds without benefit of linguistic codes.

PARADIGMATIC CHANGE IN TRANSPOSITION RESEARCH: ELIMINATING THE ONE STEP METHOD

Two sets of investigations, one with monkeys (McGonigle & Jones, 1978) and one with children (Lawrenson & Bryant, 1972), put some of these fundamental hypothesis to the test. In Lawrenson and Bryant 's study, four and six year old children were trained on one of two types of double discrimination task. In one case they were to choose the same relation such as 'bigger than'; in the other the same absolute size. Lawrenson and Bryant gave half their subjects a 'far' condition as one of the trained pairs (the absolute group were required to take the stimulus nearest in size to the original trained value). Under these conditions, there was no 'far' effect for the relational group, both the near and far training conditions were equally easy for both ages of child **and** both were substantially easier to learn than the absolute condition.

By finding such successful 'far' transposition under direct training conditions, Lawrenson and Bryant had neatly demonstrated a lower competence bound for relational responding in children than previous investigators relying on spontaneous transfer. The relative difficulty of maintaining an 'absolute' response under these same training conditions, moreover, now clearly suggested a more primitive **ontological** status for relational as opposed to 'absolute' codification. However, given that the subjects were human and were capable of covert if not explicit linguistically based augmentation of the test conditions, it was left to studies such as McGonigle and Jones (1978) to break the link with language in relational encoding which couldn't be explained away as a stimulus generalisation effect.

Begun in 1969, the Edinburgh programme on relational learning also took a training stance, informed partially by experiments on cats which showed that equivalence tests seriously underestimated positive transfer when compared to savings measures based on re-acquisition procedures. Applying a training procedure which derived from Harlow's LS procedure, but critically featuring conditions where the training stimuli varied relationally in size or brightness, McGonigle and Jones (1978) trained one group of squirrel monkeys to choose objects on a relational basis, and another group to choose on an absolute one. In the size studies, initial training stimuli were three white polystyrene cubes, which differed in size from one another by just over 1 cm.[2] The three cubes were presented as pairs. These are designated BC and CD (as two further cubes were added later), where B is the largest, C the middle-sized and D the smallest. For the relational group, choice of item, B in BC and C in CD (i.e. the larger in each case) was rewarded with a peanut; for the absolute group choice of C in both cases was rewarded.

Whilst initial learning was relatively rapid for both groups, significant differences began to emerge with task changes. The first and most obvious one was the lack of stable retention shown by the 'absolute' stimulus condition. For example, removing the visible context within which the objects were viewed (the objects glowed with luminous paint) adversely affected this condition only. Whilst context, including egocentric based reference frames based on the sight of the hand and other means of calibrating size would be removed by darkness, a relational code based on object-object differences alone proved robust enough to allow the relationally trained group to transfer both from light to dark and dark to light conditions respectively.

Transferred to triads derived from the training set, furthermore, the relational group immediately chose largest, showing an elementary form of perceptual transitivity. The fact too that monkeys in the relational condition could predict on the *first trial* of any new problem which object should be selected (as distinct from forming a non-specific LS as absolute stimulus learners did), combined with the general robustness of their performance in retaining what they had learnt, led McGonigle and Jones (1978) to claim that relational encoding was a *design primitive* and quite separate in profile and thus not derived from, associative, absolute stimulus learning. Just as crucially, such relational codes did not depend on language; instead McGonigle and Jones had found a set of precursors to linguistic comparatives. But what sort of precursors? Was the simple binary, asymmetrical relational code (bigger than) merely the tip of an iceberg obscuring a whole network or system of relationships which have yet to be exhibited by new behaviour based research methods? If so exposed, moreover, would such relationships operate as an (installed) ensemble with strict rules of interdependency, reflecting a logical hierarchy of relational types? As in the domain of transitive relationships, for example, where, within a formal symbol system, A>B, B>C implies A>C ? Or in the principled, linear seriation of size? Or in hierarchical acts of classification based on the perception of similarities and differences? Or as staged expressions of relational competences derived from minimalist design primitives and forged though environmental interaction? Or as punctate competences only, operating pragmatically and unsystematically to suit local task circumstances?

These questions are of considerable contemporary interest, not only in the light of rapidly growing research on the role of the hippocampus and relational coding (Dusek & Eichenbaum, 1997) (Morris, Garrud, Rawlins, & O'Keefe, 1982), but also because the move to ground linguistic phenomena in the pragmatics of real world adaptation has also grown apace (H. H. Clark, 1973; Jackendoff, 1983). Even the most symbol driven of symbolic representationalists such as Fodor and Pylyshyn (Fodor & Pylyshyn, 1988) have been moved to suggest that a rat learning a black versus a white discrimination would be hooked into a more extensive relational system, involving, for example, a 'darker than' relation and its inverse.

Yet no criteria are provided by these authors which would enable a distinction between punctate from non-punctate relational code use in non-humans. This is hardly surprising. In animal learning, it is rare to find both experiments and procedures which enable a **concurrency** of operation of an ensemble of relational codes in the same subject at the same time. Yet this is a condition which crucially has to be met to help evaluate the role of any one relational code in the operation of another. With humans, although inevitably confounded by linguistic factors, the trajectory of acquisition of antonymic terms by children (big/bigger; small/smaller), for example, indicates a psychological hierarchy through the apparent dependency between relational codes which in a formal (logical) system would have equal status. Illustrated by the lag shown (E. Clark, 1973; Donaldson & Wales, 1970) between the comprehension and production of the unmarked (e.g. bigger) *versus* the marked term (e.g. smaller) in young children, a natural syntax of rule acquisition is suggested as Heidenheimer () has argued, where children who first understand, e.g., big(ness), next derive an intermediate stage of understanding 'not-big', before acquiring an understanding of the inverse (smaller than), an interpretation endorsed following experiments on mental comparisons using linguistic instructions by McGonigle and Chalmers (McGonigle & Chalmers, 1984).

ORIGINS: LEXICAL MARKING AND THE P SPACE

That language is merely the expression of this developing competence rather than its cause, moreover is suggested by Clark (Clark, 1970; H. H. Clark, 1973) and others as based on the structure of the environment, rather than deriving from (deep) structures of language itself. Persisting in adult performance throughout the life cycle as shown by a consistent asymmetry in the speed with which adult human subjects make decisions following comparative questions featuring marked versus unmarked terms (Clark, 1969), such 'marking' effects were once thought to have a lexical origin. However, in a seminal paper Clark (H. H. Clark, 1973) rejected his earlier

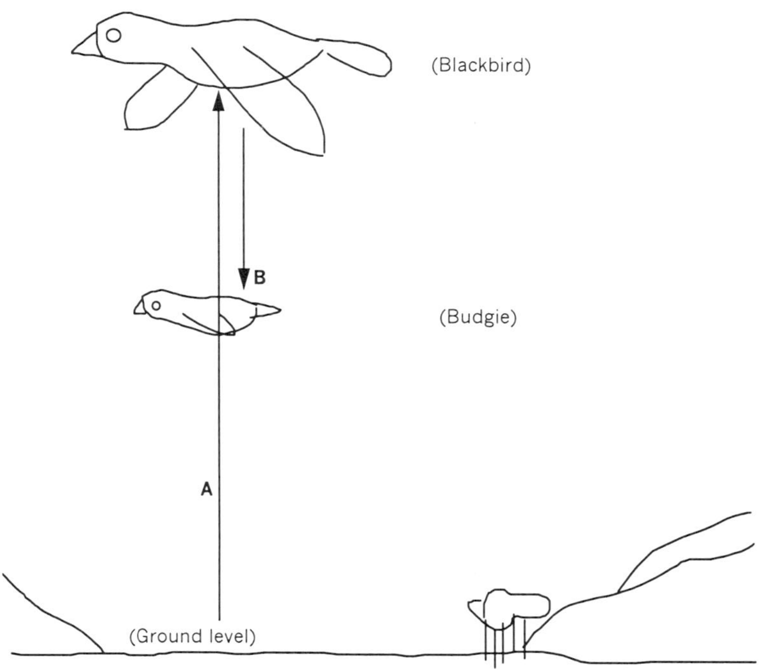

Figure 1. The perceptual origins of lexical marking (after Clark, 1973). The ground plane provides a natural reference for the primary direction A (higher). An object located along this dimension provides a secondary reference point for the derivative (marked) direction B (lower).

linguistically derived 'deep-structural' argument, arguing instead that such effects reflected instead a fundamental asymmetry in the organisation of the perceptual world—the P-space. In spatial locative terms, Clark argues, there is an egocentrically defined privilege for the upward and forwards direction as in above (below), over (under), or in front of (behind). A natural reference point for such directions is the ground—see Figure 1. However size and saliency of objects can also determine their status as natural reference points; objects on the horizon, the tallest or highest object in the visual field can form natural reference points for the judgment of size.

This example, coupled with the trajectory profile for the growth of relational comprehension, neatly demonstrates the biological answer to the symbol grounding problem (Harnad, 1987). Biological systems do not make 'abstraction' mistakes! Instead, it would seem that ecological factors extrinsic to subsequent relational coding of sets of objects anchor the ensemble to the world from the outset.

These related sets of phenomena now need to be taken to the 'proof 'stage if human psychological data on relational competence is to be used as indexical of systematic rule use by monkeys. Crucially too, as the human data is linguistico-relational, non-human competences in this sphere would unambiguously indicate a non-linguistic origin for them all. As for indicators, we shall take antonymic asymmetry as a prime 'marker' for relational dependency and system growth. And following that, a core question is the trajectory of relational acquisition. Does it have a natural syntax? If the rule for 'smaller than' derives from a negation of 'bigness', how is this expressed behaviourally? Beyond this case of possible rule dependency, furthermore, would a 'middleness' rule emerge, for example, only in subjects who had first acquired relational rules of both 'bigger' and 'smaller than'? Theoretical matters aside, how can assays of concurrent multiple rule, and thus possible system evaluation be achieved in the same subject and in the same test context without linguistic instruction? We address these matters in the next section.

RELATIONAL SYSTEMS IN THE MAKING: PRIMITIVES AND DERIVATIONS

Conventional methods of training and testing animals are inappropriate to the requirement to assess concurrent relational use in the same subject. To mimic the effects of linguistic instructions, a conditional code was needed. We (McGonigle, 1987; McGonigle & Chalmers, 1980, 1986) decided to use a colour conditional code requiring subjects to select the 'larger' of two objects if (say) all were black, and the 'smaller' of a pair if all were white.

A. 'Lexical' Marking Examined

Four squirrel monkeys (Brown, Blue, Green and White) from McGonigle and Jones (1978) were re-trained on pairwise size discriminations, using a colour conditional procedure (see Figure 2). The stimuli were wooden blocks, the same sizes as before (the sides of each face ranging from approximately 1.2 cm to 6.2 cm. with a 1.2 cm. interval difference). There were two such sets; one was painted white, the other black. For two monkeys, white denoted 'take larger', black denoted 'take smaller', and vice-versa for the other two. They were presented with the four adjacent pairs AB, BC, CD and DE in randomly presented trial blocks of five (first) and then ten, until they were 80% correct on each. In the following phase, all ten pairs deriving from the set were presented in

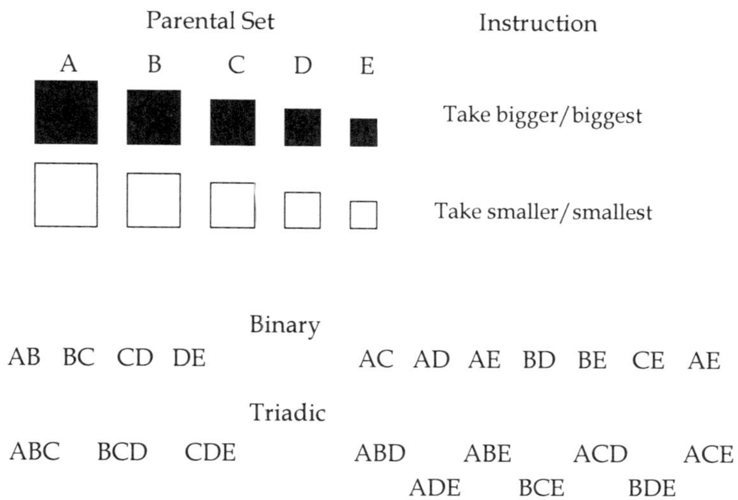

Figure 2. A colour conditional size rule training task used with monkeys (McGonigle & Chalmers, 1980, 1996).

randomised trial blocks of ten, until an 80% criterion was reached on each pair. Then the second, new instruction was learned in the same way using the other colour of stimuli.

Based on the finding from classic psychophysics that size discrimination difficulty can be measured using subjects' reaction time, we incorporated a psychophysical method of this sort, first to ensure that whatever the effects of instruction, that our RT measures would accurately reflect the relative discriminability of the pairwise combinations (10) which derived from the training sets given.

In the penultimate phase, trial blocks of 10 were given in which the instructions were alternated across blocks, until the monkeys could sustain high levels of accurate choice. Finally the instructions were randomised within trial blocks so that monkeys could predict neither instruction nor pair of objects on any trial.

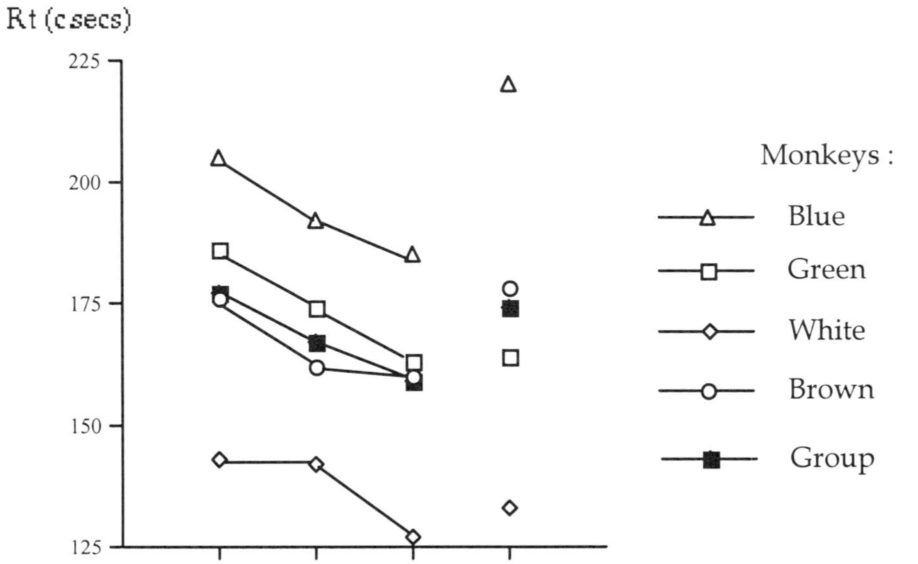

Figure 3. Psychophysical distance effects in the RTs for binary size comparisons obtained for individual monkeys, where distance is measured in steps (1 step represents the adjacent pairs).

Results

Acquisition

All monkeys completed training. Re-training on their original larger rule was relatively rapid, averaging 98 trials per pair on the initial phase. The 'smaller than' relation was learned even faster: only 39 trials were required on average. Stable performance involving both instructions in alternation was acquired with few errors and sustained at over 90% levels of accuracy for both instructions and for all comparisons.

RT Assay

A minimum of ten decision times recorded per instruction for each of the ten pairs during the final phase of testing was analysed. Analysis is based on correct choices only. There was a significant effect of psychophysical distance (measured as steps) for each subject as Figure 3 shows.

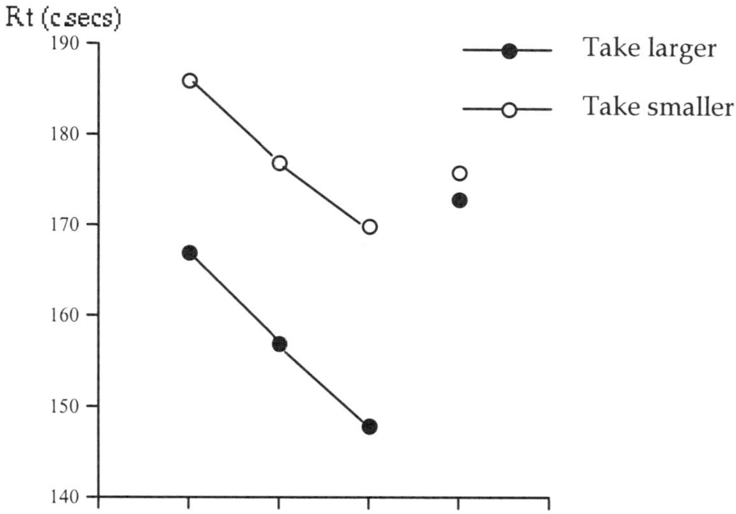

Figure 4. A marking effect in the RT data from monkeys.

As Figure 3 indicates, all monkeys show a significant inverse linear relationship between ordinal separation for all but the last item. Apart from the dramatic upturn on the last pairing (based on the two end items A and E which children also show (McGonigle & Chalmers, 1984), monkeys show profiles even at the level of the individual which are similar to those found with human subjects in memory experiments (Moyer, 1973).

Second as Figure 4 shows, there was a highly significant 'lexical' marking effect at the level of reaction times for the group as a whole and for each individual there was a significant effect of instruction in favour of the 'unmarked' instruction. This effect was unaffected either by practice or by psychophysical salience which did not interact with these variables, indicating that the marking effect is an access effect.

Third, there was an interaction between instructions and the 'ends' of the continuum found also in the memory search data for similar sorts of task with adult humans (Banks, 1977). This was indicated in the monkey data by an analysis in which the data are plotted *qua* distance functions, but from each end of the series respectively as illustrated in Figure 5. Here we can now assess the distance effects from end points only. As can be seen, the 'take larger' instruction produces a much more pronounced distance effect when plotted from the small end. Whilst the smaller than instruction also produces a more marked distance effect from the small end, they do not decrease so dramatically as distance increases. In short, the marking effect is not only an instructional effect but one which is related to a fundamental series asymmetry favoring the 'big' end of the physical continuum.

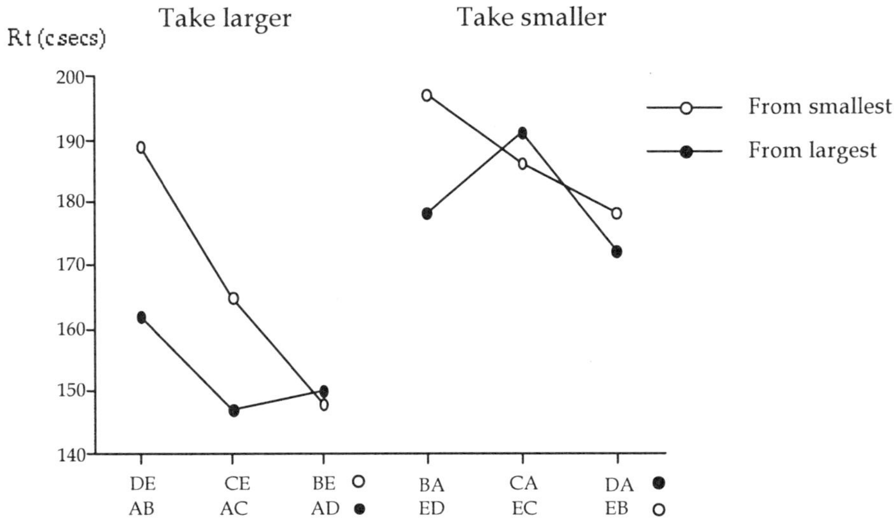

Figure 5. Asymmetrical interaction of instruction with distance effect in the RT data from monkeys' size comparisons.

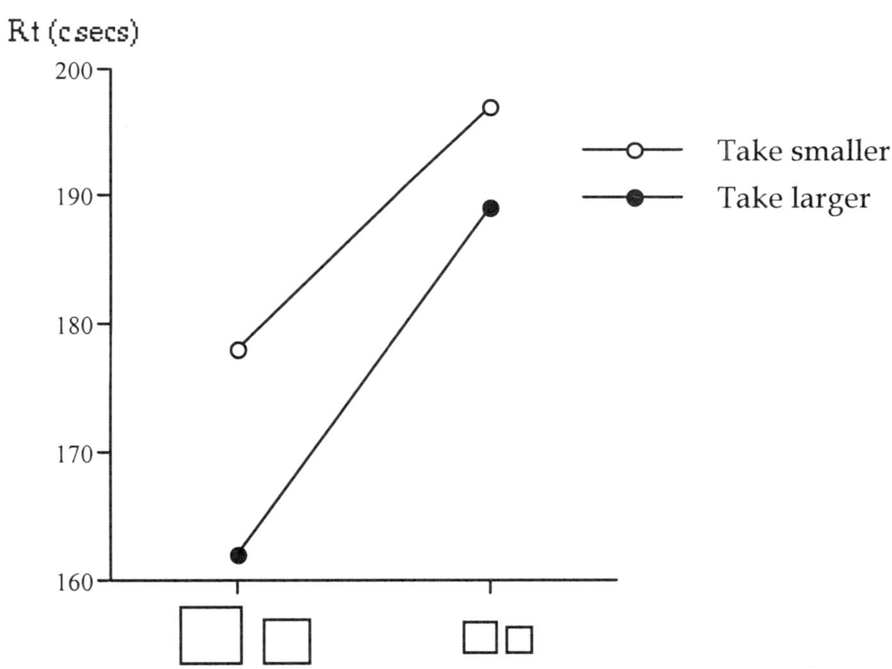

Figure 6. "Ends" by instruction asymmetry in the RT data from monkeys.

This is most clearly represented in an excerpt (Figure 6) from the data used above and which represents the variation caused by the ordinal position of test items (now taken from the extreme ends of the set).

Conclusions: 'Lexical' Marking

Without relying on acquisition data, but performatives instead, and based on steady state phases of performance at that, we can map both the macro and the micropattern of simian performance unto that of the human to an extensive degree. Crucially, we have shown not only that monkeys can operate two rules of relation concurrently in the same experiment, but that there is evidence of the marking effect used in human research to suggest a dependency between the operation of one rule and the operation of its inverse. The marking effect is clear and is undiminished either by practice or psychophysical salience. Showing impressive and parallel linear functions at the macro level, however, the micro-assays show how one end of the series anchors all decisions, such that the further away test stimuli are from the big end of the set, the slower choice is for both instructions.

We interpret these effects of instruction in the monkey data therefore, as strong evidence of asymmetrical, and unidirectional processing of the size dimension, similar to those we have obtained with children as old as nine years in memory experiments when asked comparative size questions about real world objects (McGonigle and Chalmers, 1984). Nevertheless, the implementation by the monkey of the 'smaller than' instruction was accurate and consistent. Encouraged, therefore, by the success of the conditional learning procedures, together with the gain in information obtained when more than one relation could be tested using a concurrent procedure, we next extended the method to 3 rules (middle) then to a set of 5 rules, where every size in the set we used had to have an independent code.

B. Introducing Middle-Sized

First, triads were presented, derived from the training set used above, and all subjects tested on bigger/biggest; smaller/smallest using the same instruction codes. As in the McGonigle and Jones (1978) study, monkeys generalised to the triads readily, showing once again elementary perceptual transitivity for both directions of instruction. Then a third instruction was introduced. The colour red now signified the operation of the 'middle' rule. Training on middle sized was lengthy. That monkeys required around 100 trials per triad before reaching criterion, shows that it was relatively costly for the monkeys to acquire this internal rule. However, and crucially, the monkeys were all successful, suggesting that their

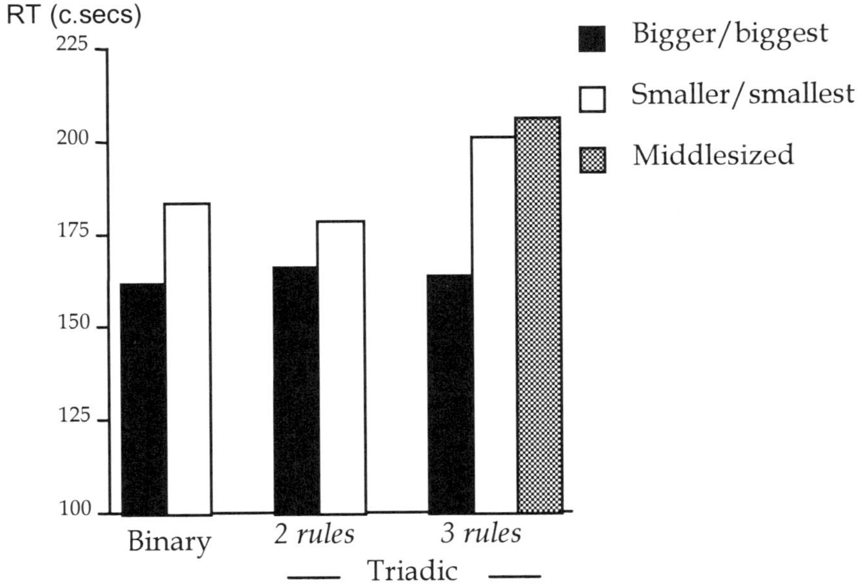

Figure 7. Mean RTs for monkeys as a function of instruction and the number of response categories required by a size comparison task.

previous failure in the McGonigle and Jones (1978) study was because they had lacked prior training on both comparatives 'bigger' and 'smaller'.

As before, a steady state phase was then introduced for purposes of measuring RTs in the context of correct performance. Figure 7 below shows these for the triplet instruction set and compares them with those obtained under double instructions from the earlier two stimulus phase and the later triadic phases.

The first finding to emerge from the comparisons depicted in Figure 7 is that whereas the triadic condition *per se* did not itself produce any increase in mean RT, the triple instruction did have an effect on RT selective to the smallest and middle-sized rules. In the latter condition, each of these showed a significant increase on their previous level. By virtue of this fact, the marking effect also increased under the triple instruction condition indicating that it is rule rather than stimulus competition which is a key factor here. Consistent, moreover, with the strong role of the 'big end' of the continuum, this rule competition seems to be between 'middle' and 'small' instructions only, leaving the 'larger/largest' instruction unaffected.

Conclusions: Middle-Sized

So far, the syntax of size rule learning by monkey is fully consistent with memory based encoding and retrieval models from human adult cognition. In their experiments on human memory and internal psychophysics, Moyer and Bayer (Moyer & Bayer, 1976) also found that, within a given size range, faster RTs were recorded at the large end of the continuum (even) where the original perceived differences are less psychophysically salient than at the small end. The memorial process, they argued, requires that the subject searches from the end of the continuum suggested by the instruction (in their case take larger) and begins a self-terminating search until one of the two items is found.

It is also has features in common with Parkman's model for digit (size) comparisons (Parkman, 1971). This suggests that when asked which is the larger of two digits, the human subject will mentally scan upward from zero until one of the digits is found. The larger digit is then deemed to be the 'other' one and is retrieved accordingly. This model was devised to explain the 'min' and 'split' effects in mental comparison experiments, where the Split effect refers to the distance effect (after Moyer (1973), as described above) and the Min effect refers to the fact that the nearer the smaller of the two items is to the small anchor, the faster the RT. For Parkman, the Split effect is seen to arise because of the statistical likelihood of one of the items being nearer to the Min as distance increases, and the Min effect itself derives from the fact that numbers are learned from zero upwards, providing the system with a fundamental directionality and a fixed starting point for any comparison. Whatever the particular search mechanisms, therefore, both models suggest, in common with Clark's p-space predictions, that binary judgments depend upon the establishment of a primary direction of relational codification.

Still consistent with Parkman's search model is the next stage of rule acquisition (smaller than) for, on this upwards or scanning model, the secondary direction is computed as a derivative only after the forward scanning has taken place. Consequently all 'smaller than' comparisons will be slower than all 'larger than' ones, where the scan procedure first finds the larger item and uses it as a reference point to locate the smaller one in a downward search.

The emergence of such processing effects is clearly evident in the monkey RT data. Thus applying Parkman's model to the question of how a series of ordinal values are acquired within a set, enables the prediction that specific rule differentiation might proceed somewhat paradoxically from the secondary reference point as it is only in the derivative downward direction that items are specifically tagged with respect to the reference point; in the upward direction, they attract an undifferentiated label 'larger than'. This

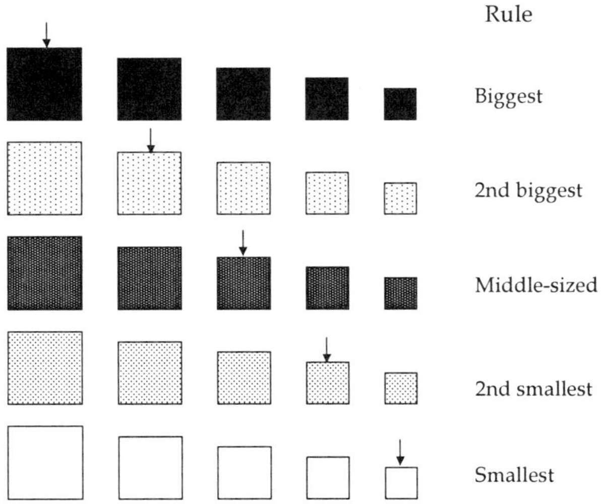

Figure 8. A colour conditional paradigm for training monkeys on five size rules (Chalmers & McGonigle, 1994; McGonigle & Chalmers, 1996).

model is well fitted to the monkey data by the fact that the introduction of middle-sized caused an apparent increase in the uncertainty attaching to 'smallest'. If this denotes the emergence of smallest as a secondary reference point (for computing middle-sized), then it should also follow that the next rule to be acquired would be the one adjacent to this. In a three item set, this would be middle-sized; in a five item set, however, it would be second smallest, not middle-sized. The next stage of our study allowed us to test this prediction.

C. Five Size Rule Learning

Now using a very large WGTA that enabled the subject to view five objects simultaneously, monkeys were adapted to new procedures using five differently coloured sets each of five sizes (Chalmers, 1994; McGonigle & Chalmers, 1996). Three new colours were added to the colour conditional discrimination task, producing the five by five design as shown in Figure 8. As before, on a given trial all stimuli were of one colour and only one size rule rewarded. The spatial configuration of the items on the tray varied randomly from trial to trial.

Training was conducted using initially trial blocks of 5 trials per rule, in random alternation with the constraint that no rule was repeated until the set of five had been exhausted. Five such trial blocks were typically run within a single session. Training continued until a performance criterion was met of 80% correct across 50 trials (10 per rule), with no one rule at or below 60%. This was repeated with trial blocks of two trials per rule. Finally the five colours were presented in single random alternation across trials, but no colour was repeated until the full set had been exhausted. Two different criteria were applied at this stage. The first was the same (80%) one as before. Training continued after this point, however, to establish a stable level of performance against which subsequent transfer behaviour could be measured and required that each and every rule could be operated (simultaneously) at the 95% confidence level. This was set at 70% correct overall across 16 consecutive sessions (80 observations per rule), with no rule less than 60% overall. Finally, an overtraining phase was given to establish, what, if any, further improvement could be sustained over a second set of 16 consecutive sessions.

To ensure that subjects were learning to make set-referenced ordinal judgments rather than ones based on absolute size, we introduced a second size range, overlapping with the first and tested size rule transfer.

Results

All subjects learnt all five rules, and operated these concurrently in the final phase of the experiment. Figure 9 illustrates the choice profiles obtained and compares them with those we obtained from four year old children following similar training. However, in the case of the children, all five size rules were trained concurrently from the outset. Unlike the initial end-point asymmetry we found under binary and triadic conditions of testing, the data now show **both** end points superior to the learning of codes denoting internal set values, especially following the attainment of criterion levels of training. Of equal interest, is the ranking of the internal ordinal rules as shown in Figure 9. From this figure, it can be seen that all monkeys now show a skewed choice distribution in favour of the 'small' end-point, from the earliest phase of acquisition. This item was not only the most accurately identified; it also attracted the fewest incorrect choices during performance on other rules. The biggest item was generally the second most accurately identified throughout, but it attracted many more incorrect choices than the smallest item, principally from rule competition with the second biggest stimulus rule. Of the three inner items—and crucial to our predictions concerning rule acquisition—second smallest was the best identified from the outset, followed by middle, followed by second biggest. Overall improvement with practice in the monkey is better characterized as a change in the pattern of error as a function of ordinal position, such that error distributions become more symmetrically distributed across the series.

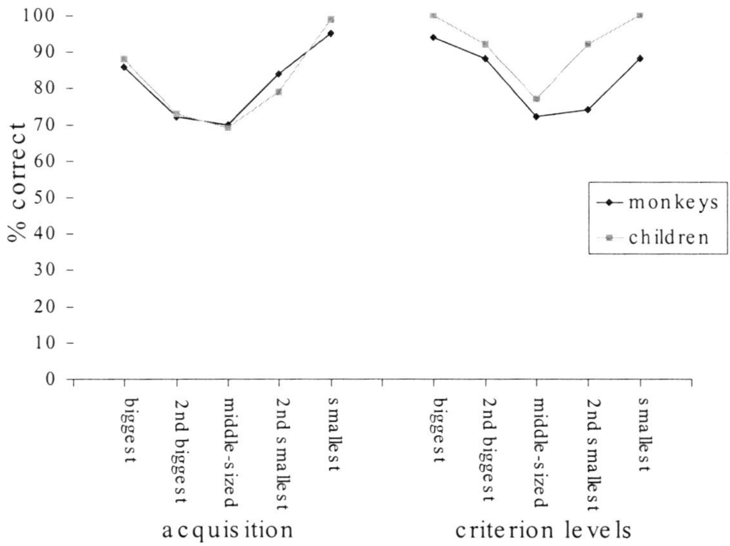

Figure 9. Pattern of size rule acquisition for monkeys and four year old children; data were taken from trial blocks during the later phases of acquisition and when the subjects had reached criterion.

In the monkey, furthermore, whereas choosing the smallest stimulus during binary tests was accompanied by relatively high costs, as reflected in the decision times (Figure 7), these costs now seem to shift to items in the centre of the set when multiple codes must be computed. This is illustrated in Figure 10.

Transferred to a second size range, designed to ensure that subjects computed 'set relative' as distinct from absolute size values, all subjects were able to sustain high levels of performance and recovered criterion levels within four sessions. The strong generalization of codes is shown in Figure 11.

Conclusions

There is now strong warrant for the belief that monkeys are adept not only at size relational codification, but that they can use such multiple relationships concurrently and systematically, following a profile of acquisition and, later, performance which suggests a very similar one to that expressed both in human development and adult human use of linguistic comparatives. That the codes operate in both sets indicates, furthermore, that

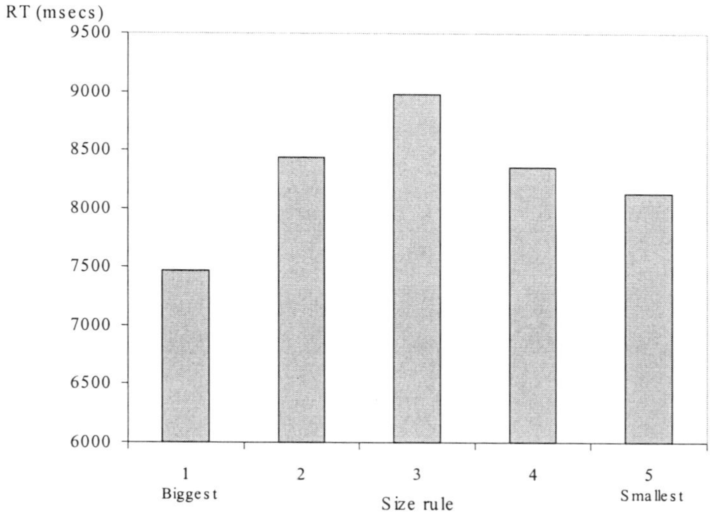

Figure 10. Reaction time distributions for monkeys during steady-state performance on five size rules.

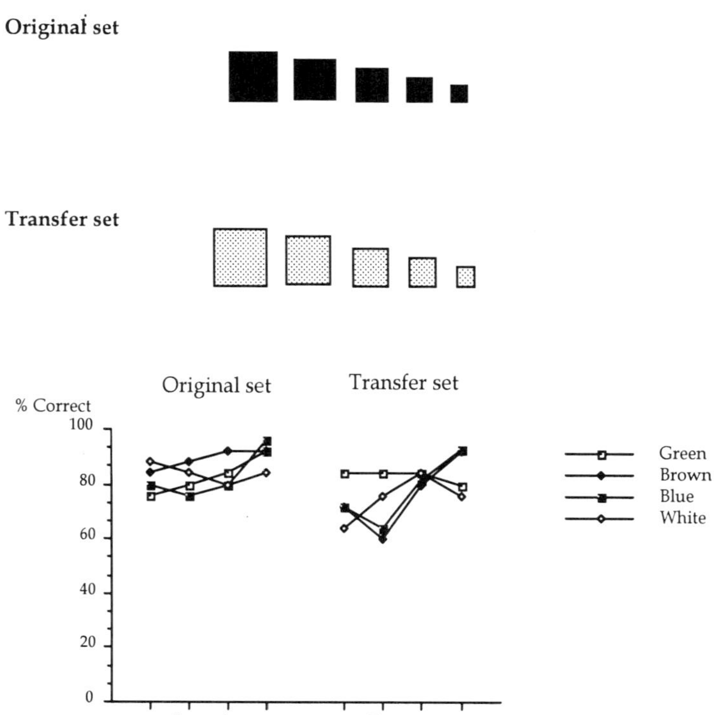

Figure 11. Transfer condition run following acquisition of five size rules and results for all monkeys for the 8 transitional sessions.

the monkey can learn to determine at least four distinct relational codes in a five item set—the 5th and 'middle 'one may only be a default procedure. In this context, it is clear that the 'assembly logic' which enables the individuation of a full set of items is not based on a co-ordinate logic which constructs a third position from two binary codes. Instead, the empirical evidence dictates that a second anchor based on the 'other' end of the range provides a (secondary) basis from which further ordinal specification takes place—see also (H. H. Clark, 1973; Parkman, 1971).

Now a symmetrical profile, the selective uptake of the relational rules, and the syntax of their acquisition indicates a **trajectory** based change in coding as the subjects moved from single to binary to a three valued and then to a five valued set of relational codes. Of course it could be argued that, initially at any rate, monkeys were taught one relational rule at a time; so some serial uncovering of the whole system was inevitable, i.e. that the rules existed all the time, only their 'existence' proof depended on staged procedures. This is a difficult point to answer in almost any domain of learning as the pragmatics of testing often requires a staged or parsed testing procedure first inducing the subject into something easy and tractable before moving to more complex matters. Nonetheless, we would argue, that the selective lag in acquisition of different rules, even under conditions of concurrent testing (five of them), followed by the durable symmetrical profiles recorded post acquisition with RT data, suggests a principled computational logic through which the whole apparatus is seeded from minimalist, inherently asymmetrical relational connections, then altered in a dynamic interaction reflecting greater demands for object individuation and calibration requiring both ends of the stimulus range. The consequence of this transition appears to be a migration from strong directional asymmetry based on one end-anchor to a more symmetrical two-anchor based search procedure.

Certainly the RT-based 'signatures' for performance when subjects operate with only two or even three relational rules is different than in cases where the same rules operate within a context where all ordinal codes must be individuated. This provides a 'look-up' table where the degree of series symmetry may usefully indicate the level of ordinal code specification within the set. And this suggests in turn a diagnostic use for such signatures in circumstances where it is difficult to disambiguate different possible control mechanisms underwriting choice. We now apply them to the assessment of the mechanisms that underwrite relational transitivity: one of the most central competences to be investigated in the domain of cognitive growth and human development.

TRANSITIVITY AND THE SYMBOLIC DISTANCE EFFECT

In the search for high level cognitive mechanisms requiring complex relational calculation, there is hardly a more seductive task than the linear inference task (Halford, 1993). At core, it provides the subjects with minimally two logical arguments as in Piaget's celebrated version, 'Edith is fairer than Lilli, Edith is darker than Susanne' (Piaget, 1928). For Piaget, a critical solution procedure indexical of abstract though was in the subject's co-ordination of the relationship between Edith is fairer than and Edith is darker than so that Edith could be located ordinally as a middle value between Lilli and Suzanne. When children failed (as they invariably did below the age of eleven or so), Piaget attributed to a failure to seriate mentally and to create three ordinal positions, the superlatives, biggest and smallest together with the new middle position, from the binary codes as given in the premises.

However, in more recent developments, memory factors have been implicated as an important restriction on performance in such tasks and a variety of procedures have been developed to ensure that young subjects remember the connecting premises at the time of transitivity testing. In this way, memory failures can be distinguished from a failure to integrate the pairwise information into a series.

Adapting the classic reasoning task for use with young children, Bryant and Trabasso (1971) developed what is known as a 5-term series task into a training paradigm. Formally represented by the predicates A>B/B<A; B>C/C<B; C>D/D<C; D>E/E<D followed by pairwise tests such as B? D, in this version, the child saw a pair of rods protruding from a box, where one was e.g. yellow (A), the other blue (B). To control for memory failure, the child was trained to remember that e.g. the blue one was bigger than the yellow and that yellow was smaller than blue. The child was similarly trained on the three other connecting pairs from the series without ever seeing the sticks varying in size.

As is always the case in transitivity tasks (Youniss & Murray, 1970), care had to be taken that simple end-point labeling strategies were not the basis for solution. The crucial comparison afforded by using five rather than three item therefore was B v D where both these items have been double coded as larger and smaller than. In the first report using this procedure, Bryant and Trabasso produced evidence that that when memory can be eliminated as factor, then children make transitive choices very well, even on the crucial B D comparison. They also supported the view that children were integrating the pairs by showing that the level of transitive choice obtained could be predicted on the basis of the Cartesian product of remembering both connecting pairs.

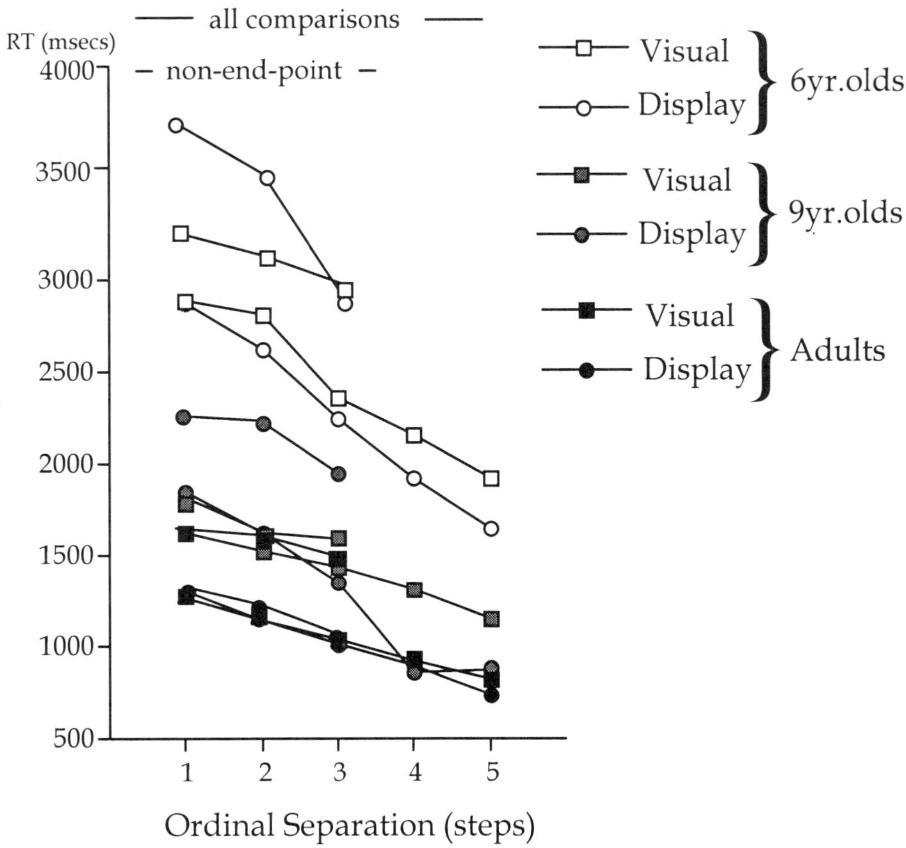

Figure 12. A sample of the SDEs (Symbolic Distance Effect) reported by Trabasso (1977) for different ages and conditions.

In a further development, Trabasso extended the task to a six term one, testing subjects of six and nine years of age as well as adults (Trabasso, 1977; Trabasso, Riley, & Wilson, 1975). In this ground-breaking series of experiments Trabasso and his colleagues also varied the medium of presentation from the quasi-concrete versions used by Bryant and Trabasso to directly perceptual ones where the subjects could either see the size values during training or could see a display of items representing the full series being tested.

With virtually all subjects and all conditions, Trabasso et al. found that transitivity of choice was accompanied by the Symbolic Distance Effect interpreted in the domain of internal psychophysics, as we have reviewed, as indicating a mental series representation of the items in a linear order. A sample of their findings are depicted in Figure 12.

MONKEY TRANSITIVITY

Given that the more recent research on transitivity in children suggested that the lower bound of a transitivity/seriation competence might be much earlier than was at first believed, there were implications for non-humans which we were the first to implement with squirrel monkeys (McGonigle & Chalmers, 1977); replicated in part with chimpanzees by Gillan (1981), pigeons (Fersen, Wynne, Delius, & Staddon, 1991) and rats (Roberts & Phelps, 1994) and now extended into the neuroscience of memory by Eichenbaum and his associates (Dusek & Eichenbaum, 1997). Our experiments with squirrel monkeys (*Saimiri scuireus*) (McGonigle & Chalmers, 1977, 1992) were based on similar conditions of training and test to those of Bryant & Trabasso (1971). Instead of sticks, however, we used coloured tins which differed in weight (not a mediate property of objects) and trained and tested as illustrated in Figure 13. Only two weight values were used throughout, however, so that no simple material scale could be used based on a weight series mapped onto the colour values. In our second study (McGonigle & Chalmers, 1992), RTs were measured during binary tests.

Training

Trained initially to a 90% accuracy on a random sequence of training pairs, all subjects received an overtraining phase on the '92 study in anticipation of extended use of RT data for an analysis of the SDE. This ensured that a minimum level of 90% success could be maintained to each pair over a block of 80 trials of post acquisition criterion testing.

Transitivity and High Retention Levels

The results of both the 1977 and the 1992 study show that choice transitivity was highly significant, and for all monkeys, including the non-end anchored comparisons B v D. High levels of retention were also recorded for training pairs, which was 99.5% correct (McGonigle & Chalmers, 1992).

THE FIRST SYMBOLIC DISTANCE EFFECTS IN MONKEYS

Uniquely, each individual was examined for evidence of an SDE. For each 100 trial test block, subjects were subjected to analysis of variance to see there was any significant variation across test pairs. If there was no such

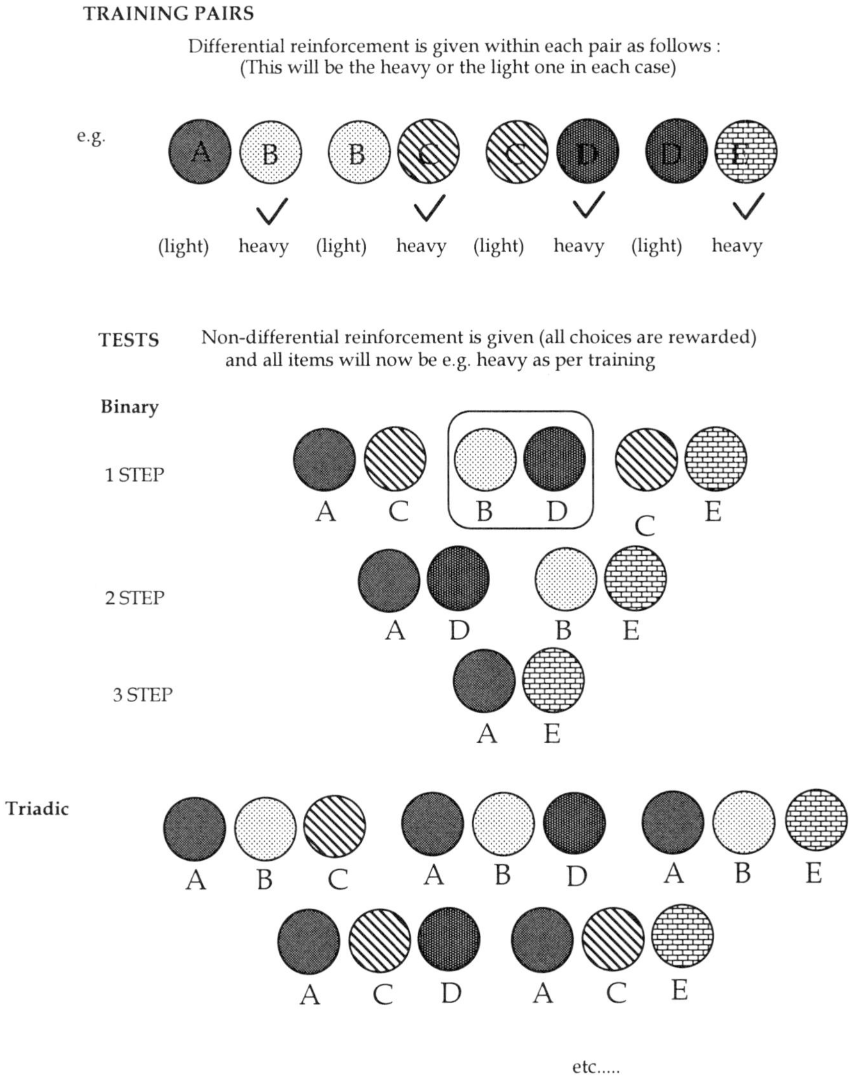

Figure 13. The transitivity paradigm used with monkeys by McGonigle & Chalmers (1977; 1992).

variation, monkeys went on to a second 100 trial test module. Similarly if pair variation was not accompanied by a significant effect of ordinal separation, or if any effect of ordinal separation was not significantly linear, the monkeys were maintained on testing, for a further module of ten (by ten) observations, and so on.

All monkeys recorded an SDE by the third ten by ten test module: one (Blue) showed the effect by the end of the first module, two (Green and

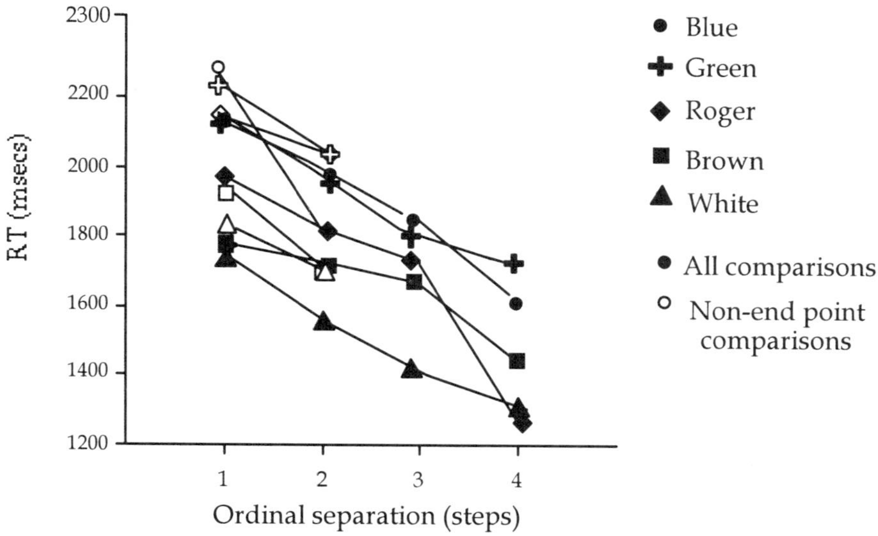

Figure 14. Symbolic Distance Effects obtained for individual monkeys.

Roger) by the end of the second and two (Brown and White) by the end of the third. No other significant RT variation was recorded. As with the marking effects we report earlier, practice did not abolish the effect. Instead, it sharpened the SDE in individuals most probably due to the reduction in 'rogue' variance or noise. All effects were also accompanied by a Symbolic Distance Effect for the non-end point comparisons which was highly significant for the group as a whole as Figure 14 illustrates.

Symbolic Distance Effects and Their Implications

In a long theoretical analysis of his team's research on the SDE, Trabasso (1977) entertains many different models all of which make similar predictions but which have very different processing implications. Trabasso tends, however, to favour an imagery account whereby a spatial layout is used 'in the mind's eye', then scanned in memory. Based on 'think aloud' protocols where some subjects at least report the use of spatial paralogical device enabling test items to be read off as if physically present, Trabasso (1977) envisages the read-off a serial search from one end of the series:

> "...suppose that the child represents the coloured sticks as an ordered list (1,2,3,4,5,6). When a question on the relationship between a pair of sticks is posed, he begins by scanning the list from the end mentioned in the question. If "longer," then he starts with 6; if "shorter," he starts with 1. Suppose the question is on the

> relationship between Stick 2 and Stick 4. If the question contains "shorter?" then he must scan two members, 1 and 2, to find 2, and he can terminate at this point. If the question contains "longer", he will scan three members, 6, 5 and 4, to find 4. On average 2 ½ members are scanned. For a question on Sticks 2 and 3, however, the respective number of members scanned are two and four, or an average of three members." (page 346/347).

A modification of that view which he then proposes, because it produces a better fit with the empirical data, is that the search is self-terminating after only one item is found. Both versions of the search model, however, are based on the idea that RT reflects distance from an end-point rather than inter-item distance *per se*. This is very compatible with the Parkman's search model for digit comparisons and the mechanisms suggested above for ordinal rule learning by monkey.[1] However, the SDE as it stands does not reveal the nature of the search mechanisms, nor how advanced search need be to produce transitive choice and the distance effect. Is it, for example with – or without – benefit of symmetrical series representation that would allow **both** ends of the series to be used as reference points?

Extended Analysis of the Symbolic Distance Effect

Given the unquestionable importance of RT data as a means of arbitrating between competing theoretical accounts, McGonigle and Chalmers (1992) undertook an extensive sub-analysis of the SDE in monkeys. With an new affluent data base per subject, McGonigle and Chalmers (1992) were able to evaluate selective directional effects, and compare the RT signatures obtained with those as reported earlier in the size relational conditions. As with our relational data, we now assessed the role of different ends of the series in the SDE, plotting decision times from each pole separately. This revealed for both the group as a whole, and for each subject bar one (who showed a symmetrical effect—the same one, White, who showed symmetry in the binary size condition) a highly **asymmetric** profile as the wedge function illustrated in Figure 15 reveals. The RT data were found to assort into two populations. One, a population of fast scores, was insensitive to distance effects. The second population was derived from slower RTs and these did show a strong distance function and internal ranking as a function of proximity to the slow end-point.

---

[1]Parkman's model suggests that zero is the reference point for digits; for size, the monkey and child data suggest it is the largest stimulus in the field.

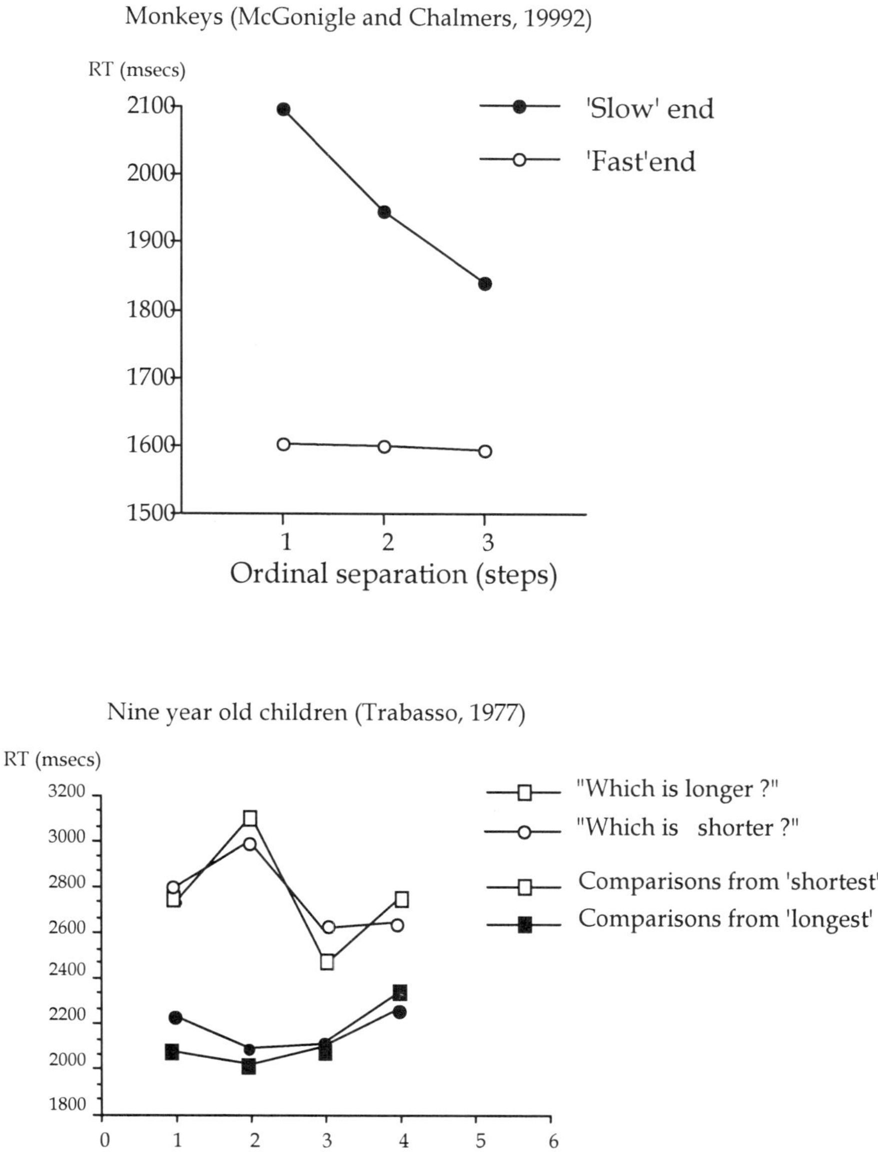

Figure 15. The SDE for monkeys and nine year old children (Trabasso, 1977) plotted separately from each end-point of the series.

These data are certainly counter-indicative of a 'read-off' from a Euclidean mental array and suggest, instead, that subjects' decision times are determined largely by proximity of the test items to a "single privileged anchor point." As monkeys, are, however, effectively only 'asked' to use one type of comparative (heavier **or** lighter denotes the direction of training), it is difficult to predict whether or not they would show symmetrical distance effects when trained with the equivalent of both comparatives. However, a re-analysis of data from Trabasso's child subjects suggests that children as old as nine (in contradistinction to adults) do **not** in fact show any 'cross-over' when the instruction changes from longer to shorter. As Figure 15 shows, there is similar evidence for series asymmetry in the child data, independently of the question asked.

Conclusions

Using our own size rule ‘signatures’ described earlier as a basis for interpretation of the levels of control implied here, we can now locate the transitivity mechanisms deployed by monkeys somewhere between an ‘early’ stage of open set control, lacking the closed, or bounded set symmetry of processing these same subjects expressed early in the five rule training and a more principled search procedure allowing search from both ends. That RTs may be ranked as a function of proximity to the 'slow' end of the series—giving rise to the SDE—suggests that monkeys are using a conjunction of a relational tag (e.g. heavy) and its negation (not heavy). Comparisons involving the ' not heavy' end-point or items associated with this anchor through proximity (Breslow, 1981), add to the RT through the operation of negation. The more likely it is that a decision can be made without negation—by using the positive, (e.g. heavy) reference point or one close to it, the faster the decision. This would conform to Heidenhemer's model for the use of antonymic terms early in development, and to one we have devised for quite independent studies of children's mental comparisons (McGonigle & Chalmers, 1984). In the present context, such a model was give a formal warrant in a production systems model devised by Harris and McGonigle (1994) and informed by our squirrel monkey data. Based on production systems each item (in the transitivity series A- to E+) was represented as a condition-action rule such as 'if E then select E' and negation-based rules such as 'if A then avoid A'. Each rule had a place in an overall stack. Thus 'if E then E (select)', could be followed in the stack by 'if A avoid A', etc. An example would be

1. If E select E
2. If A avoid A
3. if D select D
4. If C select C

A crucial property of this model is that *all* stacks that perform correctly on the adjacent pairs also perform correctly on the remote pairs without making further assumptions. This contrasts with other formal models such as the rank tensor product model of Halford (Halford, Wilson, & Phillips, 1998), in which a transitive decision about a transitive pairing is higher-level and more complex than a decisions concerning adjacent pairings.

A main implication of these converging analysis both formal and empirical is that transitive choice, for primates at least, may occur early within a trajectory, based on primitive forms of relational judgment. These leave open many possible new developments, not the least of which is the migration to finite set, symmetrical ends inward processing using two comparatives rather than one and its negation.

Triads

Some confirmation of this within the transitivity paradigm has been provided by our use of triads. As in the size conditions described above, Chalmers and McGonigle (1984) and McGonigle and Chalmers (1977, 1992) followed binary tests of transitivity with triadic tests derived permutatively from the same training set. Whilst choices were significantly transitive under these circumstances for both monkeys and children, there was an initial drop in the strength of the transitive profile for both species as Figure 16 illustrates.

Triads, where there is no perceivable solution, do, of course, require more finely tuned codes than do binary choices. This is formally shown in the Harris and McGonigle (1994) model which show that certain triads, at least, require a deeper search to disambiguate the 'winning' item. This further constrains the set of rule stacks which can sustain high levels of transitive choice in triadic situations. Take the case of the rule stacks 'if E, select E' followed by 'if A, avoid A' and *vice-versa*. In either case, the first rule reduces overall uncertainty by 40% by accounting for 40% of all pairs; the second rule by a further 30%. In the case of the triads, however, the 'avoid' rule leaves the residual uncertainty concerning which of the remaining two items to select in every triad. Thus, 'Avoid A' now has 0% uncertainty reduction, therefore, whether in first or in second place in the stack. The first useful constraint that would thus become apparent within triads is the need to proceed monotonically on the basis of a **affirmative** selection strategy. Stacking 'If E, select E', followed by 'if D, select D' and so on, produces the highest rate of information gain measured as uncertainty reduction. In the binary case, there would be no such advantage for a monotonically ordered search (McGonigle & Chalmers, 1998).

That response competition might motivate the beginnings of what might be genuine linear search was apparent from spontaneous changes in the monkeys' triadic response distributions as a function of test exposure.

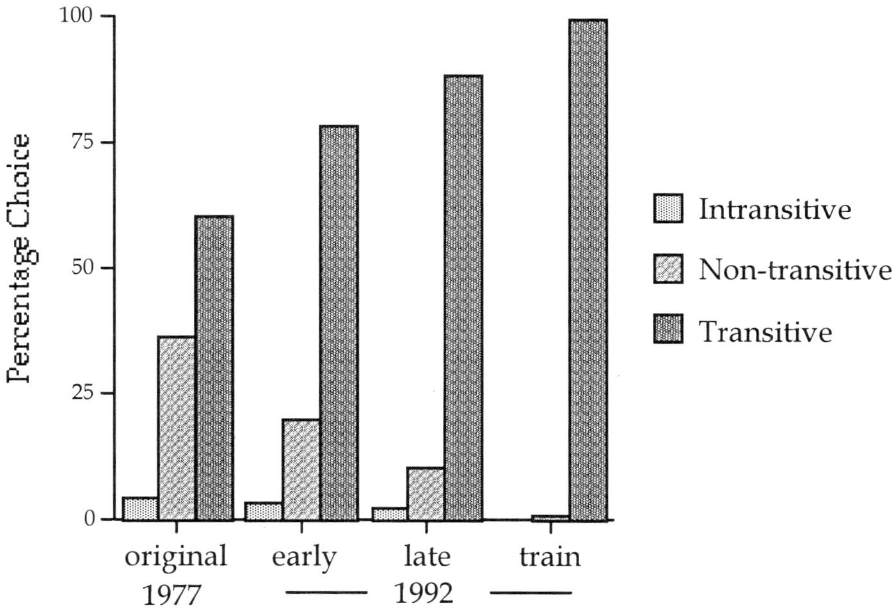

Figure 16. The spontaneous drift in triadic choice distributions for monkeys towards 100% transitivity.

Although there was an initial drop in overall levels of transitivity, even though these remained at a highly significant level, McGonigle and Chalmers (1992) found that with protracted exposure to triadic tests, monkeys achieved a level of 92%, revealing a shift in choice behaviour such that later triads revealed a level of transitive choices almost as high as in the binary case. This shift is illustrated in Figure 16, and occurred without differential reinforcement within triads.

Triads and Seriation

The crucial advance brought about by establishing a bounded set with two reference points is, as we have seen from relational rule learning the formation of a scale on which objects can be differentiated not just by the direction of comparison (bigger versus smaller), but also by establishing a finite range within which stimuli can be both calibrated and searched from both ends in accordance with the response demands of the task. In the case of object relations, it was necessary with the primate to make calibration explicit by requiring multiple responses to be sustained concurrently within the same set of values.

In the case of transitivity, however, only inferences can be drawn about the underlying calibration of the test series. And the signatures from the RT assays suggest that transitive choice is the result of low level relational encoding dominated by a single salient reference point. Whilst the triadic conditions increase item competition and induce more fine tuning, they neither motivate, nor do they require extensive ordinal codification or the extensive search that a five term series could **in principle** support.

Conclusions

Given that young children show similar patterns of binary and triadic transitive choice to those of monkeys (Chalmers & McGonigle, 1984) and similar RT profiles during test (Trabasso, 1977), the conclusion which we must inevitably draw from this is that a core rational mechanism demands rather less ordinal computation and explicit search than was once thought. Given the adaptive importance of the mechanism, furthermore, this is perhaps as it should be! It is initially, on this view, a 'small world' phenomenon adapted for small sets of alternatives, a limited and bounded rational mechanism where the ranking of test object codes is based at best on their derived distance from one or two stable reference markers, and where the search is non-exhaustive, terminating when the target item is found. Such mechanisms may indeed form the basis of judgments in real space involving deictic codification of the relative position of environmental features with respect to a common fixed source. Could this be the reason perhaps that Dusek and Eichenbaum (1997) find hippocampal substrates underwriting transitive choice in rodents in ways that Morris (1982) and others do for spatial codification? The 'landmark' logic may be similar.

Seriation

If transitivity initially reflects a limited search based ordering procedure, as we suggest, where ranking of items internal to the set derives from their relative distance from a common end anchor, it is perhaps not surprising that explicit tests of seriation which require the individuation of every item in the set *vis-a-vis* every other one of the set, emerges later in human development than the earliest demonstrations of transitivity. Achieved normatively by children around seven years, this classic seriation task as devised by Piaget (Piaget & Szeminska, 1952), requires subjects to explicitly order a large number of objects (8-10) differing in size, placing them in row. Here no inference is required that ordering of all the items has occurred as the behaviour is overt and public—especially when the child orders in a principled way (selecting and placing the rods in a monotonic sequence without trial and error), as in what Piaget terms 'operational' seriation.

In such a sequencing task, however, the use of a finite or bounded set procedure to achieve full ordinal specification by computing for example, biggest, second biggest, third smallest, etc. as a preamble to seriating them would be a cumbersome and almost certainly inaccurate procedure for sets larger than a few items—and the classic version conventionally employs 8 - 10. Instead, it would seem much more likely that the executive demands of serial ordering items within a single production (as distinct from successive relational judgments made in simple choice situations), demand much more streamlined mechanisms designed to support rapid serial control as in piano playing or typing (Lashley, 1951). However behaviours of this sort cannot be studied readily using conventional choice mechanisms as in transitivity where behaviour 'halts' following a single decision to choose one object over another. Yet conventional seriation tasks are impossible to administer to non-humans given the serious manipulative restrictions imposed by their motor control systems. These embargo the adoption of the 'select and seriate' tasks used by Piaget. To eschew these problems, we turned to touch screen based versions of seriation, developing a family of both supervised (trained) and free search based ordering mechanisms that enable us to establish a common currency of measurement in comparisons between human and non-human species.

TOUCH SCREENS AND EXECUTIVE MEMORY PARADIGMS

With touchscreens, we can now help deliver on a long-standing ambition of Lashley (Lashley, 1951), when he urged the study of serial ordering as a key to understanding the nervous system and a key player in adaptive behaviours. For few behaviours, whether at the level of language or at the level of action, make much adaptive sense without evidence of serial control, be it based on simple finite state transitions or higher order dependencies as in word strings.

However, until recently, the learning of serial orders has rarely been demonstrated (Terrace & McGonigle, 1994; Terrace, 1987; McGonigle, 1987). In simple cases of serial behaviour occurring in natural contexts, instinct rules have been posited; in complex linguistic constructions, generative grammars have a strong influence. As for child development, of which the development of seriation is one of its most robust indicants, the dynamic constructivist position of Piaget has viewed learning as a poor relation—the consequence and not the cause of cognitive growth.

A key feature of our new procedures in this context, therefore is that we can train our subjects to seriate, manipulating key independent variables such as string length, and the stimulus characteristics of the test set. These

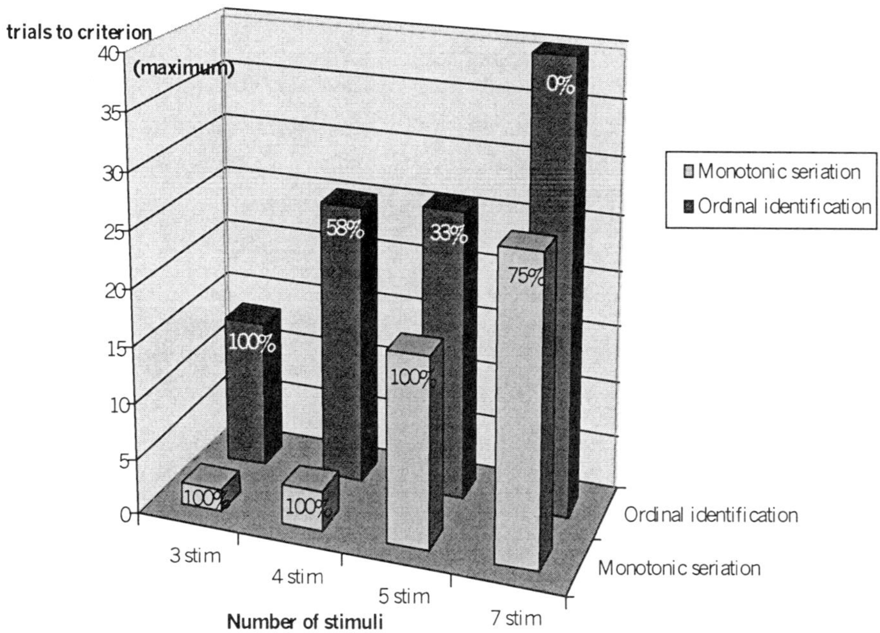

Figure 17. The relationship between learning ordinal identification for five sizes and learning to monotonically order those sizes by five year old children. The percentages on the bars represent the number of children who succeeded in reaching criterion within 40 trials (per series and per size rule).

techniques also allow subjects free search on the touch screen, enabling them to devise their own means of keeping a track of the items they have already selected. Our first experiments of this sort were of the supervised variety carried out with children of five and seven years of age (Chalmers & McGonigle, 1997). Required, for example, to touch five items (rectangles) arranged in a random linear array on the screen in the order from biggest to smallest, children as young as five were able to learn to do so using error free search within a single session or so. Whilst their executive control is clearly inferior to that of seven year olds under these circumstances, this skill is nevertheless much more easily trained in young subjects than the concurrent learning of set-referenced ordinal codes for each and every item in the set as measured by both by a matching to sample procedure and by the five rule colour conditional procedure described for actual objects earlier. This relationship is illustrated in Figure 17.

Interpreting these findings in terms of an economy-based model of serial organization (McGonigle & Chalmers, 1998; McGonigle & Chalmers,

2001), we have argued that linear monotonic size seriation is an example of a highly adaptive form of on-line memory organization, where the principled iteration of an elementary code allows a large number of items to be searched uni-directionally and rapidly on-line—without requiring the costly off-line codes of ordinal position that locate each item within a set bounded at both ends of the range.

LINEAR AND HIERARCHICAL CONTROL IN EXECUTIVE MEMORY

One of the general implications of this is that higher order cognitive organisation in humans, such as linear ordering will be inevitable on the grounds that it reflects an optimising procedure where solutions are selected both to minimize memory cost and to maximise adaptive utility (Anderson, 1990; McGonigle & Chalmers, 1998). Linear search procedures aside, classification procedures which segment the serial production into manageable chunks or units should also be selected on this view for their data reducing memory effects if these can be shown to have utility value.

In the final study we report here, based on seven *Cebus apella,* we have combined both linear size and class based ordering procedures in the same experiment run concurrently To induce concurrent classification and linear size seriation we used a novel design which first trained the subject to order three different shapes. Each shape stood for a category of icons which ultimately varied in size. Thus, from the first simple string, we generated longer sequences featuring multiple exemplars of each shape category, also varying in size. At first only three levels of size were used throughout the series. Thus the same three sizes recurred from category to category, and the subjects had to order both the categories on a shape basis, and the exemplars within each on a size basis. The rationale here was to assess the extent to which the subject could combine linear size and class based information to order a full set of nine items. Here, shape information could support only three ordinal positions, and size could only support three. So to sustain the nine item sequence, subjects had to combine shape and size information, nested hierarchically such that shape provided an ordinal segment (of three items per segment) individuated by size seriation, so that nine independent values were possible (see Figure 18).

Finally, we required the *Cebus* to order all nine items along a linear size series of nine different sizes. This was done under two main conditions run concurrently. In one all sizes came from either category A or B or C. In another, the linear series was overlaid with three categories with the first three sizes of the series going to category A, the next three to category B,

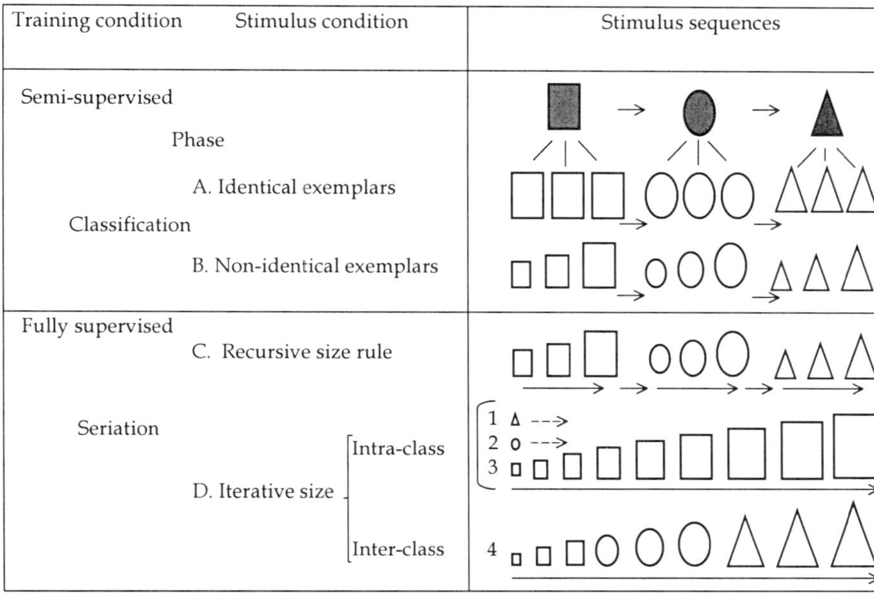

Figure 18. The design of a longitudinal study of monkey classification and seriation carried out with *Cebus apella.*

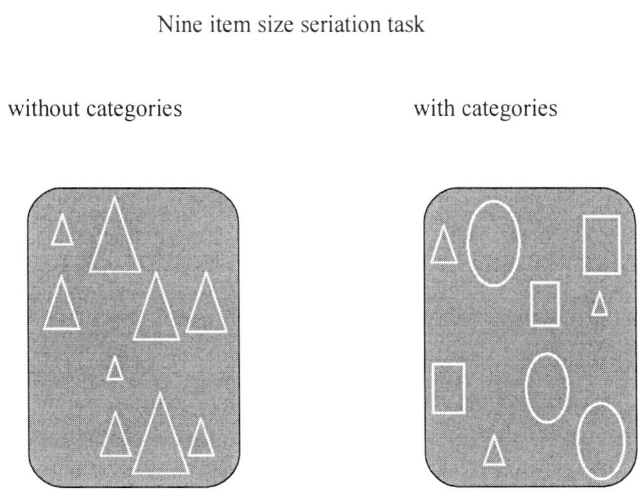

Figure 19. Examples of nine-item seriation tasks given to monkeys as they appear on the touchscreen.

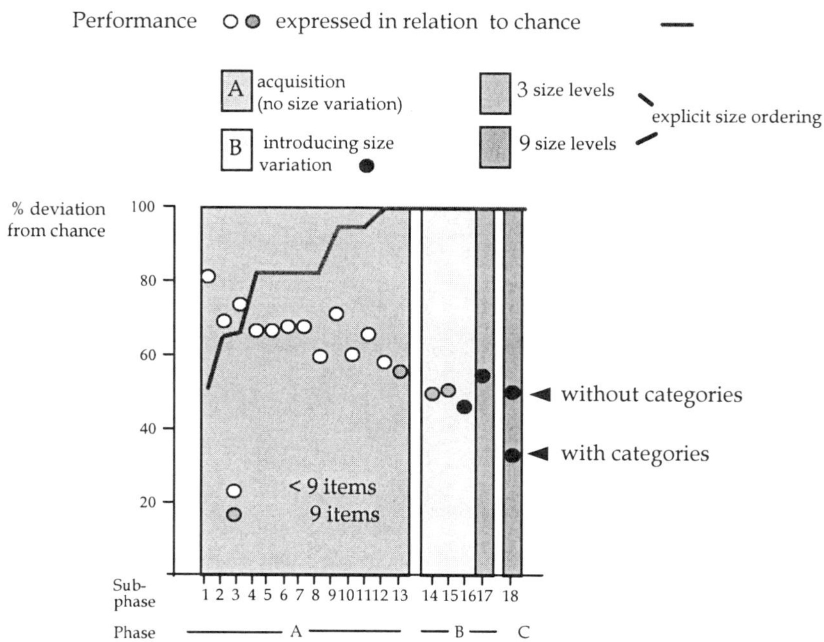

Figure 20. Learning-to-learn profile from *Cebus apella* as they proceeded through stages of the nine-item classification and seriation study. Errors committed during each sub-stage of training are expressed as a deviation from the errors expected on chance assumptions. The final points show the comparison between nine item size seriation with, and without, opportunities for classification.

and the final three to category C. The rationale here was to determine the extent to which adding category differences to a linear size series led to improved performance. Figure 19 shows the two types of sequence deployed as they appear on the screen.

Results

Over a lengthy series of stages, in an experiment lasting several years, monkeys achieved nine item seriation. As Figure 20 shows, monkeys adapted to progressive increases both in sequence length and the levels of specification demanded of the constituents of the sequences. Early stages of sequence production could be controlled, furthermore, by coarse chunks of three classes each containing identical exemplars and individuated only by their relative position to one another within the test array. Later sequences, however, required the differentiation and then principled ordering of both exemplars and categories.

In this way, the tasks can be ordered in hierarchies of difficulty. Apart from greater constituent specification in later tasks, the progressive increase in the number of constituents *per se* renders later problems subject to serious combinatorial (explosion) effects. That linear nine item size seriation is also solved, therefore, with no disproportionate increases in error, as Figure 20 also shows, indicates to us at least the high utility value of unidirectional size coding where a series can be generated, and successor items anticipated prospectively with a low cost search procedure. A further demonstration of this is provided by the enhanced utility value which categorical seriation has over the purely linear case. Figure 20 shows that at the final nine item size seriation stage, performance is uniformly better on the multiple classification condition, although both sorts of seriation return high levels of performance.

Conclusions

Nine item size seriation is regarded in human development as a benchmark of cognitive growth. Whilst the methods reported here required that monkeys were trained to do what children appear to do spontaneously at around the age of seven, their attainment of this goal is nevertheless the first of its kind to be reported, and at an operational level executed as a principled selection procedure on every successful trial. As such, it represents the most complex form of organising multiple elements in executive memory in the non-human literature. The most recent claims (Brannon & Terrace, 1998) for nine item ordering (based on numerosity) by two rhesus monkeys whilst fully consistent with these findings are based on data obtained from overlapping sequences. The objective measure of the utility of classification within a sequencing task, furthermore, gives strong warrant for the belief that the classifier system is being actively used.

Now that the objective utilities of different sorts of search procedure can be evaluated over and above successful performance, we are now addressing the question of the extent to which subjects themselves can detect and discriminate these relative utilities and self regulate executive control in the context of unsupervised free search tasks (McGonigle & Chalmers, 1998, 2001). As such tasks do not presume the way in which such strategies might be expressed (by imposing a 'right' answer on the subject), these techniques are particularly suitable in certain clinical applications as we review briefly below.

CLINICAL APPLICATIONS AND ANIMAL MODELS

There now seems little to constrain comparisons between non-primate and primate cognitive functioning at almost all levels of functioning. Executive control, spontaneous strategic elaboration, classification, phrasing, search routines and the use of state based externalised feedback to further reduce memory load (McGonigle & Chalmers, 2001)—are all now possible. In addition, the measurement of performance in real time enables on-line processing effects to be more clearly separated from off-line codification in long-term memory. For human clinical conditions, where cognitive dysfunction is under review, impairment can now be evaluated using a rich possible degradation space which enables the clinician to distinguish between meaningful (relational) and associative memory, active versus passive memory deficits, and the relative failure of patients to construct appropriate strategies in the light of progressive and well calibrated task demands which impose an ever increasing memorial burden on the subject. In the context of Alzheimer's research, for example, recent work at Edinburgh and Newcastle Infirmary has been heavily reliant on our free search tasks as described above in helping determine the sorts of strategies, if any, patients with dementia can still furnish to minimise memory loss. In addition, with a mix of supervised and unsupervised (free search) procedures, we are in a position to evaluate the extent to which strategies can be re-acquired or taught *de novo*—a remediation issue in this domain.

Another application of this work to which we have recently turned our attention concerns the characterisation of cognitive deficit associated with autism and the sex-chromosome linked condition, Fragile X syndrome. The cognitive deficit associated with such disorders are usually established through psychometric testing (especially using the Kaufman Assessment Battery for Children), and in the case of Fragile X are already thought to be as sequential in nature (Dykens, Hodapp, & Leckman, 1994). But with such testing devices the cognitive diagnosis of the condition is unlikely to advance further. Apart from the shallow, one-off nature of psychometric test items, they also confound linguistic and educational retardation with core deficits which might cause such retardation in the first place. It seemed to us that methods such as those we have described remove many of these problems at a stroke. Highly motivating and essentially non-verbal in nature they require low levels of manual dexterity, minimal communication with the experimenter, presume on no particular taught or world knowledge, and can be geared to very reduced levels of competence such as a 2-item sequence. But above all, in being capable of fractionation into component skills, they allow us to specify learning pathologies with a level of precision not available from standard techniques (Chalmers, 1998). An example is

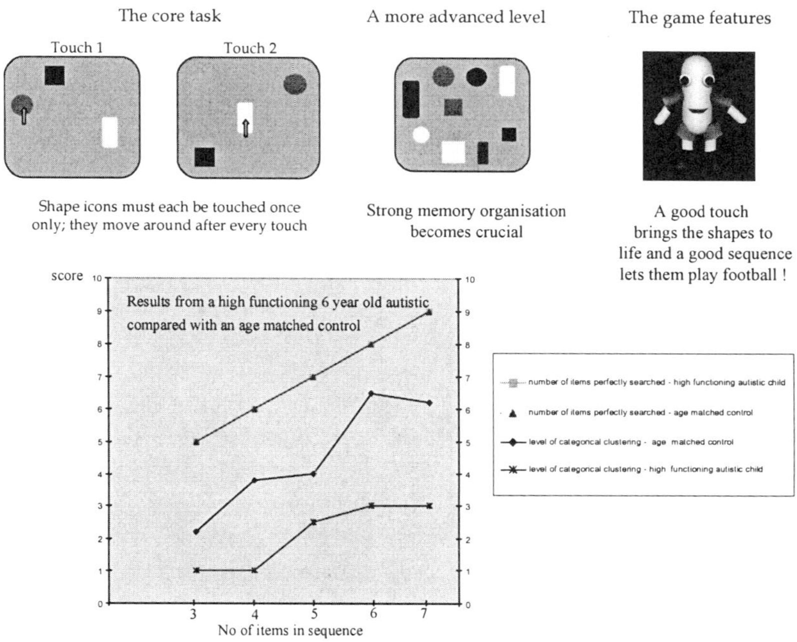

Figure 21. Example of a free search paradigm and results from an autistic subject showing perfect sequence learning for up to 7 items, but without the degree of organisational control found in a normal age matched control.

given in Figure 21, depicting a free search task as given recently by us to autistic children (Chalmers, In preparation) and children with Fragile X (Chalmers, In preparation), together with a learning profile from a high functioning autistic child, revealing that, whilst sequential learning was as good as that found with an age matched unaffected child, the signatures for categorical clustering were absent in the autistic child.

The obverse has also been found using our tasks, where an apparent failure to learn to make an exhaustive search through a set of four items by a Fragile X subject revealed a 'hidden' ability to show organisational constraints in sections of his performance similar to those exhibited by a normal (mental) age matched control subject. In both these cases we can now pursue the implications of spared and/or absent competences by explicit training or further task variations. For example, our autistic subject is to be re-tested, using new highly flexible testing programs on a task that requires exhaustive on-line search as before, but that will disallow (by varying the items from trial to trial), the characteristic autistic strategy of rote memorising a specific list of items. This should reveal, what, if any, volitional control such an autistic subject might show when confronted with visual input that demands flexibility within short-term memory, by tailored experimental techniques still relatively rare in the area of childhood cognitive dysfunction.

CONCLUSIONS AND THE FUTURE

We have offered a sample of experiments from within our programme because we believe that it is time for a new agenda to help make possible a full behaviour-based analysis of cognitive complexity without requiring language.

In this biomedical context, we are only too aware that a persistent problem in the neurosciences concerned with human cognitive function and its degradation as in the amnesias, is to find animal (species) and paradigms which will enable the neuroscientist to explore brain-behaviour relationships in animals crucially homologous with higher order functions in man. Although similarities in design logic as assessed at the behavioural level do not necessarily imply a similar form of implementation at the 'wetware' levels, it does at least create a climate of competence fractionation and evaluation enabling neuroscientists to asks more accurate questions of those adaptive competences brains (of whatever species) have evolved to deliver.

Certainly a rich space of measurement enables us now to converge comparative research with clinical and neuroscience research areas, where a major problem has been to establish task scenarios which enable both long term assays of performance longitudinally and which keep the subject's interest. With our latest generation of software we can both create intrinsically interesting interactive tasks which keep even very young subjects of 2.5 years engaged, and for long periods. And we have tested primate subjects for years in longitudinal studies in these test environments. Now being developed as an intelligent tutoring system which automatically adjusts task type, task difficulty and training conditions based on samples of a subject's ongoing performance, our new system is conceived both as a laboratory instrument for cognitive research and as a portable prototype to support a wide variety of clinical applications.

REFERENCES

Anderson, J., R. (1990). *The adaptive character of thought.* Hillsdale: Lawrence Erlbaum.

Banks, W. P. (1977). Encoding and processing of symbolic information in comparative judgments. In G. H. Bower (Ed.), *The psychology of learning and motivation* (Vol. II). New York: Academic Press.

Boakes, R. (1984). *From Darwin to behaviorism. Psychology and the minds of animals.* Cambridge: Cambridge University Press.

Brannon, E. M., & Terrace, H. (1998). Ordering of the numerosities 1-9 by monkeys. *Science,* **282**, 746-749.

Breslow, L. (1981). Re-evaluation of the literature on the development of transitive inference. *Psychological Bulletin,* **89**(2), 325-351.

Bryant, P. E. (1974). *Perception and understanding in young children*. London: Methuen.

Bunsey, M. & Eichenbaum, H. (1996). Conservation of hippocampal memory function in rats and humans. *Nature*, **379**, 255-257.

Chalmers, M. (1998). *Identifying core causes of complex cognitive deficit in Fragile X syndrome.* Paper presented at the 8th. Annual meeting of the Society for the Study of Behavioural Phenotypes, Johns Hopkins University , Baltimore.

Chalmers, M., & McGonigle, B. (1984). Are children any more logical than monkeys on the five term series problem? *Journal of Experimental Child Psychology,* **37**, 355-377.

Chalmers, M., & McGonigle, B. (1997). Capturing dynamic structuralism in the laboratory. In L.Smith & J.Dockrell & P.Tomlinson (Eds.), *Piaget, Vygotsky and Beyond* (pp. 183-200). London: Routledge.

Chomsky, N. (1980). *Rules and representations*. Oxford: Blackwell.

Clark, E. (1973). What's in a word? In T. E. Moore (Ed.), *Cognitive development and the acquisition of language*. New York: Academic Press.

Clark, H. H. (1969). Linguistic processes in deductive reasoning. *Psychological Review,* **76**, 387-404.

Clark, H. H. (1970). The primitive nature of children's relational concepts. In J. R. Hayes (Ed.), *Cognition and the development of language.* New York: Wiley.

Clark, H. H. (1973). Space, time, semantics and the child. In T. E. Moore (Ed.), *Cognitive development and the acquisition of language* (pp. 27-63): New York: Academic Press.

Donaldson, M., & Wales, R. J. (1970). On the acquisition of some relational terms. In J. R. Hayes (Ed.), *Cognition and the development of language*. New York: Wiley.

Dusek, J. A., & Eichenbaum, H. (1997). The hippocampus and memory for orderly stimulus relations. *Proceedings of the national academy of science, USA,* **94**, 7109-7114.

Dykens, E. M., Hodapp, R. M., & Leckman, J. F. (1994). *Behaviour and development in Fragile X Syndrome*. Thousand Oaks: Sage.

Heidenheimer, P. (1975) The strategy of negation and the learning of antonymic relations. *Developmental Psychology*, ii, **6**, 757-762.

Fersen, L. v., Wynne, C. D. L., Delius, J. D., & Staddon, J. E. R. (1991). Transitive inference formation in pigeons. *Journal of Experimental Psychology: Animal BehaviourPprocesses,* **17**, 334-341.

Fischer, K. W., & Canfield, R. L. (1986). The ambiguity of stage and structure in behavior: person and environment in the development of psychological structures. In I.Levin (Ed.), *Stage and Structure: re-opening the debate* (pp. 246-267). Norwood, N.J.: Ablex.

Fodor, J. A. (1983). *Modularity of Mind.* Cambridge: MIT Press.

Fodor, J. A., & Pylyshyn, Z. W. (1988). Connectionism and cognitive architecture. A critical analysis. *Cognition,* **28**, 3-71.

Gallistel, C. R. (1995). The replacement of general purpose theories with adaptive specializations. In M. S. Gazzaniga (Ed.), *The Cognitive Neurosciences* (pp. 1255-1267). Cambridge: MIT Press.

Gazzaniga, M. S., Ivry, R. B., & Mangun, G. R. (1998). *Cognitive neuroscience. The biology of the mind.* New York: Norton.

Gentner, D. (1983). Structure-mapping : A theoretical framework for analogy. *Cognitive Science,* **7**, 155-170.

Halford, G. (1993). *Children's understanding. The development of mental models.* Hillsdale: Lawrence Erlbaum Associates.

Halford, G., Wilson, W. H., & Phillips, S. (1998). Processing capacity defined by relational complexity : Implications for comparative, developmental and cognitive psychology. *Behavioral and Brain Sciences,* **21**.

Fein, G.G. (1972) The effect of within-pair variation and instructions on the transposition behavior of kindergarten and third grade children. *Journal of Experimental Child Psychology*, **14**, 379-397

Harlow, H. (1949). The formation of learning sets. *Psychological Review,* **56**, 51-65.

Harnad, S. (1987). *Categorical perception; the groundwork of cognition.* Cambridge: Cambridge University Press.

Hummel, J. E., & Holyoak, K. J. (2001). A process model of human transitive inference. In M. Gattis (Ed.), *Spatial schemas and abstract thought.* London: MIT Press.

Inhelder, B., & Piaget, J. (1964). *The early growth of logic in the child.* London: Routledge and Kegan-Paul.

Jackendoff, R. (1983). *Semantics and cognition.* Cambridge: MIT Press.

James, W. (1891). *The principles of psychology. Vol II.* London: Macmillan.

Johnson-Laird, P. N. (1983). *Mental models.* Harvard University Press.

Karmiloff-Smith, A. (1992). *Beyond modularity.* London: MIT Press.

Kuenne, M. R. (1946). Experimental investigation of the relation of language to transposition. *Journal of Experimental Psychology,* **36**, 471-490.

Lashley, K. S. (1951). The problem of serial order in behaviour. In L.A.Jeffries (Ed.), *Cerebral mechanisms in behaviour*. New York: John Wiley.

Lawrenson, W., & Bryant, P. E. (1972). Absolute and relative codes in young children. *Journal of Child Psychology and Psychiatry,* **13**, 25-35.

Luger, G. (1994). *Cognitive Science. The science of intelligent systems*. San Diego: Academic Press.

McGonigle, B. (1987). Non-verbal thinking by animals. *Nature,* **325**, 110-112.

McGonigle, B., & Chalmers, M. (1977). Are monkeys logical ? *Nature,* **267**, 694-697.

McGonigle, B., & Chalmers, M. (1980). On the genesis of relational terms : A comparative study of monkeys and human children. *Antroplogia Contemporanea,* **3**, 236.

McGonigle, B., & Chalmers, M. (1984). The selective impact of question form and input mode on the symbolic distance effect in children. *Journal of Experimental Child psychology,* **37**, 525-554.

McGonigle, B., & Chalmers, M. (1986). Representations and strategies during inference. In T.Myers & K.Brown (Eds.), *Reasoning and Discourse Processes*. London: Academic Press.

McGonigle, B., & Chalmers, M. (1992). Monkeys are rational ! *The Quarterly Journal of Experimental Psychology,* **45B**(3), 189-228.

McGonigle, B., & Chalmers, M. (1996). The ontology of order. In L. Smith (Ed.), *Critical readings on Piaget*. London: Routledge.

McGonigle, B., & Chalmers, M. (1998). Rationality as optimized cognitive self-regulation. In M. Oaksford & N.Chater (Eds.), *Rational models of cognition* (pp. 501-529). Oxford: Oxford University Press.

McGonigle, B., & Chalmers, M. (2001). Spatial representation as cause and effect : Circular causality comes to cognition. In M. Gattis (Ed.), *Spatial schemas and abstract thought* (pp. 247-277). London: MIT Press.

McGonigle, B., & Jones, B. (1978). Levels of stimulus processing by the squirrel monkey: Relative and absolute judgments compared. *Perception,* **7**, 635-659.

Morris, R. G. M., Garrud, P., Rawlins, J. N. P., & O'Keefe, J. (1982). Place navigation impaired in rats with hippocampal lesions. *Nature,* **297**, 681-683.

Moyer, R. S. (1973). Comparing objects in memory : Evidence suggesting an internal psychophysics. *Perception and Psychophysics,* **13**, 180-184.

Moyer, R. S., & Bayer, R. H. (1976). Mental comparison and the symbolic distance effect. *Cognitive Psychology*, **8**, 228-246.

Parkman, J. M. (1971). Temporal aspects of digit and letter inequality judgements. *Journal of Experimental Psychology*, **91**, 191-605.

Piaget, J. (1928). *Judgement and reasoning in the child*. London: Routledge and Kegan Paul.

Piaget, J. (1971). *Biology and Knowledge*. Edinburgh: Edinburgh University Press.

Piaget, J., & Szeminska, A. (1952). *The child's conception of number*. London: Routledge and Kegan Paul.

Pylyshyn, Z. W. (1981). The imagery debate. Analogue media versus tacit knowledge. *Psychological Review*, **88**, 16-45.

Reese, H. W. (1968). *The perception of stimulus relations*. New York: Academic Press.

Roberts, W. A., & Phelps, M. T. (1994). Transitive inference in rats: A test of the spatial coding hypothesis. *Psychological Science*, **5**, 368-374.

Russell. (1983). The child's appreciation of necessary truth and necessary falseness of propositions. *British Journal of Psychology*, **73**, 256-266.

Smith, L. (1993). *Necessary knowledge. Piagetian perspectives on constructivism*. Hove: Lawrence Erlbaum.

Spence, K. W. (1937). The differential response in animals varying within a single dimension. *Psychological Review*, **44**, 430-444.

Terrace, H., & McGonigle, B. (1994). Memory and representation of serial order by children. monkeys and pigeons. *Current Directions in Psychological Science*, **3**, 180-189.

Terrace, H. S. (1987). Chunking by a pigeon in a serial learning task. *Nature*, **325**, 149-151.

Trabasso, T. (1977). The role of memory as a system in making transitive inferences. In R. V. k. a. J. W. Hagen (Ed.), *Perspectives on the development of memory and cognition*. Hillsdale: Erlbaum.

Trabasso, T., Riley, C. A., & Wilson, F. G. (1975). The representation of linear order and spatial strategies in reasoning. In R. Falmagne (Ed.), *Reasoning, representation and process*. Hillsdale: Erlbaum.

Wills, S., & Mackintosh, N. J. (1998). Peak shift on an artificial dimension. *the quarterly journal of experimental psychology*, **51B**(1), 1-31.

Wood, D. J. (1978). The nature and development of strategies. In G.Underwood (Ed.), *Strategies for Information Processing.*: Academic Press.

Youniss, J., & Murray, J. P. (1970). Transitive inference with non-transitive solutions controlled. *Developmental Psychology*, **2.2**, 169-175.

Part 5: Conclusions

13

ANIMAL COGNITION TODAY AND TOMORROW
Diverse Themes

Stephen B. Fountain and Michael Bunsey
Kent State University

In reflecting on the Eighth Kent Psychological Forum and this resulting volume, we felt it appropriate to provide a brief review of the preceding chapters along with an account of some of the ideas discussed in and outside the formal Forum sessions. In so doing, we will try to highlight some common themes from the meeting and this volume, but we also freely interject our own opinions. We hope this treatment may give some idea as to the direction the study of sequential and temporal behaviors is taking. As will be readily noted, the volume begins with an introductory chapter in which Stewart Hulse sets the stage for the remainder of the volume by providing both a historical background for and a current evaluation of the field. Later sections focus on current approaches to the study of sequential learning and memory, counting and timing behaviors, and cognitive structure. We consider each section in turn.

HISTORY, SYSTEMS, ADMONITIONS, AND WARNINGS

In Chapter 1 of this volume, Stewart Hulse provides an historical overview of the study of serial behavior. This seems appropriate given Hulse's central importance in many of the developments in the area. Hulse traced the early history of the study of sequential behavior through Lashley, who in 1951 most persuasively argued for the inadequacy of associative chain models in explaining sequential behavior. In considering cases such as a pianist's finger movements, Lashley concluded that such movements occur far too rapidly to be explicable on the basis of associations of successive movements, but *could* be plausibly explained on the basis of centrally-mediated, hierarchically organized rules.

This type of framework was increasingly applied to nonhumans in the 1970's. At the Dalhousie conference on animal behavior of 1976 Hulse himself described what we now recognize as pioneering work on a cognitive approach to animal serial pattern behavior. This work showed, among other things, that rats more rapidly learn sequences of rewards which are formally simple than those which are formally complex. Findings such as these suggested that Lashley's notions of centrally-mediated, hierarchically organized rules were relevant for animal and human behavior alike—a point amply illustrated by the frequency of references to Lashley (1951) in comparative volumes such as this. However, examination of this volume also makes clear that these rule-based frameworks are not universally accepted and disputes persist concerning even these most fundamental questions about sequential behavior. The conflict is abundantly clear in comparing, for example, the contributions of Swartz, Capaldi, and Fountain et al. (Chapters 3, 5, and 6, respectively).

Hulse also considered three issues that he sees as being of importance to the future of comparative cognition, points which echo throughout this volume. First, is the issue of the comparative nature of comparative cognition: Hulse argued that too often we restrict our studies to one or two "canonical" species chosen primarily for historical reasons. We might be better served making more direct comparisons across meaningfully chosen species, taking into account their evolutionary histories. Actually, the contributions in this volume are not immune from Hulse's criticism, as the preponderance of work discussed here is on pigeons, rats, and monkeys. However, there are also several exceptions to the rule: McGonigle and Chalmers (Chapter 12) adopt a a truly comparative approach in their study of transitive inference in monkeys and human children. Terrace (Chapter 2), although using "standard" species, makes what can certainly be called meaningful comparisons between monkeys and pigeons in a simultaneous chaining task. This work has yielded evidence suggesting important differences between these species in the mechanisms underlying performance. In each of the latter cases, what is interesting is not any species difference in ability to learn, but rather that different species succeed in learning by using different mechanisms.

The second issue Hulse raised concerns the drift towards an increasingly neurobiological treatment of cognition. Although most (including at least some of those contributing to this volume) would agree that this is not an inherent problem, there is a problem to the extent that pure animal behavior research suffers—and in terms of available grant money there can be no doubt that it has suffered. As Hulse notes in Chapter 1, the irony is that the end result of current funding trends is that increasing resources are being funneled into the study of the brain while the study of what the brain does is gradually abandoned. The cause for concern can be illustrated by considering research into long-term potentiation (LTP), a candidate neurobiological mechanism of memory formation. Increasing amounts of money are directed into LTP research in an attempt to understand the biological bases of memory at the same time that behavioral work is making clear the limits of LTP as a memory model (Gallistel, 1990). How does LTP—with it's premise that learning is comprised of the acquisition of associations among stimuli that co-occur—explain the paradigms discussed by Roberts (Chapter 7), Bunsey (Chapter 10), and McGonigle and Chalmers (Chapter 12), in which subjects use learned associations transitively? What is the relevance of LTP to the memory for temporal duration [see Meck (Chapter 8)]? If we were to abandon behavioral research in our turn toward more neuroscience-oriented approaches, we might develop a near-comprehensive understanding of LTP without knowing how, or whether, LTP relates to memory.

Finally, Hulse warned of the dangers of anthropomorphism. This theme was echoed and elaborated upon by some of the other contributors. For example, Wright (Chapter 4) emphasized the significance of the fact that humans come to the lab with predispositions built up from years of experience which lead us to adopt different strategies than those used by animals during certain behavioral tests. On the whole, then, this volume's contributors give evidence that Hulse's warnings are being heeded.

SEQUENTIAL LEARNING AND MEMORY—CHARACTERIZING THE REPRESENTATION

Part 2 of this volume focused on current approaches to sequential learning and memory. Both Terrace (Chapter 2) and Swartz (Chapter 3) discussed acquisition and performance of a simultaneous chaining task. Wright (Chapter 4) asked a different type of question, assessing dynamic changes in the serial position effect in monkeys. Finally, Capaldi and Fountain et al. (Chapters 5 and 6, respectively) both discussed sequence learning in rodents. Together, these chapters provide a nice comparative view, with their involvement of several species trained on related sequential learning tasks. In some instances, this approach has yielded evidence for important species differences in aspects of serial learning, for example, in primates versus pigeons (reported by Terrace). In other cases, apparent species differences may have less to do with species than with task: A task in which monkeys sequentially respond to a series of four simultaneously presented visual stimuli (as in Terrace and Swartz) is quite different than one in which running times are measured in rats presented sequences of different reward magnitudes (as in Capaldi) or in which rats respond on a circular array of eight levers (as in Fountain). Perhaps we should not be surprised if these situations recruit quite different processes, regardless of the species involved.

In Chapters 2 and 3, Terrace and Swartz considered the learning of sequential responses to a simultaneously presented array of pictures by Rhesus monkeys and, in the case of Terrace, pigeons. When they are trained on an initial list of this sort, monkeys appear to learn a strategy that they are able to apply when tested on later, novel lists. This is manifested in the rapid establishment of learning set (Swartz et al., 1991). The strategy adopted during early lists seems to involve learning the first *n-1* responses, with responding to the last list item occurring by default. Probe trials have given important additional insight into the mechanisms underlying monkeys' performance in this task. For example, when items from different lists are combined into a new list, subjects show substantial positive transfer as long as the items retain their ordinal position across lists. Results such as these suggest that subjects encode information regarding ordinal position of items rather than simply learning associations among adjacent items, providing evidence for relatively high level processing in monkeys.

The use of the same type of task in pigeons has revealed a much different behavioral pattern in this species (Terrace, 1984). Pigeons differ from monkeys in their apparent inability to form a learning set across lists, and, in fact, pigeons have trouble learning a four-item list at all, showing consistent difficulty with the middle items. Tests using subsets of list items indicate that pigeons adopt a response pattern quite different than that of monkeys, responding appropriately to first and last list items and responding to middle items by default, a pattern which would put a four item list near the maximum learnable. Even when they do learn a four-item list, pigeons do so via different mechanisms than monkeys. For example, examination of response latencies in these situations indicate that monkeys, but not pigeons, seem to form a linear representation of the lists.

Interestingly, pigeons do show similarities to primates in terms of "chunking." That is, they can rapidly learn a list of even five items if trained on the list with two types of items that are segregated into chunks. Pigeons do this by applying the same response rules to the two separate chunks in turn. This is intriguing since chunks, although only sketchily defined in the human literature, are generally defined in verbal terms; we thus might not expect chunking to occur in nonhumans. Terrace suggests that the type of chunking seen in this pigeon work is substantially different than that studied in humans. Specifically, chunking in humans is usually measured during encoding in short-term memory tasks, whereas the extended training used with pigeons clearly implicate long-term memory processes. The chunking going on here can be thought of as "output chunking"—that is, the organization of responses at the time of retrieval from long-term memory. Interestingly, humans performing a 6-item list task pause regularly during output of these lists, seemingly reflecting self-imposed organization, or chunking. During each pause, it is thought, a chunk is retrieved from LTM and held in working memory until the subject has responded to the relevant items. This model recalls Lashley (1951), whom had long ago emphasized the significance of output chunks in his analysis of skilled movement sequences. This idea is also discussed by McGonigle and Chalmers (Chapter 12) who reason that this type of chunk is in fact necessary for sequential responding to occur.

In Chapter 4, Wright addressed a very different aspect of sequence learning, namely, the serial position effect—the U-shaped function describing how subjects remember stimuli from the beginning and end of a list better than stimuli from the middle of the list. Previously, attempts to develop animal models to study the serial position effect have been hampered by the apparently poor list learning ability of nonhumans. Wright found that this poor performance in fact reflects interference effects which are exaggerated by traditional task designs. Monkeys can learn lists of visual stimuli quite well if interference effects are minimized by the use of many distinct stimuli (Sands and Wright, 1980). Using such tasks, Wright has been able to examine serial position in monkeys and has found, surprisingly, that the classic U-shaped serial position function apparently represents a temporary transitional stage: With visual stimuli, primacy and recency effects are both seen only with an intermediate retention interval. With an immediate test, the primacy effect is absent, whereas following a long retention interval the recency effect disappears.

Serendipitously, Wright and colleagues have found that the opposite pattern obtains if auditory cues are used instead of visual cues; that is, recency is not seen with an immediate test and primacy disappears after a long retention interval (Wright, Shyan, & Jitsumori, 1990). Together, these surprising results pose a serious challenge to existing models based on the human data. For example, if information must go sequentially from STM to LTM, as outlined in the so-called modal model (Atkinson & Shiffrin, 1968) how can there ever be a primacy effect in the absence of a recency effect? Wright's data suggest that new models must include more dynamic memory and forgetting processes in order to accommodate such results.

As it turns out, these results also point to an advantage in the use of animal over human subjects. Specifically, humans come to any learning task with strategies, one of which is cross-modal coding of to-be-remembered information (e.g., verbal coding of visual material). Such dual coding of stimuli presumably obscures the modality patterns Wright has detected with monkey subjects—humans can compensate for a missing recency effect in one modality by taking advantage of the intact recency in another.

In Chapter 5 Capaldi took a much different view of sequential learning than the previous authors, arguing that such learning *can* be explained by relatively simple associative mechanisms. In Capaldi's test situation, rats are trained to respond for reward, with reward magnitude varying in an ordered, sequential manner. Learning in this situation is inferred when a subject's running speed tracks reward magnitude, with larger and smaller rewards producing faster and slower running times, respectively. Capaldi argued that this learning can be explained on the basis of memory for single items: a rat need only associate a large reward with a preceding small reward to develop an appropriate expectancy. By this framework animals need not learn ordinal positions (*a la* Terrace and Swartz) or formal rules (*a la* Hulse and Fountain) in order to learn a sequence. Capaldi argued that response speed is also likely to be affected by at least two other factors. First, a large reward on Trial n will have excitatory properties independent of anticipation of Trial *n+1*, and this factor can either diminish or potentiate the effects of expectancies. A second important factor is the discriminability of reward magnitudes in series. Specifically, if successive trials consist of 0 pellets of food (nonreward) on some trials versus 10 pellets on others, subjects will learn more rapidly than if trials consist of less discriminable quantities, for example, 5 versus 7 pellets.

Capaldi argues that these factors together can comprehensively explain reward magnitude sequence learning. Capaldi and associates have illustrated how this framework might be compatible with results derived from longer sequences in Hulse's runway tasks, though because of the heuristic nature of their theory they have been forced to adopt the weaker position that the theory "cannot be excluded" (Capaldi, Verry, & Davidson, 1980). Hulse has argued that this type of framework cannot extend to longer sequences (Hulse, 1980), and this view is widely accepted in recent textbooks. However, Fountain et al. argue in Chapter 6 that this question remains largely unresolved.

Fountain et al. (Chapter 6) offer evidence to support the view that animals may use both simpler associative mechanisms and rule-learning mechanisms in complex sequential tasks. They report evidence from behavioral studies and computational modeling to argue that a surprising number of sequential phenomena—many aspects of reward magnitude pattern learning and phrasing effects—can be explained by simple associative mechanisms. However, they also report evidence from behavioral studies and brain manipulations indicating that for now simple associative views alone fail to account for all the behavioral and psychobiological data.

Fountain et al. discuss behavioral evidence that phrasing cues facilitate sequential learning and performance in at least two ways—by emphasizing the structure of the sequence and by cueing subsequent responses. Evidence from cue-removal studies reported by Fountain and colleagues (e.g., Stempowski, Carman, & Fountain, 1999) indicate that phrasing cues have many of the properties of discriminative cues. However, earlier evidence shows that phrasing cues that do not occur at structural boundaries actually impair performance and that under some conditions training with phrasing cues can result in savings after cue-removal (Fountain et al., 1984). Both of these results are inconsistent with hypotheses that cues merely break down a sequence into smaller chunks. Importantly, both are also inconsistent with the notion that pauses act only as discriminative cues, as might be expected from Capaldi's framework (Chapter 5). If a cueing effect were the only mechanism involved, we might expect facilitation regardless of where the cue occurred and a deficit after cue-removal. The phrase-induced impairment and savings after cue-removal are thus critical to the interpretation that subjects create an organized representation of the sequence. It should be noted that this is a case in which procedural differences might cause different results in different labs or between tasks. Breaking a sequence into small arbitrary chunks might be more helpful in tasks such as that used by Swartz and Terrace in which a hierarchical rule does not guide performance.

Fountain et al. also illustrate how computational modeling and neurobiological manipulations can shed new light on old theoretical problems. On the one hand, Fountain et al. show how a computational model endowed with mechanisms for item association and generalization can account for many of the classic rule-learning phenomena in reward magnitude pattern learning. These data might seem to argue against rule-learning in rat serial learning, but other behavioral phenomena, such as the phrasing phenomena just described, are not likely to be as easily described by their computational model. Furthermore, hippocampal dysfunction brought on by MK-801, an N-methyl-D-aspartate (NMDA) antagonist drug, suggests that nonhippocampal mechanisms may be involved in learning about pattern structure whereas hippocampus mediates learning about "exceptions-to-the-rule." Although the debate over the existence, nature, and importance of rule-learning mechanisms versus widely-accepted associative mechanisms in animal learning is far from resolved, one can hope that these new approaches promise to elevate and broaden the discourse.

It will be important to try to further extend the associative models of Capaldi and Fountain et al. to other sequential learning and memory phenomena and tasks like those discussed in previous chapters. In addition to having difficulty explaining some rule-learning phenomena, these models may also have particular difficulty explaining findings such as those of Terrace and Swartz (Chapters 2 and 3, respectively) showing that animal subjects use ordinal position to guide responding in recombined lists. On the other hand, as noted by Fountain (Chapter 6), we should be wary of too readily passing over simple associative mechanisms in explaining complex human-like cognitive performance in any given task.

COUNTING AND TIMING—COMPARATIVE INSIGHTS AND NEUROBIOLOGICAL MODELS

Part 3 of this volume addressed a variety of issues related to counting and timing behaviors in nonhumans and humans. Roberts (Chapter 7) gave an historical perspective on their attempts to assess and characterize counting behavior in pigeons and rats. Meck and Kesner (Chapters 8 and 9, respectively), in contrast, focused on the neural basis of timing and temporal processing.

As Roberts noted, at least some components of counting behavior appear to be quite fundamental. Several species have been shown capable of relative numerosity judgments, and monkeys and rats have been shown capable of discriminating on the basis of absolute number (Roberts & Mitchell, 1994; Davis & Albert, 1985). Several species also seem capable of ordinality, demonstrating the capacity for transitivity. It should be noted that these are difficult studies to conduct cleanly and there has been controversy regarding whether animals are showing truly transitive behavior or whether their choices can be more parsimoniously explained on the basis of the development of simple response biases. Robert's lab has addressed these concerns and their recent work seems to support ordinality explanations (e.g., Roberts & Phelps, 1994).

Roberts has suggested two other general areas in which animals fail to achieve human-like levels of performance. One is in the active use of counting. This is akin to the claims that one of the critical differences between human and chimp language is in the spontaneous generation of sentences. Second, animals have not yet been shown to demonstrate a suitably abstract ability to count, that is, transferring numerical discrimination across situations. With respect to both of these issues, though, more research is needed before we can make any definitive claims.

Meck also addressed timing behaviors in Chapter 8, though with a much more neurobiological slant. Meck outlined a model mapping the different components of a quantitative theory of scalar timing in memory onto neurobiological structures. This approach represents a nice model system in its combination of a neurobiological approach with a well-developed model from the behavioral literature. Using a discrete trials peak interval procedure in which a rat's "estimate" of stimulus duration can be assessed, Meck's work has demonstrated a role for a frontal system in storage of information regarding the expected time of an event and a role for the hippocampal system in storage of information about duration from a single trial. These dissociations between hippocampal and frontal systems are noteworthy since in most cases in the literature, lesions of these two systems have similar effects.

In recent work, Meck and colleagues have been investigating the role that the cholinergic system plays in these areas in timing behavior. They have found that treatments which increase cholinergic function cause underestimation of time of reinforcement while those that decrease cholinergic function have the opposite effect. Such results illustrate the power of this model system—with such a linear and quantifiable behavior, subtle, bidirectional drug effects can be measured. This is a far cry from showing a crude, unidirectional memory impairment after drug injection, for example. In terms of practical application, these results are also interesting in that cholinergic dysfunction is implicated in aging. Meck has now demonstrated changes in frontal cholinergic markers and time estimation in aged rats, and has shown that treatments that reduce the cholinergic changes also ameliorate the temporal overestimation. Nonetheless, much work remains to be done, as even the causal direction of these relationships is unclear.

This model has also provided evidence of cholinergic correlates of the *direction* of discrepancies in time estimation. Driving the frontal theta rhythm to faster rates decreases remembered time of reinforcement, and this relationship is proportional. This intriguing result illustrates clearly how a relatively complex behavior (viz., timing) which has no environmental referent can arise from an animal using an internal cue or structure, in this case the theta rhythm. This is similar to the type of mechanism hypothesized by Roitblat in his later chapter. It is interesting to see a potential neurobiological mechanism compatible with Roitblat's notion that complex behaviors can arise without complex cognitive structures provided organisms can take advantage of internal structure.

Kesner has put forth an attribute model of memory formation according to which specific brain regions are posited to mediate formation of distinct memory attributes. In Chapter, 9 Kesner specifically focused on memory for time. Working with humans and rats, Kesner has found evidence for evolutionary conservation in the neurobiological substrates for various types of timing behavior. In both rats and humans, memory for duration seems dependent on the hippocampus but not medial prefrontal cortex. Importantly, the ability to detect duration seems intact in hippocampals—it is specifically *memory* of duration which is impaired. Performance on temporal order tasks and on frequency discriminations, in contrast, seem dependent on both hippocampal and frontal structures. However, only so-called data-based temporal order tasks, in which task information varies from trial to trial, are dependent on both structures: If temporal order of items is constant across trials, only the prefrontal cortex is critical (Chiba et al. 1992, 1994).

Kesner's findings tie in nicely with Meck's timing model. Both authors provide evidence that hippocampal and frontal areas are critical to temporal processing. At a more specific level, although they use different terminology, both attribute temporal working memory functions to the hippocampal system and temporal reference memory functions to the frontal system. However, the dramatically different approaches the authors take make further comparison somewhat difficult. Whereas Meck has used multiple techniques to assess many aspects of a single type of timing behavior, Kesner, in keeping with his more general attribute model approach, has assessed many different types of temporal processing following ablation of various brain structures. What these authors do share (as does Bunsey in Chapter 10) is an emphasis on conservation of function across species. This bias to focus on shared processes stands in stark contrast to the stance of many of the contributors here, a discrepancy which reflects the different perspectives of comparative versus physiological psychologists, more generally.

CONSTRAINING THE COMPLEXITY OF EVENTS IN TIME: COGNITIVE STRUCTURE IN SEQUENTIAL PROCESSES

The chapters by Meck and Kesner provide a nice segue into Part 4 of the volume which focused on cognitive structure in the sequential and temporal organization of behavior. Mike Bunsey (Chapter 10) addressed neurobiological substrates for such behaviors, whereas Herb Roitblat

(Chapter 11) and Brendan McGonigle and Margaret Chalmers (Chapter 12) discussed the relevance of constraints on such behaviors.

In Chapter 10, Mike Bunsey addressed the more general role of the hippocampal system in memory formation. A key part of this work has involved investigating the neurobiological substrates underlying a form of transitivity, namely, associative transitivity. In associative transitivity, subjects trained on paired associates which share elements (e.g., A-B and B-C) may infer relationships between novel pairings (A-C), based on the shared items. Bunsey and Eichenbaum (1996) have shown that specific hippocampal lesions do not affect learning of the premise pairs in this situation but block associative transitivity, suggesting that the hippocampus is critical in forming a memory representation which can be used transitively.

More generally, these results are consistent with the notion that the hippocampus is responsible for relational memory, the memory for multiple sensory-sensory associations which are coded in such a way that they can be used flexibly, that is, in new ways or under novel conditions (Eichenbaum, Otto, & Cohen, 1992). Associative transitivity is seen as being emblematic of relational memory as it requires a novel use of information, but it is only one example of this type of memory; the same relational memory processes are hypothesized to underlie memory for spatial locations, as well as human declarative memory, each of which is dependent on the hippocampus. The evidence thus indicates that the hippocampus subserves a similar (and complex) role in memory processing across species. This emphasis on conservation of function across species is fundamental to this theoretical framework, much as it is in Kesner's model.

The specific task used by Bunsey is important, too, as associative transitivity is distinctly different than the transitive inference discussed in previous chapters, and cannot be explained by the simple associative mechanisms mentioned earlier. Work with a related sensory preconditioning task suggests, instead, that in this type of transitivity presentation of item A may activate a representation of item B, which then biases choice behavior to C, with which B is associated (Matzel et al., 1983). According to this account, associative transitivity is not far removed from simple association, going beyond only in making explicit a role of a stimulus representation. As such this type of task may represent another useful tool for bridging the gap, noted in Chapter 14 by McGonigle and Chalmers, between simple associative and cognitive approaches. Currently, there is abundant effort being directed toward understanding the neurobiological mechanisms of simple conditioning. Bunsey's effort represents one way behavioral neuroscience can move beyond simple conditioning, examining a task which is only slightly more complicated but which involves the more cognitive notions of stimulus representation.

Incidentally, as with Meck and Kesner, the intriguing results seen after borrowing a task from the purely behavioral literature nicely illustrates the potential for interactions across the neuroscience and more behavioral fields.

In Chapter 11, Herb Roitblat addressed the fact that traditional "reflex" approaches to behavior fail to deal with many aspects of animal behavior, including most of those discussed in this volume. For example, the ability of animals to engage in timing behaviors, such as those described by Meck and Fetterman, poses problems for traditional approaches, as there is no cue in the environment corresponding to time. Likewise, performance in Hulse's runway task, in which subjects learn to run more slowly as they near the end of a sequence of decreasing reward sizes, is difficult to explain in reflex terms. For example, so-called "wildcard experiments" indicate that rats respond on the basis of overall sequence structure, as performance is unaffected on probe trials in which one element of the sequence of rewards is replaced by a 0-reward wildcard (Roitblat, Pologe, & Scopatz, 1983). These results suggest that rats have an abstract, hierarchical representation of the sequences.

How is it possible that seemingly simple organisms such as rats or pigeons can engage in such complex, structured learning? Roitblat argued that it is misleading to assume that complex, structured learning implies complex cognition. Rather structured learning may result from the use of more peripherally located structure to organize information. After this peripheral organization, learning is simple. By way of example, Roitblat discussed the ability of dolphins to use echolocation to derive complex information regarding objects in the environment. The physical structure of the dolphin's cochlea make it tantamount to a bank of frequency filters: Time-extended auditory information about an object is transformed into a time-independent representation on the cochlea. The dolphin now has a simpler code with which to do further processing. Thus, much of the complex representational work is done before information even reaches the brain. What is the relevance of this to the types of behavior discussed in this volume? Meck has shown how timing behavior may be accomplished through the use of an internal oscillator. Timing a stimulus becomes trivial if the number of oscillations during the stimulus is recorded. Although the oscillator cannot be considered peripheral, the principle is similar. It is perhaps not too far a stretch to suggest the same principle is reflected in Robert's notion that transitive behavior depends on a mapping of stimuli onto a spatial cognitive structure.

In Chapter 12, McGonigle and Chalmers, though using quite different approaches, made a similar point—that constraints render the learning of complex structure simpler than might be suspected. McGonigle and Chalmers also emphasized the inadequacy of simple traditional learning theory to explain complex behavior, using transitivity to make his case. An advantage to transitivity as a model system is that, although a complex behavior, it does not require linguistic ability. As such, it may be a tool with which to bridge the gap between the simple associative approach of learning theory and what McGonigle and Chalmers call "the abstract, symbolic level ridden currency" of the traditional human cognitive approach.

McGonigle and Chalmers outlined the history of their work with transitivity which initially suggested that transitivity could be explained on the basis of an abstract spatial array upon which items were ordered in a linear fashion (see also Roberts). A key piece of evidence here was the symbolic distance effect, by which reaction times increase the greater the ordinal distance between tested items. More recent work, however, suggests that at least under certain conditions the symbolic distance effect is an artifact (McGonigle & Chalmers, 1996). These latter findings have inspired McGonigle and Chalmers to adopt a different approach, modeling transitivity with a stack of production rules through which subjects proceed in a fixed order. The appropriate rule stack is capable of modeling many aspects of primate transitive behavior, including some aspects which cannot be explained by the spatial array model.

This model of transitive behavior also seems to resolve an apparent paradox in the human literature: Children of an age capable of the abstract ordering inherent in transitivity, are incapable of ordering ten sticks from tallest to shortest when the sticks are placed directly in front of them. According to McGonigle and Chalmers' framework, this is due to the fact that in this situation there are simply too many possible combinations for children to handle. In standard transitivity tasks, on the other hand, although the comparison is more abstract, subjects only face a 2-item choice. In McGonigle and Chalmers' rule stack framework, it is the number of items, and thereby the number of possible combinations, that determines complexity. Children get better at high number seriation tasks over development because they become able to detect and exploit serial constraints in ordering tasks, functionally reducing the number of combinations involved. By using these constraints they thus make the tasks easier for themselves. This latter type of finding has led McGonigle and Chalmers to more basic tasks than transitivity, and their group is now using simple search tasks to examine self-regulation and the development of strategies. These experiments assess, for example, the ability of subjects to spontaneously use category types to "chunk" information during search, somewhat akin to the output chunking discussed by Terrace in Chapter 2. Like Terrace and Roitblat, then, McGonigle and Chalmers are increasingly emphasizing the organism's use of organization to render unwieldy problems solvable.

CLOSING THOUGHTS

Sequential behavior, including counting and timing behavior, can sometimes appear to be so complex that it would defy analysis and, therefore, it seems to demand reduction. On the other hand, behavior is undeniably sequential in nature and, as many investigators from Lashley (1951) to Terrace (Chapter 2, this volume) have noted, it may not be possible to properly understand much of human and animal behavior if behavioral sequences are conceived as nonsequential and are reduced for study to independent behavioral events. This volume describes some of the most successful ongoing programs of research attempting to grapple with the problems inherent in viewing ongoing behavior as fundamentally sequential. The work reported in this volume illustrates how much of sequential behavior can be described in traditional associative terms, but that there are clear limitations to this view of sequential behavior. The challenge for the future is to recognize the strengths of Lashley's (1951) approach to the problem of sequential behavior. His method was to marshal evidence from every possible source—behavioral analysis, neurobiological manipulations, and, we may add today, computational modeling (Church, 1997, 2001)—to constrain and extend psychological theories of sequential behavior.

REFERENCES

Atkinson, R.C., & Shiffrin, R.M. (1968). Human memory: A proposed system and its control processes. In K.W. Spence and J.T. Spence (Eds.), *The Psychology of Learning and Motivation*, Vol. **2** (pp.89-105). New York: Academic Press.

Capaldi, E.J., Verry, D.R., & Davidson, T. (1980). Memory, serial anticipation pattern learning, and transfer in rats. *Animal Learning & Behavior,* **8**, 575-585.

Chiba, A.A., Johnson, D.L., & Kesner, R.P. (1992). The effects of lesions of the dorsal hippocampus or the ventral hippocampus on performance of a spatial location order recognition task. *Society for Neuroscience Abstracts,* **18**, 1422.

Chiba, A.A., Kesner, R.P., & Reynolds, A.M. (1994). Memory for spatial location as a function of temporal lag in rats: Role of hippocampus and medial prefrontal cortex. *Behavioral and Neural Biology,* **61**, 123-131.

Church, R.M. (1997). Quantitative models of animal learning and cognition. *Journal of Experimental Psychology: Animal Behavior Processes,* **23**, 379-389.

Church, R.M. (2001). A Turing test for computational and associative theories of learning. *Current Directions in Psychological Science*, **10**, 132-136.

Davis, H., & Albert, M. (1986). Numerical discrimination by rats using sequential auditory stimuli. *Animal Learning & Behavior*, **14**, 57-59.

Eichenbaum, H.E., Otto, T., & Cohen, N.J. (1992). The hippocampus: What does it do? *Behavioral and Neural Biology*, **57**, 2-36.

Fountain, S.B., Henne, D.R., & Hulse, S.H. (1984). Phrasing cues and hierarchical organization in serial pattern learning by rats. *Journal of Experimental Psychology: Animal Behavior Processes,* **10**, 30-45.

Gallistel, C.R. (1990). *The Organization of Learning*. Cambridge, MA: The MIT Press.

Hulse, S.H. (1980). The case of the missing rule: Memory for reward vs. formal structure in serial-pattern learning by rats. *Animal Learning & Behavior*, **8**, 689-690.

Lashley, K.S. (1951). The problem of serial order in behavior. In L.A. Jeffress (Ed.), *Cerebral Mechanisms in Behavior* (pp. 112-146). New York: Wiley.

McGonigle, B., & Chalmers, M. (1996). The ontology of order. In L. Smith (Ed.), *Critical Readings in Piaget.* Routledge: London.

Matzel, L.D., Held, F.P., & Miller, R.R. (1983). Information and expression of simultaneous and backwards associations: Implications for contiguity theory. *Learning and Motivation*, **19**, 317-344.

Roberts, W.A., & Mitchell, S. (1994). Can a pigeon simultaneously process temporal and numerical information? *Journal of Experimental Psychology: Animal Behavior Processes*, **20**, 66-78.

Roberts, W.A., & Phelps, M.T. (1994). Transitive inference in rats: A test of the spatial code hypothesis. *Psychological Science*, **5**, 368-374.

Roitblat, H.L., Pologe, B., & Scopatz, R.A. (1983). The representation of items in serial position. *Animal Learning & Behavior*, **11**, 489-498.

Sands, S.F., & Wright, A.A (1980) Primate memory: retention of serial list items by a rhesus monkey. *Science*, **209**, 938-940.

Stempowski, N.K., Carman, H.M., & Fountain, S.B. (1999). Temporal phrasing and overshadowing in rat serial-pattern learning. *Learning and Motivation*, **30**, 74-100.

Swartz, K.B., Chen, S., & Terrace, H.S. (1991). Serial learning by rhesus monkeys: I. Acquisition and retention of multiple four-item lists. *Journal of Experimental Psychology: Animal Behavior Processes*, **17**, 396-410.

Terrace, H.S. (1984). Simultaneous chaining: The problem it poses for traditional chaining theory. In M.L. Commons, R.J. Herrnstein, & A.R.

Wagner. (Eds.) *Quantitative Analyses of Behavior: Discrimination Processes* (pp. 115-138). Cambridge, MA: Ballinger Publishing Co.

Wright, A.A., Shyan, M.R., & Jitsumori, M. (1990) Auditory same/different concept learning by monkeys. *Animal Learning & Behavior*, **18**, 287-294.

Index